Space and the American Imagination

SPACE

and the
American
Imagination

SECOND EDITION

Howard E. McCurdy

The Johns Hopkins University Press
Baltimore

© 2011 The Johns Hopkins University Press
All rights reserved. Published 2011
Printed in the United States of America on acid-free paper
9 8 7 6 5 4 3 2 1

The Johns Hopkins University Press
2715 North Charles Street
Baltimore, Maryland 21218-4363
www.press.jhu.edu

Library of Congress Cataloging-in-Publication Data

McCurdy, Howard E.
  Space and the American imagination / Howard E. McCurdy.
— 2nd ed.
    p. cm.
  Includes bibliographical references and index.
  ISBN-13: 978-0-8018-9867-9 (hardcover : alk. paper)
  ISBN-10: 0-8018-9867-6 (hardcover : alk. paper)
  ISBN-13: 978-0-8018-9868-6 (pbk. : alk. paper)
  ISBN-10: 0-8018-9868-4 (pbk. : alk. paper)
  1. Astronautics—United States—Public opinion. 2. Mass
media—United States—Influence. 3. Astronautics and state—
United States. 4. Popular culture—United States. I. Title.
  TL789.8.U5M338 2011
  387.80973—dc22   2010019456

A catalog record for this book is available from the British Library.

*Special discounts are available for bulk purchases of this book. For
more information, please contact Special Sales at 410-516-6936 or
specialsales@press.jhu.edu.*

The Johns Hopkins University Press uses environmentally
friendly book materials, including recycled text paper that is
composed of at least 30 percent post-consumer waste, whenever
possible. All of our book papers are acid-free, and our jackets
and covers are printed on paper with recycled content.

To my classmates at Queen Anne High School

Seattle, Washington

Class of 1959

Contents

Acknowledgments

When I was not yet in my teens, a relative of mine, Grace Warren, wrote me a series of letters on the planets and extraterrestrial life. Her husband, Dana Warren, was a theoretical physicist at the Lawrence Livermore National Laboratory in California and an expert on cosmic rays. Together, they excited my imagination with extraterrestrial facts and fantasies. I visited them in the summer of 1956. The following year, the Soviet Union launched *Sputnik 1* and *Sputnik 2*. The excitement surrounding that event encouraged my classmates at Queen Anne High School in Seattle to form a rocket club (unauthorized by school authorities). I joined, along with one of my best friends, Joel Farley. Together we conducted crude but delightful experiments with homemade rockets. Appropriately motivated, I decided to pursue a career in science. I chose chemistry and went to Oregon State University to study the subject. William E. Caldwell, who had been my father's professor twenty years earlier, welcomed me into the science program. It took one year to discover that I was as interested in social science as I was in physical science. My professors at Oregon State, the University of Washington, and Cornell University made public policy understandable. Cornell professor Paul P. Van Riper gave me my first research assignment involving the civil space program, an insignificant but nonetheless stimulating responsibility.

Starting out as a young professor, I was encouraged by the dean of public administration scholars, Dwight Waldo, to explore the relationship between imagination and public policy. He had written occasionally about fiction and public administration, and I did the same. The more I wrote, the more I learned about the ways in which imagination affected the course of public affairs. I was impressed by the work of scholars such as Roderick Nash and Joseph Corn, who described how imagination shaped public attitudes toward conservation policy and aviation.

This book has allowed me to combine my interests in science, imagination, and public policy, permitting me to revisit those wonderful images that entertained me in my youth and investigate in a scholarly way the achievements to emerge from them. I could not have completed this work without the inspiration of the people who encouraged these interests in the past.

Special thanks go to Roger Launius, who now works as a space historian at the National Air and Space Museum, and Sylvia Kraemer. As director of the History Office at the National Aeronautics and Space Administration (NASA) Headquarters, Dr. Kramer encouraged me to pursue my interest is space policy and supported my work. She was succeeded by Dr. Launius, who oversaw the History Office grant that allowed me to complete the first edition of this book. Much of the material used in the original study was gathered from the archives of the NASA History Office, masterfully assembled by the irreplaceable Lee Saegesser. Colleagues at American University and the University of Washington, on whose faculties I serve, listened and critiqued as the book evolved. A new NASA chief historian, Steven J. Dick, organized a series of symposia that provided opportunities to test ideas for the current edition.

I completed the manuscript of the first edition at the University of Washington during the 1995–96 academic year and finished the second edition while working there in 2009. I am grateful to the deans of the Evans School of Public Affairs, Sandra O. Archibald and Margaret T. Gordon, who allowed me to teach and write in my hometown. At American University, Neil Kerwin, William LeoGrande, Meg Weekes, Bernard Ross, Claire Felbinger, Ann Ferrin, Robert Cleary, Robert Boynton, David Rosenbloom, and Robert Durant provided encouragement and support. Professor Richard Berendzen, who provided frequent lessons in astronomy and astrophysics, deserves special thanks. Graduate students at both institutions checked sources and located materials; I particularly want to recognize the assistance of Suzanne Roosen, Margaret Bolton, Vince Talucci, Mary Huston, Jennifer Dale, Marcela Onyango, and Richard Faust. To those who read parts or all of the manuscript, especially James R. Hansen and Robert Wohl, I extend my appreciation.

Preparation of this study required a wide-ranging knowledge of popular culture, space science, rocket technology, and public policy. Any errors that may have crept into the work in spite of repeated cross-checks are my own. Or they might even be the result of an overambitious imagination.

Space and the American Imagination

Imagination

The ability to imagine is the largest part of what you call intelligence. You think the ability to imagine is merely a useful step on the way to solving a problem or making something happen. But imagining it is what makes it happen.
                                        —Michael Crichton, *Sphere*, 1987

Humans possess an astounding capability. They can imagine themselves living in circumstances very different from the ones in which they currently exist. They can communicate those visions in such a way as to motivate other people to desire them. Using tools and technology, humans can then reconstruct their circumstances so that the visions they imagine become real.

In few areas is this power more dramatically illustrated than in the realm of space exploration. The vision of space exploration is both simple and astounding. Humans imagine that they—or entities like themselves—will leave their home planet, fly in space, and reach other destinations.

In America, as the technology permitting space flight became possible, the advocates of space exploration offered a straightforward explanation of how this would occur. They imagined themselves building winged spaceships, defying gravity, and accelerating those machines to velocities capable of orbiting the

Earth. They imagined themselves building large, Earth-orbiting space stations. They imagined flying to the Moon, landing and returning to Earth, and eventually establishing lunar research stations. They imagined their ability to construct robotic spacecraft that could fly to other planets and, without the advantages of humans on board, orbit those spheres, land on them, and traverse their surfaces. They imagined large, orbiting space telescopes. They imagined humans establishing colonies on other planets and inventing methods for the investigation of Earth-like objects around nearby stars. They imagined the discovery of extraterrestrial life. Exploration advocates imagined ways in which they could use current events and inspiring images to enlist support for the achievement of these goals.

In fifty years, just a half century after the launch of the first small, Earth-orbiting satellite, humans achieved much of what they had dreamed. Given enough time, humans may realize even more. The human capacity for turning vision into reality could be as unlimited as the realm into which those efforts intrude.

This book deals with space exploration and the manner in which it has been imagined, principally in the United States of America. It recounts what people thought, how they sought to convert vision into reality, and the obstacles they encountered. It concludes with some observations about the process by which visions inspire public action and some comparisons to other public activities in which imagination has played a significant role. It attempts to explain why the spacefaring vision proved so powerful and why humans confronted with adversity chose to bend reality to fit their visions rather than abandon their beliefs. Many books have been written about space exploration—some about decisions such as the one to go to the Moon, others about the space race between the United States and the Soviet Union, and more about individual projects and the machinery that flies in space.[1] This is a book about vision, imagination, popular culture, and its translation into achievements that make dreams come true.

Imagination involves a process of forming mental images of events or processes that are not actually present. A mental image produced by the imagination may be called a vision, although the latter typically refers to an idea of considerable scope that broadly depicts situations that have not yet occurred. Images and visions are transmitted in various ways. Some reach their audience through works of fiction, which depict characters and events that are not real.

Others are proclaimed in the form of forecasts by advocates seeking to popularize new ideas. In the realm of space exploration, the underlying vision has been transmitted through the media of fiction, popular science, books, movies, television, radio, newspapers, magazines, and paintings.

The most pervasive images enlarge the ideas, customs, and beliefs held by the public at large—what is generally known as popular culture. Persistent ideas become part of the popular culture. They become part of the stories that communities of people tell about themselves, stories that help to define who they are and the things in which they believe.

We know that visions have a transformational effect, encouraging private action and public change. Transformations often begin when humans alter their subjective interpretations of the world. Subjective interpretations consist of the assumptions that people make about the circumstances in which they live. Humans once assumed illness to be the byproduct of demonic possession or a punishment for sins. Healers set aside that subjective belief by imagining that humans had the power to cure the ill. Changes in objective reality followed, including the discovery of cures. In another case, social reformers attacked the view that poverty was "nature's way" and thereby helped to create the intellectual foundation for the modern welfare state. Conservationists convinced the public—which previously had viewed wild areas as savage and unfriendly places—that tranquillity and spiritual renewal could be found in the wilderness. The national park movement was born. Aviation enthusiasts imagined a future in which people could fly, promised that it would change the world, and made it occur.[2]

History repeatedly shows that humans, having changed their subjective interpretations of the world, can change objective reality as well. Arthur C. Clarke, the great science fiction writer, once observed that any sufficiently advanced alteration in objective reality was "indistinguishable from magic." A change in objective reality—or technology—need not violate the laws of physics in order to occur. It must merely do that which was once thought to be impossible. Clarke's thoughts on the subject became known as Clarke's laws. "When a distinguished but elderly scientist states that something is possible, he is almost certainly right. When he states that something is impossible, he is very probably wrong." The great diversity of national cultures around the world demonstrates that humans are capable of creating widely different subjective interpretations of the world. Clarke insisted that humans could change

The grand vision of space exploration drew considerable strength from its capacity to excite and entertain. Advocates stirred public interest with images of space stations, exotic spacecraft, alien life-forms, lunar bases, and expeditions to Mars. This painting by Chesley Bonestell, which appeared in a 1954 issue of *Collier's*, depicts preparations for the return of the first human expedition to Mars. (Reproduced courtesy of Bonestell LLC)

objective realities as well. "The only way of discovering the limits of the possible," Clarke announced, "is to venture a little way past them into the impossible."[3] What once seems like magic becomes real.

In a dramatic way, space exploration tests the connection between culture and technology. The visions associated with it stretch credulity and assault physical barriers. Standing alone, the vision of space exploration is one of the most fascinating that humans have conjured. How could anyone have the foresight to suggest, with the aviation era barely under way, that humans might fly from the Earth to other spheres? Who could have predicted, just a few years after the first humans learned how to fly in the atmosphere, that barely a half century later humans would land on the Moon? As acts of imagination, these are fantastic. As achievements, they are magical.

The vision of space exploration is impressive on its own terms. Yet it is made even more imposing by the manner in which it connects to other social traditions. The vision of space exploration owes much of its power to its ability to attach itself to other cultural traditions that define the human experience.

Consider the search for extraterrestrial life. For centuries, humans have debated whether living entities might inhabit other worlds. Early Greek writers supported the notion; Aristotle denied it. Many modern scientists favor the idea; others suggest that conditions supporting the development of complex life forms such as those on Earth are extremely rare.[4] To gather evidence, scientists have dispatched spacecraft to investigate conditions on Venus and Mars. Other persons have established listening stations to search for electromagnetic signals that alien civilizations might be broadcasting into space.

Through four decades of watching and listening, no signs of extraterrestrial life have been found. The surface of Venus is too hot, the exterior of Mars too sterile, and no messages from outer space have been received. Rather than deter further investigation, these findings have simply promoted calls for more. Part of this perseverance draws its force from the scientific method, in which theories are constantly tested and refined. Yet a large part of it arises from a series of events that have created a collective social predisposition that favors the expectation of life on other worlds.

The whole history of human exploration on Earth supports the presumption that expeditions dispatched to distant places will discover strange life forms in the places explored. Medieval travelers returned from foreign lands with stories of exotic creatures. Sea captains and the naturalists who accompanied them returned from long voyages with sketches and samples of biological wonders. Modern scientists have discovered life forms on the bottom of the ocean that need no sunlight to survive. Space exploration offers an opportunity to extend the exploration process into new realms, and that line of extension supports the notion that human discoveries in the cosmos will resemble those on the Earth. Although the metaphor may be misplaced—space exploration may not be like terrestrial exploration—the idea that it could be kindles the desire to explore.

Through such metaphors, humans have come to view space exploration as an extension of the most dramatic events shaping human history. To advocates of space exploration, the venture offers an opportunity to extend the frontier story that characterized the American experience for four hundred years; for military officers, space is an extension of the "high ground" that gives nations enjoying it a security advantage. Metaphors can make a complex subject easy to explain to an often-inattentive public. Rather than explain the intricacies of cosmic radiation or toxic atmospheres, it is easier to say that space exploration will be "like the frontier."

Through metaphors and association, space activities interlock with the most important characteristics of the American experience. The relationship gives the space exploration vision a level of desirability far beyond what it would receive if it had to stand on its own. The exploration of space promises to maintain the spirit of innovation and discovery that has made American strong. It connects to the corporate experience in a nation that has grown rich through business firms. It expands the experience with aviation in a nation that invented heavier-than-air flight. It affirms the idea that progress occurs through science. It has helped to define the conservation movement and is associated, in an odd sort of way, with the American agonies over slavery and servitude. The associations are so strong that Americans would want to believe in space travel even if it was not true. The associations give space travel a faith-like quality, encouraging belief even in the face of doubt and adversity.

The means by which humans communicate this vision further reinforce its believability. Communities of imagination come into being when people who might be physically separated from one another acquire a means for sharing common beliefs. Just as the printing press allowed new religious denominations to arise and as committees of correspondence (along with delivery of the mail) encouraged fresh nations to form, so the electronic and print media help people to believe in the desirability of space travel. Magazines printed paintings of spaceships and extraterrestrial bodies before humans and machines ventured there. Science fiction writers told fantastic tales. Movies and television shows promoted the vision of fast space ships, large space stations, and extraterrestrial life. Aliens may not live among us in fact, but they inhabit the mass media in large numbers, in movies such as *E.T.: The Extra-Terrestrial* and in television shows such as the *Star Trek* series.

The images entertain. Stories fictional and true dealing with space exploration are among the most widely consumed in American culture. They exalt explorers who investigate new worlds, honor scientists who attack the mysteries of the universe, and praise machines that follow the intentions of their creators. In an era of television and mass communication, the capacity of ideas to excite and entertain is crucial to their survival in the marketplace of imagination. Ideas that do not excite and entertain are replaced by those that do. In this respect, space exploration has proved to be one of the most remarkably persistent stories that Americans tell about themselves—a defining characteristic of national culture. The initial vision was transmitted through works of fiction, enthralling readers with stories of voyages to the planets and stars. Through a deliberately organized public relations campaign, using the medium of popular science, space enthusiasts convinced Americans that those stories could come true. When the actual space program arrived, reporters treated it as one of the greatest news stories of all time.[5] Commentators elevated the place of space exploration by treating it as important. By displaying the story, they communicated a central message to the general public. This subject is important. These achievements deserve your attention.

The cultural significance of the spacefaring story gives it a special place in American society, well above the official priority accorded space exploration within the national policy agenda. The National Aeronautics and Space Administration receives less than 1 percent of federal tax revenues, yet its activities rank among the most visible of American achievements. The NASA administrator heads a small, independent agency that does not entitle its leader to sit as a member of the president's cabinet, yet space accidents precipitate periods of deep national introspection.

When the line between policy and culture blurs, difficulties inevitably arise. To make space exploration entertaining, storytellers often bend events to fit salable story lines. They may simplify or alter reality. The resulting "infotainment" can attract large audiences, but it can also fail to capture the challenges that lie between culture and realization. Docudramas that twist facts to create interesting tales may mislead the public into believing that imagined events

The reality of space exploration, during the early years of the venture, differed considerably from the romantic vision offered by advocates of cosmic flight. Although much was accomplished, the results often surprised. Robotic expeditions to Mars, such as this 1997 *Pathfinder* mission, did not reveal the lush and habitable planet of imaginative lore, but a dry and frigid sphere. (NASA)

actually occurred. Stories that utilize stereotypes and familiar narratives may weaken the public's ability to make intelligent decisions about real challenges and genuine events.[6]

Through this cycle the vision of space exploration moves. In the beginning, the vision seems familiar, feasible, and desirable. It interlocks well with the American experience. The truth or validity of the vision is largely irrelevant to its undertaking, because no one knows for certain at that time how its pursuit will turn out. The power of the vision to enlist public support is determined largely through its compatibility with established cultural beliefs and its ability to attract a large audience. Entertainment and the collective presumptions about space exploration found in historical metaphors guarantee a receptive audience for the undertaking. They also assure that gaps between imagination and reality will arise. The truth often turns out to be different from what people imagine it to be.

One might think that the discomforting presence of reality intruding upon a wonderful vision would disappoint its creators. In fact, the opposite occurs. Activities in the modern world constructed on imaginary foundations do not spontaneously disappear when facts intrude in discomforting ways. If the space exploration experience is a guide, then efforts constructed on a sufficiently solid foundation of subjective understanding will persist even in the presence of adversity.

Some fifty years into the venture, the reality of space exploration differs considerably from the anticipating vision. Many objectives have been achieved, but not all in the anticipated way. Visionaries anticipated that winged spaceships would provide inexpensive and relatively safe access to space; NASA's reusable space shuttle failed to meet those goals. Dreamers prepared plans for large, rotating space stations that could serve as launching pads to the cosmos; the International Space Station is a microgravity research laboratory with a skeleton crew. Scientists hoped to find plants or primitive life forms on Mars; the first machines found a cold and sterile soil. The vision of space exploration proved harder to achieve than its promoters imagined; its pursuit led to discoveries that few people anticipated. Few engineers, as space travel began, anticipated the staggering cost and technical difficulties of building orbital space stations and flying reusable spacecraft. Few scientists foresaw the astonishing differences between the Earth, Venus, and Mars. Practically no one anticipated the capabilities that space robots would achieve after a half century of innovation. To paraphrase J. B. S. Haldane, space exploration turned out to be not only stranger than we imagined but sometimes stranger than we could imagine.[7]

Yet this reality has not deterred advocates of exploration. Not in space. Humans possess an impressive capacity for reshaping their lives and an astonishing capability for imagining how it might occur. Confronted by unforeseen difficulties, humans are prone to increase their efforts. Rather than abandon their beliefs, visionaries modify them.[8] If nature imposes cosmic speed limits on the velocity of propelled devices, then humans envision shortcuts through space and time. Perhaps wormholes will do. If Mars proves too cold and airless for human habitation, then humans envision methods for changing the local atmosphere. The process, called terraforming, is achieved by pumping greenhouse gases into the air.[9] Humans believe they have altered the atmosphere of their own planet, so why not achieve the same results on a nearby sphere?

Reconsideration and abandonment of efforts require an opposing sense of reality and a countervailing vision around which the opposition can rally. Alternatives to the dominant narrative of space exploration in America have been ill formed and their advocates poorly organized. In the absence of a persuasive alternative, space enthusiasts confronted with findings that do not fit the underlying vision tend to reinterpret the vision or modify it in nondestructive ways. The faithful do not abandon their beliefs just because their prophecies fail. Rather than acknowledge defeat, they work harder to make the original vision come true and recommit themselves to their original aspirations. In such ways, they move toward their vision. If new worlds do not fit old dreams, it is the worlds that tend to change, not the dreams.

The retelling of this fascinating process through the subject of space exploration demonstrates the power of human beings to reshape not only their subjective interpretations of reality but also the objective worlds on which humans live.

1

The Vision

Where there is no vision, the people perish.

—Proverbs 29:18

I n the spring of 1959, less than eight months after the creation of the National Aeronautics and Space Administration, an internal NASA committee chaired by Harry J. Goett met to chart a long-range plan for the human exploration of space. Committee members agreed on the overarching purpose of their endeavor. "The ultimate objective of space exploration," members concurred, "is manned travel to and from other planets."[1] They also agreed on the steps by which this would occur: the development of spacecraft capable of orbiting Earth; a laboratory or station in space; human expeditions to the Moon; and precursor missions to the planets conducted by robots in advance of human beings.

Later that year, NASA officials ensconced the recommendations of the Goett committee within the agency's first comprehensive long-range plan. While the document embraced a number of robotic and remotely controlled missions, the vision for the ensuing decades remained clear: "Manned exploration of the moon and the nearer planets must remain as the major goals."[2] Seventeen months later, President John F. Kennedy approved one of the most significant steps in this long-range plan with his 1961 decision to go to the Moon. Advocates of space exploration faithfully pursued the vision in the decades that followed. In 1989 and again in 2004, presidents George H. W. Bush and George W. Bush endorsed efforts that would return astronauts to the Moon and eventually send humans to Mars. In 2010, after a far-reaching review of the national space program, President Barack Obama reaffirmed the national commitment for a Martian expedition.

This approach to space travel, with its heavy reliance on human-piloted spacecraft, dominated the public vision. It became synonymous with space travel, so much so that other alternatives not only were dismissed but, in some cases, could not be imagined. To the dominant group of advocates promoting space exploration, space travel meant human flight. Advocates of alternative means contemplated a program focused on scientific investigation, carried out by machines or robots that would do the tough work of exploration while humans stayed behind. Their alternative was neither as well organized nor as adroitly presented as the dominant vision. Other alternatives were so exotic that they received no serious attention at all.[3] If technological civilizations engage in space exploration for sufficiently long periods of time (say millions of years), they may do so in forms that are unrecognizable to species just beginning the venture. Yet for earthlings contemplating the possibility of space travel in its initial stages, the vision of human travel to and from the Moon and inner planets dominated their collective imagination.

An impressive program of scientific discovery could have been organized around satellite technology, robotic probes, instruments in space, instruments on the ground, or the emerging science of remote sensing. Granted, such technologies were not well understood at the beginning of the space age, but neither were rocketry and human space travel. To conceive of space exploration in any other form required an alternative act of imagination. In this respect, the dominant vision of human space flight possessed one great advantage. It was, given the history of contemplation and the cultural traditions upon which that history drew, easier to imagine.

Proponents communicated the adopted vision of space travel so effectively that any other approach to cosmic investigation seemed inconceivable. They created a set of expectations so wonderful and exciting that the expectations became cultural norms—assumptions so deeply embedded in American society that their selection hardly needed validation.

For as long as humans have been able to conceive of heavenly bodies other than Earth, people have dreamed about traveling to alternative worlds and exploring them. Lucian of Samosata, one of the few Greek thinkers who believed in the existence of other worlds, describes an odyssey in which his characters travel to the Moon and encounter lunar inhabitants who move about on large, three-headed vultures. Seventeenth-century astronomer and mathematician Johannes Kepler wrote a fanciful work in which a journey to the Moon reveals that its craters are artificial constructions, built by the local inhabitants in order to provide shaded places from which they can escape the rays of the Sun. Public interest in stories about extraterrestrial travel flourished after Galileo Galilei revealed in 1610 that the traveling points of light in the night sky were planets like the Earth, though different in size. Acceptance of the plurality-of-worlds doctrine naturally encouraged people to ponder what humans might find if they journeyed to and explored these Earth-like realms.[4]

Rather than agonizing long over the means of transport to and from these extraterrestrial bodies, people who wrote such stories surmounted that obstacle with little more than an active imagination. Lucian transported the crew of his sailing ship via the power of a giant whirlwind. In his 1638 story *The Man in the Moon*, Domingo Gonsales (pseudonym of the English bishop Francis Godwin) relied upon a flock of wild swans capable of carrying large loads to transport him to the lunar surface. Cyrano de Bergerac wrote two seventeenth-century novels describing yet another voyage to the lunar surface. One of his methods of transportation depended on bottles filled with early morning dew, Bergerac observing that dew seemed to rise upward when exposed to the rays of the morning Sun. In 1775 Louis-Guillaume de la Follie produced levitation through electric power, which the hero of his novel generated by turning cranks and gears attached to two large sulfur balls. Joseph Atterley (actually American professor George Tucker) relied on an antigravity substance called *lunarium* to transport his crew to the Moon in 1827, as did H. G. Wells in his 1901 novel *First Men in the Moon*. The centrifugal force produced by a giant flywheel hurled the components of an Earth-orbiting station into space in Edward Everett Hale's 1869 novel *The Brick Moon*. In an influential 1897 novel, *Auf Zwei Planeten*, German author Kurt Lasswitz likewise employed a gravity-defying material to transport Martian explorers to Earth. Suspended in space at a point directly above the North Pole, the Martians' descent to the Earth's surface was expedited by the erroneous observation that descent would prove effortless because the planet does not move at the point of its poles.[5]

The most fantastic schemes for extraterrestrial travel, avoiding physical devices altogether, transported humans by telepathic means. In the fantastic *Somnium*, Kepler falls asleep and dreams that spirits can speed humans to the lunar surface through shadows of the Earth and Moon during eclipses of each. The dream method of transportation was also favored by Edgar Rice Burroughs, who in the early twentieth century thrilled a generation of young readers with an adventurous sequence of novels set on the planet Barsoom (the local name for Mars). The hero of his novels, John Carter, falls into a trance outside the mouth of an Arizona cave, awakening a moment later on that distant sphere.[6]

Such methods of transport required readers to suspend disbelief. Having overcome the limits of reason, readers could enjoy the contemplation of conditions on unreachable lands. With transport problems resolved, the main purpose of the story could proceed. Fortunately for the authors of such stories, a broader collection of travelers' tales had already set the precedent for fanciful methods of conveyance. The earthbound fantastic voyage had been a central part of Western literature ever since Homer wrote the *Iliad* and the *Odyssey* in the eighth century B.C. Stories depicting epic journeys on Earth provided a far more sturdy foundation for the contemplation of space travel than science fiction alone.

Never mind that many transportation schemes required a suspension of the laws of physics. Miraculous events are a staple fare in travelers' tales, in both the methods of transport and the nature of local conditions in the wondrous places the travelers visit. In Shakespeare's *The Tempest*, the exiled Duke of Milan uses magical powers to create a terrible storm that blows the King of Naples and his ship's crew onto the island where the play is set. The executor of this scheme, an airy spirit named Ariel, culls the king's ship out of its fleet and brings the vessel unharmed into a quiet cove. This magical tale provided the inspiration for the classic 1956 science fiction film *Forbidden Planet*.[7] Because Shakespeare's play challenges the audience's imagination, the audience is encouraged to contemplate the possibility that humans using their powers of imagination might learn how to control the weather and perform other seemingly miraculous events.

Like the characters in *The Tempest*, Lemuel Gulliver in *Gulliver's Travels* and the famous Robinson Crusoe both reach their destinations through terrible storms. In his classic children's story *The Lion, the Witch and the Wardrobe*, C. S. Lewis, who was also a writer of science fiction, has his characters find the land of Narnia by climbing through the back of a enormous wardrobe in a professor's country home. Alice travels by means of a rabbit hole and looking glass, and Dorothy commutes from her Kansas home to Oz in a ferocious tornado.[8]

Because no one has ever been to the places depicted in these tales, readers are allowed to exercise their imaginations and speculate on the plausibility of the wonders described within. No one has ever visited the mythical islands described by Homer in the *Odyssey* or the land of the Lilliputians in *Gulliver's Travels*. So far as we know, no one-eyed giants or diminutive people inhabit the islands of the Earth, but who can say with certainty that stranger creatures do not exist in places as yet unexplored?

Not constrained by the requirements of reality, the authors of such tales are free to embellish their stories with highly romantic themes. In a typical exploration fantasy, an earthly traveler is called upon to resolve a struggle between good and evil, or at least to demonstrate intellectual superiority among misguided people. After saving the Lilliputians from a naval invasion by neighboring Blefuscu, Lemuel Gulliver proceeds to rescue the conquered nation from slavery at the hands of the Lilliputian emperor. Four human children help rescue the inhabitants of Narnia from the tyrannical rule of the evil White Witch in *The Lion, the Witch and the Wardrobe*, a story full of magic, including the resurrection of the good lion Aslan from the dead. Like explorers on real terrestrial expeditions, Odysseus is called upon to overcome unimaginable hardships on his ten-year journey home in the *Odyssey*. For centuries, explorers in fact and fiction have been treated like outstanding athletes, capable of superhuman acts of strength and bravery. The romantic quality of these stories invites readers to hope that at least some of the elements in travelers' tales might be true.

During the late nineteenth and early twentieth centuries, a number of individuals, motivated by stories such as these, sought to take romantic fantasy and make it happen. Sometimes these individuals prepared additional works of imagination as a means of promoting their cause. Until the mid-nineteenth century, nearly all extraterrestrial tales employed exploration methods that were physically impossible. Inevitably, science and invention exposed the more fanciful assertions as the falsehoods they were. Scientific observations, for example, confirmed that the Moon had no atmosphere, upsetting the plot lines in a number of travelers' tales. For the most part, this did not bother the audience for such stories, as most readers were pleased simply to be entertained. A growing number of individuals, however, conscious of advances in science and technology, found these shortcomings troublesome, as standing in the way of actually completing the fantastic voyages described in the tales. Fascinated by the prospect of space flight, they identified inaccuracies and attempted to correct them.

The so-called fathers of modern rocketry—Robert Goddard, Hermann Oberth, and Konstantin Tsiolkovsky—were all inspired by errors of imagination. "My interest in space travel was first aroused by the famous writer of fantasies Jules Verne," noted Tsiolkovsky. "Curiosity was followed by serious thought."[9] Among the outpouring of fictional works on space travel during the mid-nineteenth century, the work of the French writer Jules Verne was distinct. In his early twenties, struggling to establish himself as a successful writer in mid-nineteenth-century Paris, Verne developed the technique of inserting scientific explanations into simply told travelers' tales, enabling readers to acquire scientific information while enjoying an otherwise romantic adventure. By the time he released his first spacefaring story, Verne had already produced novels on crossing Africa by balloon, journeying to the center of the Earth, and exploring the North Pole. In 1865 he published *De la terre à la lune* (*From the Earth to the Moon*), a sequel, *Autour de la lune* (*Round the Moon*), appearing five years later. Verne took special care to make his stories appear as plausible as possible. In the two lunar adventures, he relied upon mathematical formulas to calculate the velocities and transit requirements necessary to carry his spacefarers toward the Moon. He provided them with devices for the replenishment and purification of air and even stumbled across the use of rocket power as a means of correcting the trajectory of the space capsule during the voyage. The actual means of departing the Earth, however, stumped Verne. Lacking an appropriate technology to propel his travelers to an escape velocity, Verne fell back upon the entertaining but physically implausible scheme of shooting the space capsule out of an enormous cannon.[10]

By 1875 Verne's lunar adventures had been translated into Russian. Konstantin Tsiolkovsky, born in 1857, had by this time moved from his village home to Moscow, where he struggled to educate himself at the Chertovsky Library. An earlier bout with scarlet fever had left him nearly deaf, a condition that caused Tsiolkovsky to withdraw from formal schooling and isolate himself in a world of books and dreams. "The idea of travelling into outer space

French novelist Jules Verne helped inspire serious investigations of space flight by injecting scientific information into otherwise romantic adventure tales. In his 1870 story *Autour de la lune* (*Round the Moon*) he inadvertently stumbled upon the use of rocket power as a means for controlling the trajectory of his space capsule on its route toward the Moon. (NASA)

constantly pursued me," he later wrote. Like many people of his day, Tsiolkovsky struggled to identify a realistic method of space travel. He toyed with a number of propulsion schemes, including centrifugal force, perpetual motion, and antigravity devices, and wrote science fiction stories in an effort to popularize his ideas. After repeated mathematical calculations, Tsiolkovsky settled on the use of multistage rockets, a conclusion explained in his "Investigation of Universal Space by Means of Reactive Devices," the first parts of which appeared in 1903. The treatise was a serious work of imagination itself, introducing readers who could obtain copies of this obscure work to orbiting space stations and methods for colonizing the cosmos.[11]

Hermann Oberth was born more than a generation later, in 1894, in Transylvania, a site more famous for its vampire tales than space fantasies. While attending the Bischof-Teutsch Gymnasium in his hometown of Schässburg, the young Oberth also discovered the novels of Jules Verne. He read Verne's lunar adventures "at least five or six times and, finally, knew [them] by heart."[12] Abandoning his original intent to follow his father into medicine, Oberth instead pursued his boyhood interest, earning advanced degrees in mathematics and physics and writing as his doctoral dissertation a treatise on the problem of extraterrestrial flight. Like Tsiolkovsky, Oberth adopted rocket propulsion as a substitute for Verne's cannonball approach. He arranged for publication of his dissertation by Rudolf Oldenbourg, a Munich publishing house. Once past the lengthy presentations on launch angles and trajectories, propellants, and valves, readers of *Die Rakete zu den Planetenräumen* (The Rocket into Planetary Space) were treated to an imaginative discussion of a human expedition and establishment of an Earth-orbiting space station. The book was an immediate success, moving through three editions between 1923 and 1929. To help promote his ideas, Oberth traveled to Berlin to provide scientific advice for a film on space flight being produced by Fritz Lang. The 1929 movie, *Frau im Mond* (released in the United States as *By Rocket to the Moon*) is generally considered to be the first realistic treatment of space travel in cinematic form. It depicts a number of principles set forth in Oberth's book and utilized in actual flight forty years later, including the rollout of a multistage rocket ship from a large vehicle-assembly building.[13] The movie allowed Oberth and his associates to spread the gospel of space flight among the German intelligentsia. As one of Oberth's disciples explained, "The first showing of a Fritz Lang film was something for which there was no equivalent anywhere. . . . It is not an exaggeration to say that a sudden collapse of the theater building during a Fritz Lang

Works of imagination often motivated the pioneers of space flight to undertake their studies. The Russian space advocate Konstantin Tsiolkovsky wrote science fiction stories, and German rocketeer Hermann Oberth, pictured here, read Jules Verne's space adventures so frequently that he knew their details and deficiencies "by heart." Tsiolkovsky and Oberth both settled on the use of rocket power as the most reliable means of propelling humans beyond Earth. (NASA)

premiere would have deprived Germany of much of its intellectual leadership at one blow."[14]

Like his European counterparts, Robert Goddard's interest in rocketry was motivated by works of imagination. According to his wife, Goddard read *From the Earth to the Moon* several times and absorbed himself by penning comments and corrections in the margins.[15] As a young person, Goddard read a serialized version of *War of the Worlds* in the *Boston Post*, which, he said, "gripped my imagination tremendously."[16] In his autobiography, Goddard recalls climbing a backyard cherry tree and dreaming of a voyage to Mars, an event he quietly celebrated thereafter as his "anniversary day."[17]

Goddard believed that in order to escape the inevitable alteration of the Sun, humans would eventually leave Earth and migrate to planets in other solar systems. "The navigation of interplanetary space must be effected to ensure the continuance of the race," he wrote at the age of thirty. He further believed that the Moon could be used as a launching pad for trips into interplanetary space and contemplated the construction of space factories on the Moon and planets designed to produce rocket fuels.[18] Although imagination inspired him to think wondrous ideas, it rarely caused him to promote them. Owing to personal shyness and a discomforting encounter with the national press, Goddard was notoriously reluctant to advance his vision of space flight in public.

In 1919 Goddard, then thirty-seven years old, published *A Method of Reaching Extreme Altitudes*, a treatise on the use of rockets to propel objects into space. Sickly and reclusive, he had spent most of his life in Worcester, Massachusetts, where he earned three college degrees, accepted a teaching position, and experimented with rocket propulsion. The pamphlet, published by the Smithsonian Institution, is a rather sober discussion of the mass required to propel objects to altitudes beyond those attained by high-altitude balloons. Toward the end of the pamphlet, Goddard advanced a proposal from an earlier paper he had prepared on "The Navigation of Interplanetary Space" while recuperating from a bout with tuberculosis in 1913. He suggested that a small rocket of sufficient power (without humans on board) could be launched from Earth's surface and travel to the Moon, where its collision with that body would be confirmed by the ignition of a pyrotechnic device bright enough to be viewed through telescopes on Earth.[19]

Public reaction to *A Method of Reaching Extreme Altitudes* caused Goddard to recoil from further promotional ideas. His proposal for a "moon rocket" was lampooned in the national press, where it was front-page news. Editors at the *New York Times* pointed out (incorrectly) that a rocket would not work in a vacuum because it would have nothing "against which to react." Goddard knew this to be false, having already produced rocket thrust in vacuum tubes. The sensational national publicity distressed Goddard, and, as historian Frank Winter has observed, he "became even more press-shy and secretive."[20] His future writings concentrated on turbines, pumps, stabilization, and fuels. In 1926, from a field near Auburn, Massachusetts, he completed the first successful launch of a liquid fuel rocket, an event of abundant importance. Goddard requested that the Smithsonian Institution, his main source of financial support, not publicize the results. Four years later, in an effort to avoid further publicity and seek better weather, he moved his launch operations to a remote area near Roswell, New Mexico.

Tsiolkovsky, Goddard, and Oberth all turned to rocket power as the most efficient means of producing energy sufficient to achieve the velocities re-

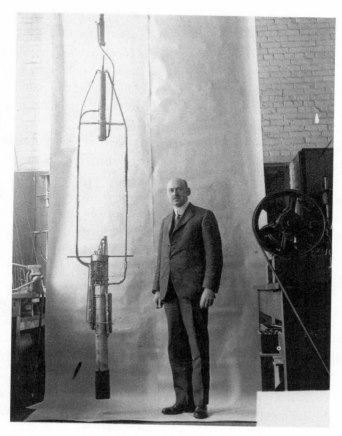

The writings of Jules Verne and H. G. Wells inspired America's leading rocket scientist, Robert Goddard, to undertake his work. In his autobiography, Goddard remembers climbing a backyard cherry tree and dreaming of a voyage to Mars, an event he subsequently celebrated as his "anniversary day." In 1926, from a field near Auburn, Massachusetts, Goddard became the first person to successfully launch a liquid-fuel rocket. (NASA)

quired to escape the gravity of Earth and steer transit craft through space. In developing the science of rocketry, each made profound contributions. In shaping popular visions of space travel in America, their contributions were most unequal.

Tsiolkovsky's work had practically no effect on the public conception of rockets and space travel in mid-twentieth-century America. He was an obscure figure within Russia, and even less well known outside. Although Goddard's work was widely recognized within U.S. scientific circles, his refusal to partici-pate in even the most elementary promotional efforts assured that his influence on popular conceptions would be nearly as slight as Tsiolkovsky's. As strange as it might seem, Oberth had the greatest influence on the public image of space travel, although it arrived in the United States indirectly, through the work of his followers in the German Rocket Society.

The absence of any eminent U.S. scientist to serve as a spokesperson for the awakening space movement left the American wing in the hands of two groups devoutly committed to the romantic version: German expatriates and science fiction fans. Both found reinforcement for their astonishing views within newly organized rocket societies. Responding to the increasing volume of extraterrestrial narratives, both fictional and nonfictional, partisans of space flight throughout Europe and the United States founded societies devoted to rockets and space travel. They were inspired by an explosion in the quantity of science fiction as well as the appearance of serious investigations into the dynamics of rocketry. Tsiolkovsky's followers founded the Society for Inter-planetary Travel in 1924, which collapsed less than one year later because of internal disputes. The Verein für Raumschiffahrt, popularly known in the West as the German Rocket Society, began its work in 1927. The American Inter-planetary Society (later renamed the American Rocket Society) was founded in 1930 in New York City, and the British Interplanetary Society (BIS) appeared in 1933. Similar groups arose in Austria (1926), Canada (1936), and France (1938).[21]

Leaders of the rocket societies steadfastly committed themselves to the at-tainment of human space flight, a dedication that went considerably beyond the general curiosity about astronomy prevalent at that time. Using the tech-nology of the telescope, astronomers and their popularizers beginning in the nineteenth century had generated substantial public interest in the nature of cosmic phenomena. Books and magazines regularly reported the latest find-ings, and by the 1930s associations devoted to the investigation of astronomi-cal phenomena were commonplace in Europe and America. The Société As-tronomique de France was formed in 1887 and the American Astronomical Society in 1899.

The rocket societies distinguished themselves from the astronomical groups by promoting human space flight as the primary means of cosmic investiga-tion. The founders of the American Interplanetary Society announced in the opening sentence of their first bulletin that their principal aim was "the promotion of interest in interplanetary exploration."[22] The official statement announcing the formation of the society set out the philosophy in more detail: "It is our intention to build this society into a national organization with financial and other resources of such importance we can offer real inducement and stimulation to American scientists . . . in the development of rockets, rocket cars and other proposed methods of traveling in space and communicating with the planets."[23] Space travel in the context of the times meant travel by human beings. Writing about the creation of the German Rocket Society, of which he was a founding member, the prolific science writer Willy Ley explained that its creators were "dedicated to promoting the idea of space travel." That "driving ambition," Ley said, served as the inspiration for their work.[24] The scientific term used to describe the investigations of the rocket societies, invented by a French science fiction writer, was *astronautics* (navigating the stars).[25]

Leaders of the rocket societies advanced a vision of remarkable power. Humans, they said, would carry out expeditions of discovery in space as ambitious as those of earlier explorers on Earth, maintaining the spirit of adventure and discovery those individuals had inspired. Scientists would build large rocket ships capable of transporting people into space, and once in space, humans would construct space stations and assemble spacecraft capable of voyages far beyond Earth. Expeditions to the Moon and nearby planets would ensue, followed by permanent bases and settlements. Explorers would probe the mysteries of the universe, locate strange creatures, and make miraculous discoveries. It was a remarkable vision, powerfully attractive to people so inclined to believe.

It was also at that time wholly fantastic. Practical experience with the science of rocketry during the 1930s was primitive at best, and the production of a workable vehicle capable of launching even the smallest object into space lay a quarter century away. The lack of practical experience, however, did not deter visionaries; to the contrary, it encouraged them to make bolder claims.

In the late 1930s members of the British Interplanetary Society formulated plans for a rocket trip that would land humans on the Moon and return them safely to Earth. The members were utterly serious and developed a number of technical devices in pursuit of their goal, such as plans for a "coelostat" to assist in navigating the spacecraft along its path, a "carapace" to protect the crew from atmospheric heating, and fall-away rocket tubes that could be jettisoned once the ship had spent its solid fuel. Various details appeared in two issues of the *Journal of the British Interplanetary Society* in 1939 and attracted widespread attention in the British press. Arthur C. Clarke, a twenty-one-year-old amateur astronomer and civil servant, calculated the escape velocities. That same year, in one of the first publications in what would be an extraordinarily prolific writing career, Clarke published an article titled "We Can Rocket to the Moon—Now!"[26]

The German Rocket Society was the most influential of the advocacy groups. Its members formed the club for the purpose of implementing the concepts contained in Oberth's book on rocket travel. Many rocket clubs conducted field tests; for the Germans, however, this was more than an exciting hobby. By 1932 members of the German Rocket Society had conducted eighty-seven flights and more than 270 static tests with real rockets. Experimenters had made impressive progress with the mechanics of liquid-fuel propulsion and the techniques of regenerative cooling. Beginning that year, the first of several members went to work for the German army, where they eventually produced the world's first large liquid-fuel launch vehicle, the V-2. After World War II, 125 members of the German rocket team emigrated to the United States, forming the nucleus of the group that built the large rockets used to propel the first Americans to the Moon.[27] Of equal importance to their technical accomplishments was their ability to promote the space flight dream. The Germans produced not only better rockets but also experts better able to communicate a technically sophisticated vision of space flight. After moving to the United States, two of the Oberth's followers, Willy Ley and Wernher von Braun, became the principal spokespersons for the spacefaring movement.

Ley was born in Berlin in 1906. His father was a wine merchant, and his mother the daughter of an official in the German Lutheran Church. As a student in Berlin and East Prussia, Ley developed a broad-ranging interest in paleontology, astronomy, and physics. In 1926 he obtained a copy of Oberth's treatise on space flight, which excited Ley so much that he gave up his plans for a career in the earth sciences and devoted himself to the promotion of space flight. Ley had an unusual proclivity for communication, a talent he put to use as a lecturer and writer of popular science books. At the age of nineteen, he rewrote Oberth's book in a style compatible with popular consumption. The following year he helped organize the German Rocket Society and accepted the position of vice-president (Hermann Oberth agreed to serve as president), from which Ley helped build the membership to more than one thousand, with sufficient funds to operate a rocket proving ground with its own staff of mechanics and engineers on the outskirts of Berlin.[28]

When the National Socialist Workers' Party took control of Germany in 1933, the work of the German Rocket Society shriveled, due in part to a lack of funds and the increasing suspicions of the new government. "The value of the sixth decimal place in the calculation of a trajectory to Venus interested us . . . little," said the army captain placed in charge of developing long-range rockets for the German military.[29] Some of the society's members migrated to the government's military rocket development program at Peenemünde, a remote island in the Baltic Sea. Ley fled to the United States. In 1935, with help from friends in the American Rocket Society (it had changed its name the previous year), Ley settled in New York City, where he eked out a living writing articles and books on zoology. As V-2 rockets began to fall on London and other targets in the fall of 1944, Ley returned to his primary interest, publishing *Rockets: The Future of Travel beyond the Stratosphere*. The book was an instant success. Ley's ability to link the development of rocketry to space travel established him as the primary popularizer of space flight in the United States at that time.

Like others before him, Wernher von Braun became involved in rocketry as a result of reading works of imagination. Recalling his reaction as a young man to one such story, von Braun said, "It filled me with a romantic urge. Interplanetary travel! Here was a task worth dedicating one's life to! Not just to stare through a telescope at the moon and the planets but to soar through the heavens and actually explore the mysterious universe! I knew how Columbus had felt."[30] Introduced to members of the German Rocket Society at the age of eighteen by Willy Ley, von Braun quickly ingratiated himself with the society's top staff. A talented engineer, von Braun was also remarkably charismatic and a relentless promoter of his own career. That summer he helped launch small, liquid-fuel rockets from the society's Raketenflugplatz testing grounds in northern Berlin. Two years later, having completed work on his bachelor's degree, von Braun went to work for the German army's new rocket program. The army captain who recruited the twenty-year-old von Braun commented on "the energy and shrewdness with which this tall, fair young student with the broad massive chin went to work."[31]

Von Braun played a key role in the development of the German rocket center at Peenemünde. He was technical director for research and development within an overall organization that by the war's end had produced more than six thousand V-2 rockets, twenty-nine hundred of which fell on targets in Great Britain and the Continent. As the Third Reich collapsed, von Braun led the nucleus of his rocket team to a small village on the Austrian border so that they could surrender to the U.S. Army. Interrogators were amazed at the assertion that this thirty-three-year-old had developed the V-2 rocket and had even prepared plans for a missile that could reach New York.[32]

As a group, the Germans were no more technically adept and no more obsessed by nationalistic concerns than their counterparts in other countries, nor did they possess a deeper tradition of science fiction. What distinguished the leading members of the German Rocket Society was their unusual aptitude for communicating the gospel of space flight. The society's top engineers and scientists were drawn from intellectual classes whose members were accustomed to communicating with social and political elites. Oberth's father was a physician, Ley grew up in middle-class Berlin, and von Braun's father was a Prussian aristocrat, the Baron Magnus von Braun, a landowner in the province of Silesia and an important public official in pre-Nazi governments. Von Braun's mother was a well-educated woman from the Swedish-German aristocracy with strong interests in biology and astronomy.[33]

Once in the United States, Ley and von Braun found themselves isolated. "No one had the slightest interest in the subject [of space travel] except science fiction magazines," Ley complained after his arrival in the United States in 1935.[34] A polyglot of science fiction writers had begun meeting some five years earlier at the Manhattan apartment of G. Edward Pendray, a reporter for the *New York Herald Tribune*. Pendray, a strange-looking man with a pronounced goatee, had come to New York from Wyoming in 1925 (he was born in Nebraska in 1901) and was something of a feature in the local literary scene, having penned a number of science fiction stories under the pseudonym of Gawain Edwards. His wife, Leatrice Gregory, was a widely syndicated newspaper columnist.

The writers who met at Pendray's apartment were active contributors to *Science Wonder Stories*, one of the cheap pulp-paper magazines that specialized in chimerical adventure tales. "The most popular theme of science fiction," Pendray explained, "was interplanetary (or interstellar) travel." Being writers rather than engineers, their "imaginations could outpace dull practical considerations with a velocity comparable to that of light."[35] At one meeting in the spring of 1930, a group of twelve attendees formed the association designed to promote their fantastic schemes. The "principal moving spirit," according to

Pendray, was David Lasser, managing editor of the magazine in which their stories appeared. A Baltimore native, Lasser had studied engineering at the Massachusetts Institute of Technology from 1920 to 1924. After a brief career as an engineer and technical writer for a variety of New York firms, he signed on as managing editor for Gernsback Publications. The following year, he helped found the American Interplanetary Society and was elected its president. Pendray became vice-president, while Ms. Gregory served for a time as the society's librarian and sewed parachutes for use in the group's experimental rocket program.[36]

As a consequence of these developments, the principal explication of space flight in the United States during the twenty-five-year period leading up to the beginning of the space age fell to people with a highly romantic view. Pendray became the principal spokesperson for the group after Lasser left Gernsback Publications in 1933 for a career in the trade-union movement. Pendray's position was gradually supplanted by that of Ley and von Braun, who added technical credibility to the fantastic schemes.

As if to merge the two trends, von Braun began work on a novel describing an imaginary mission to Mars, completing it while helping to develop rockets for the U.S. Army after World War II. The novel did not enlist a commercial publisher, but the appendix to the novel, which outlined the technical requirements for the voyage, did. The ideas contained in *The Mars Project* were part of an increasingly public campaign that pushed von Braun to the forefront of the effort to realize the dreams of lunar and planetary pioneers.[37]

One additional factor helped reinforce the magnetism of the spacefaring dream. The founding of the rocket societies and the promotion of their point of view occurred at a time when the last great era of terrestrial exploration seemed to be closing. Advocates of interplanetary flight used popular interest in terrestrial exploration, along with the myths that had grown up around it, to promote space travel. The idea that the traditions of terrestrial exploration could continue in a new realm attracted many people to the cause.

To the public, it seemed that few areas worthy of exploration remained on the surface of the Earth. The golden age of polar exploration ended, by most accounts, in 1929, with the first airplane flight over the South Pole. The search for the great Northern Passage had ended a quarter century earlier, marked by the completion of Roald Amundsen's tortuous three-year expedition through the ice-bound seas of northern latitudes, concluding in 1905. Completion of the Yellowstone expedition of 1870 and John Wesley Powell's navigation of

Because Robert Goddard was notoriously shy, promotion of space flight in America fell to science fiction fans. One of the leading promoters, Edward Pendray (*right*), was a reporter for the *New York Herald Tribune* who wrote science fiction stories on the side. A collection of science fiction writers gathered in Pendray's Manhattan apartment in the spring of 1930 to organize the American Interplanetary Society. (National Air and Space Museum)

the Grand Canyon in 1869 seemed to finish more than three hundred years of major North American expeditions. The heroic age of African exploration had drawn to a close with confirmation of the source of the Nile in 1876 and completion of the full-scale expedition by Henry Morton Stanley down the Congo River in 1877. The golden age of Alpine mountaineering had ended with the conquest of the Matterhorn in 1865. The science of discovery did not cease with the conclusion of these events, but the ability of newspaper owners and other publishers to generate popular interest in new expeditions declined.[38]

Terrestrial expeditions such as these became a form of mass entertainment in the hands of the publishing industry. Stories of exploration were used to sell newspapers. To boost circulation, the editors of the *New York Herald* dispatched the journalist Henry Morton Stanley to lead and report on an expedition to Central Africa searching for the British missionary David Livingston. Public interest in expeditions and the reports from them were also used to raise funds. Early members of the National Geographic Society, who numbered barely two hundred, decided in 1888 to issue a popular magazine in order to spur interest in their affairs. The rise of the so-called pulp magazines, generally marked by the appearance of Frank Munsey's *Argosy* in 1896, was due in large part to the general interest in adventure stories set in foreign lands. The public desire to personally, and safely, experience the great expeditions helped to create interest in travel and to promote the family vacation, a trend promoters of the national park movement and their allies in the American railroad industry effectively used.[39] Whether the last great era of exploration had actually closed was of less importance than the dwindling supply of mysterious terrestrial lands in which to set entertaining tales.

As a form of mass entertainment, expeditions were mythologized in ways that conveyed a romantic view of the world and the place of humans within it. Recounted in lectures and the popular press, exploration tales became a vehicle for expressing the beliefs of the times, validating, by comparison to practices in foreign lands, faith in the superiority of Western institutions and values. Expeditions also helped reconfirm the belief that humans were capable of making great sacrifices and enduring extreme hardships in order to achieve presumably great ends.

Proponents of space travel suggested that interplanetary travel would continue the adventure of exploration in an endless realm. Said Ley in introducing the story of space flight to readers of the first edition of *Rockets*:

It is the story of a great idea, a great dream, if you wish, which probably began many centuries ago on the islands off the coast of Greece. It has been dreamt again and again ever since, on meadows under a starry sky, behind the eyepieces of large telescopes in quiet observatories on top of a mountain in the Arizona desert or in the wooded hills near the European capitals. . . . It is the story of the idea that we possibly could, and if so should, break away from our planet and go

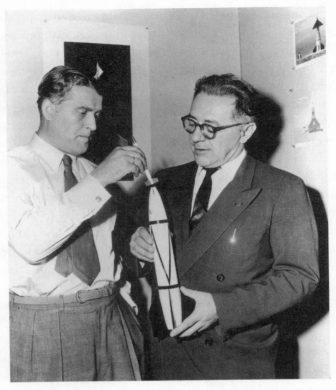

After immigrating to the United States, Willy Ley (*right*) found the American space-flight movement firmly in the hands of science fiction writers. In his native Germany, Ley had helped found the influential German Rocket Society. When he was joined after World War II by German rocket engineer Wernher von Braun (*left*), the two visionaries became the most prominent advocates in America for human space flight. (National Air and Space Museum)

exploring to others, just as thousands of years ago men broke away from their islands and went exploring to other coasts.[40]

Philip E. Cleator, founder of the British Interplanetary Society, wrote in 1936 that "the spirit of adventure, the lure of the unknown, will attract man to Mars just as surely as they caused him to penetrate into the wilds of Africa and the solitude of the Polar regions."[41]

By introducing space travel as an extension of terrestrial exploration, advocates of space flight found themselves promoting the traditional image of "small ships and brave men" sailing off into the unknown. Given the elementary state of space flight technology at that time, it was hard to imagine cosmic exploration occurring in any other way. When members of the British Interplanetary Society prepared their plans for an expedition to the Moon, they assumed that the rocket ship and its crew would not be able to communicate directly with people back on Earth, a presumption derived from the seagoing expeditions of centuries past. BIS members toyed with schemes such as a flashing light to signal the arrival of the crew on the Moon.[42] The image of spacefarers as lonely men (and, occasionally, a woman) cut off from terrestrial contact persisted through the 1950s.[43]

An essential ingredient of any motivating vision, whether terrestrial or otherwise, is the ability to imagine distant events and places without the benefit of knowing exactly what is going on there. This allows people to transfer diverse expectations to the vision, drawn from their own cultural beliefs, precisely because the real circumstances are unknown. The illusion of a Great Southern Continent, excited by Marco Polo's 1477 report of a land, rich in wealth, too far south for the Great Khan to conquer, was, in the words of Daniel Boorstin, "embellished precisely because it could not be disproved."[44] Once experience reveals the true state of nature, expectations invariably fall, but the underlying vision rarely dies. Rather, people update the vision. The dream moves on.

Myths supporting the great earthbound expeditions had fallen under the weight of frequent visitation to the places imagined. Beliefs about the Great Southern Continent collapsed after explorers discovered Antarctica to be a cold and barren land. The South Pacific was not an earthly Eden, free from Christian notions of sin, a myth that persisted well into the twentieth century, and Africa was not the "dark continent," the source of animal instincts and original sin, any more than other terrestrial locations.[45]

Popular conceptions of exploration, however, fulfill important human needs. Old myths die hard. As each of the old frontiers yielded to investigation, humans transferred their expectations to new realms. Interest in Alaska, the polar regions, and the Himalayas kept the promise of human exploration alive through part of the twentieth century. Before long, however, romantic images of even those places wore away. Outer space appeared just in time to maintain the dream. As the middle of the twentieth century approached, the vision of human exploration had nowhere to go but up.

Having formulating a dominant vision, advocates of space exploration faced a daunting task. They had to find a group of people willing to finance their schemes, and, in spite of the occasional fantasy to the contrary, that proved extremely difficult. Money that appeared in fiction vanished in the real world. (In *Frau im Mond*, industrialists hand out money for a lunar expedition in the belief that the Moon is rich in gold. The intrepid explorers do not disappoint their backers, for the expeditionaries, once on the Moon, discover a cave full of the precious metal.) Sensible advocates recognized the need for government support: only governments, with their extensive treasuries, possessed the resources necessary to implement elaborate space plans. Placing space exploration on the public policy agenda, however, was as hard as actually traveling to the Moon. In the beginning, the public did not believe that human space flight was real; political leaders knew it was expensive; and military leaders, who had the funds to lavish on rocketry, were more interested in practical applications than in exploring the galaxy. In the United States, where exploration advocates encountered especially high levels of disbelief, converting popular fantasies into real policies required a concerted, well-planned campaign.

2

Making Space Flight
Seem Real

Man will conquer space *soon*. What are we waiting for?
—*Collier's*, 1952

n late 1949 George Gallup conducted a poll in which he asked
Americans to imagine what sort of scientific developments
would take place by the year 2000. Eighty-eight percent of re-
spondents believed that a cure for cancer would be found within
the next fifty years. Sixty-three percent envisioned a future in
which trains and airplanes would be run by atomic power. When
asked whether "men in rockets will be able to reach the moon
within the next 50 years," however, doubters prevailed. In spite
of the outpouring of books and articles on space exploration
during the previous two decades, only 15 percent of those polled
believed that this reality would occur.[1]

Imagination matters when societies contemplate new ventures. People must have the ability to visualize a solution to the
phenomenon with which the society grapples and possess confidence in the attainability of the goal. In his excellent work on
the origins of the U.S. space program, Walter McDougall pays
homage to the importance of imagination in determining the

shape of America's response to the challenge of space travel. He identifies three forces that were required to get the space program under way: an economy rich enough to afford this expensive endeavor; the appropriate technology, particularly the development of rocketry; and imagination, or what McDougall calls "culture, the realm of symbolism." Once these forces started "pushing in the same direction," McDougall argues, a large, government-supported space program "was automatic."[2]

By 1949, when Gallup tested the belief in space flight, the United States hardly lacked a popular culture devoted to human space flight. Opinion leaders, however, had done little to eliminate public disbelief. Far more than in Europe, where engineers and rocket scientists played a leading role in the promotion of space travel, spacefaring in America remained the province of people interested in fantastic tales. Science fiction played such a large role in the American image of space travel that public skepticism remained high even after rocketry became a practical science. To the American public, space travel was intriguing but infeasible. Before government support for actual space missions began, advocates had to convince Americans that space flight was possible. This required feats of imagination as impressive as those contained in fictional tales.

Fantastic stories had dominated American interest in extraterrestrial phenomena for some time. In 1835 the *New York Sun* had titillated readers with reports that the famous astronomer Sir John Herschel had observed a large number of creatures on the surface of the Moon through a specially constructed telescope. Herschel was a real person, who had actually arrived in Cape Town to make astronomical observations of the southern sky. In a series of stories, which were wholly fictional, readers of the *Sun* learned that Herschel had located moon bison, bat-persons, and a unicorn.[3] Beginning in 1869, Edward Everett Hale had serialized a fictional account of an Earth-orbiting space station in the *Atlantic Monthly*. He called it "The Brick Moon," a reference to the material out of which the orbital facility was constructed.[4] By the 1870s Jules Verne's book *From the Earth to the Moon* had become available in the English press. Two decades later the English writer H. G. Wells published *War of the Worlds*, serializing the story in the American magazine *Cosmopolitan*. At the Pan-American Exposition in Buffalo in 1901, visitors could take a simulated trip to the Moon in a winged spacecraft. The ride reappeared at Coney Island the following year, where its popularity inspired promoters to build

Luna Park, a rival amusement facility.[5] By one estimate, sixty million visitors passed through the turnstiles at Luna Park during its first five years.[6]

Fans of space fantasy could expand their imaginations by reading dime-store magazines, an important medium for transmitting popular culture in early twentieth-century America. The American publisher Frank Munsey turned *Argosy* into a cheaply priced, pulp-paper magazine in 1896. *Argosy* carried adventure stories about the West, Africa, sea travel, and an increasing number of science fiction tales.[7] In 1926 Hugo Gernsback launched the first pulp magazine entirely devoted to science fiction, *Amazing Stories*. It spawned a series of competitors, including *Astounding Stories*, and ushered in what many have called the golden age of science fiction in the United States.[8]

As the market for science fiction in the pulps expanded, so did the interest of Hollywood film producers. Swashbuckling Flash Gordon jumped from the comic strips to movie matinees in 1936, followed shortly by Buck Rogers, who made a detour through the radio waves on his way to the silver screen.[9] Television followed. Producers of the new medium began offering a steady fare of space cadets to youthful viewers in 1949 with the premiere of *Captain Video*. *Buck Rogers*, *Tom Corbett*, and *Space Patrol* premiered the following year.

By 1949 science fantasy had developed a loyal following, and images of human space travel had become part of American popular culture. Much of the popular culture of space flight, however, was meant to be fanciful, not real. Visitors to Coney Island who rode the Trip to the Moon disembarked from a spaceship to be greeted by dancing Moon maidens and costumed midgets handing out samples of green cheese.[10] It is said that the producers of the wildly popular Captain Video confined their weekly budget for special effects to twenty-five dollars, producing such comical devices as the Captain's "opti-con scillometer" and "cosmic ray vibrator."[11] E. E. Smith, a food chemist and part-time writer of science fiction, created the "space opera," galactic adventures featuring fleets of starships from which space cadets used exotic technologies to enforce Judeo-Christian ethics on villainous aliens. Smith's *Skylark of Space*, which he had begun writing in 1915, appeared in *Amazing Stories* in 1928.[12] Writers of science fiction and fantasy did not represent outer space as it actually existed any more than writers of Westerns accurately sought to portray the American frontier. Space was not meant to be real. Rather, it was meant to be entertaining. Science fiction writers used space as writers of Westerns used the American frontier—as "a vast colorful backdrop against which

CRUCIBLE OF POWER   **By Jack Williamson**

Members of the public remained skeptical about the possibility of space flight, relegating it to the realm of fantasy, the domain of pulp fiction magazines such as this 1939 issue of *Astounding Science-Fiction*. Americans placed more faith in a future attended by atomic-powered airplanes and cancer cures than flights to the Moon, according to a 1949 Gallup poll. (Copyright © 1939 by Street & Smith Publications, Inc.; reprinted by permission of Dell Magazines, a division of Penny Publications) any kind of story could be told," a philosophy that dominated space fiction for most of its history.[13]

As a consequence, the principles of space travel were not well understood among the public at large. In the case of other new technologies, such as airplanes and automobiles, people could witness actual operations firsthand. Although outrageous claims were made about the future of those technologies, those claims at least had their feet on the ground. With space travel, however, excepting the Europeans upon whom military rockets fell, most people had to experience the new technology vicariously. This gave the purveyors of popular culture considerable freedom in stretching technology to fit literary needs. Fantasy overwhelmed fact, and public understanding suffered as a result.

The golden age of science fiction, with its familiar aliens and galactic travel, did little to dispel disbelief among the public at large. Early science fiction, in the main, was not very scientific. The core of devoted readers remained clubby and small. Many people were exposed to science fiction, but few were so devoted as to acquaint themselves with the scientific facts involved. The 15 percent who responded affirmatively in 1949 to George Gallup's question about lunar expeditions undoubtedly exceeded the total population of science fiction devotees.

Faced with an extraordinary level of public skepticism, space boosters decided to deliver a more realistic message. David Lasser, one of the founders of the American Interplanetary Society, published the first serious treatment of rocketry and space travel for the general U.S. public in 1931. Lasser explained how developments in the science of rocketry would make possible human travel to the Moon, Venus, and Mars. Through his book, Lasser hoped that "the mists of misunderstanding, ignorance, and prejudice that surround the 'interplanetary rocket' question may be cleared up."[14] Both Lasser and his associate G. Edward Pendray sought to awaken public interest in the possibility of space flight.[15] In 1932 the *Brooklyn Daily Eagle* announced that Pendray would broadcast a talk on rocketry and space flight over W2XAB, the Columbia Broadcasting System's experimental television station. It must have been, said one historian, "the world's first TV space report."[16]

The proponents of human space flight encountered extensive skepticism. In the 1936 volume of *Rockets through Space*, another early English-language exposition on space travel, the British writer Philip E. Cleator complained, "Most people either do not believe that interplanetary travel is possible, or else they are utterly indifferent towards it. . . . There is no indication that the situation will be any different in the immediate future."[17]

Although none of the early works promoting space travel reached an exceptionally large audience, the themes they developed did. The principal theme was straightforward: human space flight was just around the corner. The obstacles to lunar and planetary voyages were not insurmountable, and the basic technologies had already been developed. Early publications provided readers with lessons on the dynamics of rocket flight and orbital mechanics as a way of proving this point. They also laid out an order of exploration that passed to the U.S. civil space program practically unchanged.

Following the success of *Rockets: The Future of Travel beyond the Stratosphere*, which had gone through a second edition in 1947, Willy Ley continued to produce books championing space travel. In 1949 he released the award-winning *Conquest of Space*, a collaborative effort with artist Chesley Bonestell.[18] In *Conquest*, as in the other books, Ley provided plainspoken descriptions of the principles involved in orbiting the Earth, building a space station, and exploring the Moon and other planets. *Conquest* opened with a description of a rocket launch from the White Sands Missile Range in New Mexico. The book's most distinguishing features were the illustrations by Bonestell, so remarkable that Viking Press gave him top billing on the title page. The paintings gave readers astonishingly realistic views of the Earth as it might be seen by passengers in a rocket ship ascending into space. Thanks to telescopic images, Bonestell prepared equally realistic aerial views of the surface of the Moon, which looked remarkably like the photographs NASA would produce with real spacecraft more than a decade later. Having captured the attention of readers with these realistic views, Bonestell offered illustrations of the surface of Mars and other planets based on astronomical information filtered through an active imagination.

Two years later Arthur Clarke published *The Exploration of Space*, another popular treatment of modern astronautics. This was Clarke's second book, *Interplanetary Flight* having appeared in 1951. Clarke was then thirty-three years old. Following his work on the British Interplanetary Society spaceship proposal, he had served as a radar instructor in the Royal Air Force and earned a bachelor's degree in physics and mathematics. He had also distinguished himself by publishing a number of articles on electronics, including a 1945 proposal for a system of global communication satellites. Before the Second World War, space travel had been a hobby, spurred by a childhood interest in science fiction magazines. In 1951 he learned that he could support himself by writing about space travel, this being the first year his income as a writer exceeded the pay he received from his job as a scientific journal editor. Clarke had also dabbled in science fiction, producing a few short stories and a manuscript for a first novel on interplanetary flight. In *The Exploration of Space*, he argued that the possibility of human space flight "must now be regarded as a matter beyond all serious doubt. . . . The conquest of space is going to be a very difficult, dangerous and expensive task. The difficulties must not, however, be exaggerated, for the steadily rising tide of technical knowledge has a way of obliterating obstacles so effectively that what seemed impossible to one generation becomes elementary to the next."[19]

In spite of his interest in electronics, which might have caused him to anticipate the potential for remotely controlled spacecraft, Clarke envisioned no suitable alternative to the widespread desire among space enthusiasts for human ventures. If his technical background suggested alternatives, his desire to entertain invalidated them. In his chapter on the spaceship, Clarke suggested that the exploration of space would occur in seven stages, all but the first requiring human crews. The stages, with a few technical changes, set forth the long-range plan that NASA adopted nearly one decade later:

(1) Unmanned, instrument-carrying missiles will enter stable orbits round the Earth, and will travel to the Moon and planets.

(2) Manned, single-step rockets will ascend to heights of several hundred miles, landing by wings or parachutes.

(3) Multistage, manned rockets will enter circular orbits just outside the atmosphere and, after a number of revolutions, will return by rocket-braking and air resistance.

(4) Experiments will be made to refuel these ships in free orbit, so that they can break away from the Earth, make a reconnaissance of the Moon, and return to the Earth orbit.

(5) The type of ship designed for a lunar landing will be flown up from Earth or assembled in free orbit, and after refueling will descend on the Moon. The ship may then return direct to an orbit around the Earth, or it may make a rendezvous, *in an orbit round the Moon*, with tankers sent from Earth.

(6) While the exploration of the Moon is proceeding by the use of such ships and techniques, attempts will be made to refuel rockets for the journeys to Mars and Venus. . . .

(7) Finally, landings will be made on Mars and Venus.

Space artists helped arouse public interest in cosmic exploration with paintings that anticipated distant landscapes. None was more influential than Chesley Bonestell, an architect and Hollywood special-effects artist. This painting of Saturn as it might appear from Titan was published in a 1944 issue of *Life* magazine. Sixty years later the *Huygens* spacecraft landed on this distant moon. (Reproduced courtesy of Bonestell LLC)

After humans reached the nearby planets, Clarke concluded, "the first era of interplanetary flight would be ended." The next era would be devoted to the challenges of improving spacecraft efficiency; building up bases on the Moon, Mars, and Venus; accumulating stores of fuel at useful places; and preparing for more demanding journeys to the outer planets and their moons.[20]

Ultimately, Clarke believed, humans would touch intelligent life beyond the solar system: "Even if we never reach the stars by our own efforts, in the millions of years that lie ahead it is almost certain that the stars will come to us. Isolationism is neither a practical policy on the national or the cosmic scale. And when the first contact with the outer universe is made, one would like to think that Mankind played an active and not merely a passive role—that we were the discoverers, not the discovered."[21]

In spite of the demonstrations of rocket power during World War II, the initial prophecies of people such as Ley and Clarke continually struck a wall of public skepticism. Few believed that industrialists or the U.S. government could be enticed to sponsor a massive program of lunar and planetary exploration. Serious plans for space travel had been envisioned but had not produced even the authorization necessary to launch the first tiny Earth satellite.

This began to change in 1951. To broaden interest in their astronomy program, officials at the Hayden Planetarium, part of New York City's American Museum of Natural History, organized the first of three symposia on space travel. Although attended by only a few hundred persons, the products of the symposia eventually reached millions of Americans. Museum officials asked Willy Ley to coordinate a group of speakers for the first meeting. Ley's purpose in accepting the task was clearly promotional. "The time is now ripe to make the public realize that the problem of space travel is to be regarded as a serious branch of science and technology," Ley explained in his letter to potential speakers. "Invitations will be sent to institutions of learning, to professional societies and research groups, and also to the science editors of metropolitan newspapers and magazines (plus those out-of-town and foreign publications which have offices in New York)."[22] The first symposium was held on 12 October 1951. Ley, along with planetarium chair Robert R. Coles and Robert P. Haviland, scheduled an appearance on the Nancy Craig television show for the afternoon following the symposium to report on their activities.

In his address to the symposium, Ley recounted the advances in rocketry that had made space travel possible. "Thirty years ago [in 1921] all serious thought about space travel consisted of a short book by Prof. Robert H. Goddard and a few articles in professional magazines." By 1931 the serious literature had grown in volume and scope, and rocket enthusiasts had formed societies in Germany and in the United States. Small liquid-fueled rockets had reached altitudes of about three thousand feet. By 1941 rockets had risen 10 miles above the surface of the Earth, and by 1951 two-stage rockets had attained a peak altitude of 135 miles. "The obvious question," Ley said, "is what will come next." He predicted that rockets without people on board would carry "all the way to the Moon," rockets would circle the Earth in orbit, and nations would build orbital stations to house the first people to live and work in space.[23] Before Ley's remarks, participants had been treated to an imaginary trip to Mars, part of the planetarium's current demonstration on "The Conquest of Space."[24] Ley was followed on the podium by Robert P. Haviland (a rocket expert with the General Electric Company), Fred L. Whipple (chair of the Department of Astronomy at Harvard University), Heinz Haber (a professor of space medicine), and Oscar Schachter (who spoke on the legal claims to outer space).

"It is obvious to all concerned that any project of such magnitude as the conquest of space can be successful only if it enjoys the full support of the public," Coles reiterated in opening the second symposium on 13 October 1952.[25] Ley coordinated the second symposium, inviting his old colleague Wernher von Braun to deliver one of seven formal papers. Von Braun and his German rocket team had moved to Huntsville, Alabama, where the forty-year-old von Braun oversaw the U.S. Army Ordnance Guided Missile Development Group as its technical director.[26] Anxious to summon forth public funding for human space flight, von Braun called for a separate government program devoted solely to the exploration of space: "Many a serious rocket engineer, while firmly believing in the ultimate possibility of manned flight into outer space, is confident that space flight will somehow be the automatic result of all the efforts presently concentrated on the development of guided missiles and supersonic airplanes. I do not share this optimism. . . . The ultimate conquest of space by man himself is a task of too great a magnitude ever to be a mere byproduct of some other work."[27] In calling for an independent space program, von Braun raised an issue with which speakers at the second and third symposia struggled. Many agreed with von Braun that space exploration was too important be developed as a byproduct of military research or some other endeavor; they also understood the difficulty of justifying human space flight on its own commercial or scientific merits, given the staggering cost.

While introducing the speakers at the third symposium, museum official Joseph Chamberlain promised the audience, "You will hear a suggestion that we completely alter our stodgy and limited perspective concerning departure from Earth and accept a new philosophy that may permit realization of this goal in our time."[28] Chamberlain had invited Arthur C. Clarke to coordinate this final symposium, held on 4 May 1954. Clarke had published five science fiction novels in the previous two years, including the classic *Childhood's End*, making him one of the most prolific and widely read advocates of human space flight.[29] His symposium address on the history of the space flight idea was followed by a blunt assessment of the activities necessary to implement it. The commander of the Navy Bureau of Aeronautics, R. C. Truax, laid out the central challenges confronting advocates of human space flight like himself: "There is simply no overwhelming rational reason why we should try to set up a station in space, send a rocket to the moon, or take any other steps along the road towards interplanetary flight." The military utility was difficult to demonstrate, the commercial potential uncertain, and scientists were in general "a poverty-stricken lot." Eventually the United States might engage in space travel as an outgrowth of practical considerations, but "you and I," Truax noted, "would very likely not be alive to see even the beginnings." He wanted a space program soon, and he admitted that he wanted one simply for the excitement of it, because it was a great human adventure: "If the majority of the people of this country feel the same way, the arguments of immediate utility are unnecessary. One does not have to justify the manner in which he spends his own money. Ultimately every thing we do is done simply because we want to."

Government officials would be forced to support an ambitious space program if the public wanted it. But members of the public would desire it only if advocates excited their imaginations. Truax called on advocates to mount a crusade that would "fire the imaginations" of the uninformed and unconvinced. This work could not be carried out by a few informed rocket enthusiasts alone, he said, but "must be passed by newspapers and magazines, by commentators, by editors, by civic and fraternal organizations, by letter and by word of mouth from individual to individual."[30]

Members of the editorial staff at *Collier's* magazine had attended the first symposium on space travel in October 1951. *Collier's* was a weekly magazine published in New York with a circulation of 3.1 million copies, making it one of the top ten magazines in the United States at a time when millions of

Americans received information about current events through weekly and bi-weekly outlets such as *Life, Look, Collier's,* and the *Saturday Evening Post.*[31] Intrigued by statements emerging from the first symposium, *Collier's* managing editor Gordon Manning dispatched associate editor Cornelius Ryan to a less-publicized symposium on space medicine held in Albuquerque, New Mexico, early the following November. As related by Fred Whipple, who presented papers at both conferences, Ryan was initially skeptical about the possibility of human space flight and artificial satellites. Following the formal presentations at the New Mexico conference, Whipple, Wernher von Braun, and Joseph Kaplan (a professor of upper atmospheric physics at UCLA) took Ryan aside for an evening of cocktails and dinner. An impassioned discussion ensued. Whipple described the result. "Whether or not he was truly skeptical, we persevered. Von Braun, not only a prophetic engineer and top-notch administrator, was also certainly one of the best salesmen of the twentieth century. Additionally, Kaplan carried the aura of wisdom and the expertise of the archetypal learned professor, while I had learned by then to sound very convincing. The three of us worked hard at proselytizing Ryan and finally by midnight he was sold on the space program."[32]

Ryan returned to New York and met with Manning. In preparation for what would eventually become an eight-part collection of articles spanning two years, the editors initiated a series of discussions in New York with the leading advocates. The first issue to emerge from these discussions appeared on 22 March 1952 with a cover painted by Chesley Bonestell. The cover displayed von Braun's design for a 265-foot-tall winged rocket dropping its second stage as it ascended past the forty-mile mark above the Earth. Four of the speakers from the first Hayden symposium (Ley, Whipple, Haber, and Schachter) contributed articles, as did von Braun and Kaplan. The cover copy read: "Man Will Conquer Space Soon: Top Scientists Tell How in 15 Startling Pages."[33]

The eight-part series laid out what by then had become the conventionally accepted stages for the exploration of space. The 22 March issue opened with a lengthy article by von Braun describing plans for a large, Earth-orbiting space station, which he presented as the facility that would establish a human presence in space. It could be built, he said, within ten to fifteen years. To introduce von Braun's article, Bonestell contributed what would become one of the most widely reproduced images of space travel—a two-page illustration depicting the 250-foot-wide, rotating space station, co-orbiting platforms, space "taxis," and a winged space shuttle. Ley described work inside the space station, Haber

To dispel public disbelief, advocates of human space travel undertook a deliberate public relations campaign designed to convince Americans that space flight could actually occur. Advocates presented their vision of the future in an eight-part series in *Collier's* that began with the 22 March 1952 issue. Following instructions provided by Wernher von Braun, Chesley Bonestell painted the magazine cover. (Reproduced courtesy of Bonestell LLC)

discussed the problems of human survival in space, Kaplan gave a short lesson on the transition from atmosphere to space, Whipple advanced the potential for space-based astronomy, and Schachter discussed legal claims to the Moon and other celestial bodies.

The second set of articles, in the 18 October issue, described an elaborate expedition from the space station to the Moon. "We will go to the moon in the next 25 years," the editors predicted after further discussions with their expert panel.[34] Von Braun contributed an article on the technical procedures for the voyage, and Ley provided a detailed description of the passenger vehicles. The following week, Whipple and von Braun described activities on the lunar surface, and Ley provided a short description of a lunar base. Stunning illustrations by Chesley Bonestell and Rolf Klep again graced the issues.

The next three issues, 28 February, 7 March, and 14 March 1953, were devoted to the problems of selecting, training, and protecting the people who would fly into space, reemphasizing the assumption that humans would pilot the rocket ships of tomorrow. Only with the 27 June 1953 issue did the editors acknowledge the role of automation. Von Braun and Cornelius Ryan described what they called the "baby space station," a ground-controlled satellite that would collect scientific data. "What kind of scientific data do we hope to get?"[35] Von Braun and Ryan proposed collecting information on how three monkeys launched in the nose cone of a satellite would react to sixty days in orbit. Automated flight, in the view of the advocating group, existed for the purpose of determining the effects of space travel on living organisms, a prelude to human flight.

The ultimate aim of this early activity was presented in the 30 April 1954 issue of *Collier's*. Humans would explore Mars. Von Braun, who with Ryan had come to dominate the magazine's series, presented a plan for a two-and-a-half-year expedition to that mysterious, Earth-like planet. Fred Whipple introduced von Braun's article by posing the question that had fascinated astronomers and science writers for decades: was there life on Mars? "There's only one way to find out for sure," Whipple concluded, "and that's to go there."[36] Chesley Bonestell, Fred Freeman, and Rolf Klep again provided vividly realistic illustrations depicting the proposed expedition.

As part of an aggressive campaign to promote the 22 March 1952 issue, the *Collier's* staff dispatched von Braun on a media speaking tour. Von Braun had emerged from the early sessions as the most adept exponent of human space flight. Handsome and charismatic, with his slight German accent suggesting the archetypical rocket scientist, von Braun had an unusual talent for making space travel seem feasible. No single individual would attain his media status until astronomer Carl Sagan hit the cover of *Newsweek* in August 1977. Von Braun appeared on the *Camel News Caravan* with John Cameron Swayze and on Dave Garroway's *Today* show, promoting the theme that human flights into space would occur within the lifetimes of the viewing audience. Shortly after the *Collier's* series appeared, *Time* magazine characterized von Braun as "the major prophet and hero (or wild propagandist, some scientists suspect) of space travel."[37] *Life* magazine called him "the seer of space" and put him on the cover of its 18 November 1957 issue, devoted to the dawning space age.[38]

The most influential opportunity for public exposure of the space flight idea occurred in the spring of 1954 as the *Collier's* series ended. Walt Disney, who had already achieved considerable success with his animated cartoons and full-length motion pictures, had agreed to produce a one-hour weekly television program to begin on the American Broadcasting Company (ABC) television network in fall 1954. Disney was motivated by the desire to promote his Disneyland theme park, scheduled to open the following year. Disney executives had struggled to raise the money necessary to build the park—Walt observed that "dreams offer too little collateral"—and the television series offered both a source of revenue and public exposure.[39] Producers organized the program around the park's four themes: Adventureland, Frontierland, Fantasyland, and Tomorrowland. Of the four themes, Tomorrowland was the least developed. Disney asked one of his senior animators, Ward Kimball, to develop ideas for the television segment. Kimball, who had been following the *Collier's* magazine series, was very impressed that "there were these reputable scientists who actually believed that we were going out in space."[40] To assist with the show's story lines, Kimball called in Willy Ley, who in turn recruited von Braun and Heinz Haber. Ley, von Braun, and Haber, along with Kimball and Disney, appeared on the first show. It gave the space boosters access to an enormously large audience and a huge chunk of American popular culture.

Millions of Americans watched the first program, "Man in Space," which aired on Wednesday, 9 March 1955. (It was rebroadcast on 15 June and 7 September.) In opening the show, Disney noted that "one of man's oldest dreams has been the desire for space travel—to travel to other worlds. Until recently, this seemed to be an impossibility, but great new discoveries have brought us to the threshold of a new frontier—the frontier of interplanetary space."[41] Following an animated segment on the history of rocketry, Ley explained the development of the first multistage launch vehicles. Haber described the challenges of protecting space travelers from hazards such as meteorites and cosmic rays, illustrated by an amusing cartoon figure. Von Braun then presented the design for a large, four-stage rocket that could carry six humans into space and back. The show ended with an impressive animation depicting the first launch of the giant rocket ship from a small Pacific atoll and its return to Earth.

Von Braun and the others worked hard to convince viewers that the concepts presented were attainable and that much of it could occur soon. "If we were to start today on an organized and well-supported space program," von Braun asserted, "I believe a practical passenger rocket could be built and tested within ten years."[42] His prediction was not far-fetched. The first flight of von Braun's three-stage Apollo-Saturn V rocket ship with astronauts on board took place thirteen years later, on 21 December 1968.

Von Braun was more cautious in his predictions for a trip around the Moon. "Even though we now have the theoretical knowledge to make a trip to the Moon," he said in the second Disney program on space flight, "it will be many years yet before our plans can fully materialize."[43] The second program, originally titled "Man and the Moon," aired on 28 December 1955. As before, von Braun emerged as the leading spokesperson for the space scientists and engineers, delivering a lengthy lecture on the construction of an Earth-orbiting space station and the preparation of the spaceship that would make the first voyage around the Moon. Von Braun and Kimball, who presented the major segments of the show (Ley and Haber did not appear), depicted the voyage to the Moon as one of the great adventures of humankind. "Such a trip has long been the dream of many men since history began," Kimball explained in introducing a humorous cartoon segment on lunar fantasies. Later in the program, four actors portrayed the humans dispatched on the first journey around the Moon, with dramatic close-ups of the lunar surface.

Von Braun appeared only briefly in the third program titled "Mars and Beyond," which aired in late 1957. By then, he was preoccupied with efforts to launch the first U.S. Earth-orbiting satellite. He and Ernst Stuhlinger were photographed at work on plans for an interplanetary spacecraft powered by electromagnetic drive and did not speak. Most of the show consisted of animated segments and cartoons dealing with the development of life in the solar system and an imaginary trip to Mars.[44]

The series accomplished the purpose sought by the space boosters. To millions of Americans it portrayed human space flight as something real, as no longer relegated to the realm of fantasy. The Disney series promoted this idea, as did the *Collier's* articles and other vehicles, simply by describing the voyages in a spellbinding, imaginative way.

In the summer of 1955, Disney opened his Anaheim, California, theme park, a monument to American mythology and popular culture. The original Tomorrowland sought to present a picture of the future as the Disney people envisioned it thirty-one years hence, in 1986.[45] Among its attractions, Tomorrowland offered patrons the opportunity to ride an all-aluminum passenger train, cruise in pink and blue fiberglass boats, and drive miniature automobiles on an Autopia freeway. At the center of Tomorrowland, dominating the still-barren landscape of Disney's new park, rose an eighty-foot needle-nosed rocket ship, which Ley and von Braun had helped design.[46] The one-third scale model of an atom-powered rocket ship marked the entrance to Disney's Rocket to the Moon ride. As Disney himself explained, "After entering the Disneyland space port, visitors may experience the thrills that space travelers of the future will encounter when rocket trips to the Moon become a daily routine."[47]

The ride drew extensively upon the work undertaken for Disney's second television program on space. In a building behind the model rocket, visitors viewed a fifteen-minute briefing on space travel and then entered a circular chamber designed to simulate the ship's passenger cabin. Viewing screens on the floor and ceiling showed the Earth recede and the Moon appear; seats vibrated to simulate flight.[48] Disney wanted to make the ride as realistic as possible—a challenging task in 1955, when no humans had yet traveled into space.[49] In only one respect did Disney delve into a bit of fantasy: guests on the ride looked down at the ruins of an ancient lunar base as the rocket traversed the back side of the Moon.[50]

The conquest of space as von Braun and Ley described it began with the construction of an Earth-orbiting space station. Consequently, the Disney people prepared Space Station X-1, an exhibit that opened with the park in mid-1955.[51] From a revolving platform, visitors watched the surface of the Earth and the United States pass beneath them from dawn to dusk. As one Disneyland publication proclaimed, "In the future, according to scientists, space stations similar to the one at Disneyland will have full living quarters for several score of men and its own gravitational field."[52]

Disney and *Collier's* exposed millions of Americans to the possibility of human space flight. They did so, moreover, in a manner that allowed people to visualize how the venture would actually appear to the people who participated

The Disneyland theme park, which opened in 1955, featured a simulated rocket ride to the Moon. Wernher von Braun and Willy Ley helped design the eighty-foot-high model rocket ship that stood outside the entrance to the attraction. To promote his new theme park, Walt Disney asked animator Ward Kimball to produce three hour-long programs for the popular *Disneyland* television show, featuring von Braun and Ley. Through the Disneyland theme park and television series, space advocates helped convince millions of Americans that travel to the Moon and planets could happen soon. (© Disney Enterprises, Inc.)

in it. Cameras in rockets had photographed the curvature of the Earth from high altitudes, but nothing beyond. The growing interest in space flight in an increasingly visual world excited public demand for anticipatory images of the heavenly landscapes to which humans and their machines would soon depart.

Artists had begun to produce realistic images of landscapes on other worlds (independent of illustrations for works of fiction) in the late nineteenth century.[53] James Nasmyth furnished a set of pictures for the 1874 book on *The Moon* by James Carpenter.[54] Nasmyth prepared plaster models of lunar landscapes, which he photographed against black, starry backgrounds for most of his illustrations. The work of a number of space artists accompanied astronomical photographs and drawings in the 1923 book *Splendour of the Heavens*.[55] Editors of various periodicals, including science fiction pulp magazines, commissioned astronomical art. The editors of *Astounding Science-Fiction*, for example, displayed views of Mars and Saturn as seen from their moons in place of the more melodramatic illustrations promoting space fantasies.[56] Nearly sixty years after the event, Arthur Clarke still remembered "the splendid cover" of a 1928 edition of *Amazing Stories*, a painting of Jupiter with its atmospheric eddies and swirls.[57] Charles Bittinger illustrated a 1939 *National Geographic* article on recent developments in astronomy.[58] Like most other serious artists, he depicted astronomical wonders such as a terrestrial eclipse (an eclipse of the Sun covered by the Earth as seen from the Moon).

No artist had more impact on the emerging popular culture of space in America than Chesley Bonestell. Bonestell did for space what Albert Bierstadt and Thomas Moran had accomplished for the continental frontier. Like Bierstadt and Moran, Bonestell transported viewers to places they had never been before. Although the paintings were based on real sites, Bonestell used his imagination to exaggerate features in such a way as to create a sense of awe and splendor. He used light and shadow, as artists had done with western landscapes a century earlier, to portray space as a place of great spiritual beauty. Through his visual images, he stimulated the interest of a generation of Americans and showed how space travel would be accomplished. Many readers remembered the paintings of planets and spaceships more than the words in the articles that accompanied them.

Born in San Francisco in 1888, Bonestell was trained as an architect. He designed numerous buildings in the San Francisco Bay area. In 1938, at the age of fifty, he changed careers, accepting a position as a special-effects artist at

RKO studios in Hollywood. He became a highly paid specialist, painting backgrounds for such movie classics as *The Hunchback of Notre Dame* and *Citizen Kane.*

Since boyhood Bonestell had possessed an amateur's interest in astronomy. He occasionally prepared sketches of the Moon and planets in his spare time.[59] Using photomontage skills acquired from painting movie backgrounds, Bonestell prepared a series of paintings showing the planet Saturn as it would appear from five of its moons. He sold the paintings to *Life* magazine, which published them in its 29 May 1944 issue. The article contained what would become one of Bonestell's most famous illustrations—a stunning portrait of Saturn framed by snowy cliffs, as it might be seen from its giant moon Titan. Lit from behind, Saturn sat in a clear pastel blue sky, which, the editors explained, appeared blue instead of black because Titan was a moon with an atmosphere. In fact, as scientists later learned, the atmosphere of Titan is opaque. Any beings on the surface of this frozen moon would not be able to discern Saturn through the haze. That did not discouraged Bonestell or other artists from preparing such impressive scenes.[60]

In the tradition of astronomical art, Bonestell concentrated on landscapes rather than people or spaceships. He would occasionally add tiny figures to his landscapes, but these, as the editors of *Life* magazine explained, were "purely imaginary, put in to give scale."[61] Two years later, in 1946, *Life* published Bonestell's portrayal of a trip to the Moon. Again the paintings emphasized landscapes, particularly aerial views of the Earth and the Moon. Only one of the paintings, an oblique perspective, featured the winged rocket ship that would carry the explorers on the voyage.[62]

Impressed by Bonestell's capacity to excite interest in space exploration, Willy Ley approached him to collaborate on the 1949 classic *The Conquest of Space.* Later, when *Collier's* magazine first contemplated its series on space exploration, its editors recruited Bonestell to meet with Ley and von Braun and others planning the stories.

Ley and von Braun provided Bonestell with the technical information necessary to prepare detailed paintings of spacecraft, and Bonestell in turn provided the human flight advocates with an imaginative visual representation of their ideas. Von Braun was especially influential. As Bonestell explained, "Von Braun would send me sketches drawn on engineer's graph paper, which I converted into working drawings and then into perspective. The courses I had had at Columbia University [as an architecture student] enabled me to handle some very complicated problems, and my courses in structural engineering helped me to understand the mechanics of space machinery."[63]

A stunning painting of a V-2–type rocket ship sitting on the surface of the Moon with astronauts setting up experiments graced the jacket of *Conquest of Space*. Bonestell later collaborated with Ley and von Braun on *The Exploration of Mars*.[64] In addition to his usual landscapes, Bonestell contributed a series of nine paintings depicting the expedition craft and landing party. Nearly all of Bonestell's paintings for the *Collier's* series contained a spacecraft, a space base, or other sign of the underlying human theme. Bonestell's art, along with paintings of rocket ships and space station cutaways by Rolf Klep and Fred Freeman prepared for the *Collier's* series, reappeared in two books based on the magazine articles.[65] At a time when the public had not yet accepted space travel, Bonestell's images showed it to be something that humans could actually do.[66]

Space art became increasingly realistic under the influence of space advocates in the 1950s.[67] R. A. Smith, considered one of the pioneers of space hardware art, illustrated two of Arthur C. Clarke's early works on the exploration of space.[68] Jack Coggins illustrated two children's books that prominently displayed space travel technology.[69] Frank Tinsley contributed dramatic artwork that commercial firms employed in their corporate advertising. As the space race began, print advertising provided a major outlet for visual images extolling the wonders of space exploration.[70]

Public interest in the visual aspects of space encouraged American movie producers to depict human space travel in realistic ways. In 1948 Robert Heinlein, a science fiction writer and unabashed promoter of space exploration, convinced movie producer George Pal to make a film based on Heinlein's 1947 book *Rocketship Galileo*, a rather fanciful children's tale.[71] Heinlein's screenplay depicted a trip to the Moon as it might actually occur. Pal recruited Chesley Bonestell to make the movie look accurate. Bonestell's artwork for the film duplicated the cover of his and Ley's 1949 *Conquest of Space*. Bonestell chose for the movie destination the crater Harpalus, on the northern latitudes of the Moon, which allowed him to show the Earth near the horizon of his lunar landscape.[72] In one concession to dramatic imagery, he sculpted the crater walls to appear as if they had been carved by wind and rain. *Destination Moon*, released in 1950, provided large numbers of Americans with their first sense of what a lunar landing site might actually look like. It set new standards for the portrayal of space travel, returned a nice profit, and won the 1950 Academy Award for special effects.[73]

Encouraged by the popular reception to *Destination Moon*, Pal and Bonestell collaborated to produce *When Worlds Collide*, released in 1951. Bonestell again provided the dramatic artwork. Pal followed with *War of the Worlds* in 1953, for which Bonestell provided some art and technical advice. The special effects consumed six months of effort, whereas filming the actors took only forty days.[74] Buoyed by his success, Pal released *Conquest of Space* in 1955.[75] The movie was based on the Bonestell-Ley book by the same name and on von Braun's earlier attempts to produce a Mars novel. Von Braun, Ley, and Bonestell all assisted with the movie, which depicts the adventures of the crew on a multiyear voyage from von Braun's rotating space station to the surface of Mars.

In sharp contrast to the success of *Collier's* and Disney, this space booster failed. *Conquest of Space* was incredibly dull and a box office flop. Combined with huge cost overruns by the art and special effects department on the film *Forbidden Planet*, the failure of *Conquest* led moviemakers to abandon space realism for the next twelve years.[76] In 1957 the film industry returned to established formulas with clunkers such as *Attack of the Crab Monsters* and *The Amazing Colossal Man*. Bonestell's Hollywood work dried up; Pal turned to science fantasy films such as *The Time Machine* (1959) and *Atlantis: The Lost Continent* (1960).

As the message of space exploration swept through American popular culture in the 1950s, public opinion began to shift. People paid more attention to rocket technology and grew less skeptical of those promoting space exploration. By early 1955 the proportion of Americans who believed that "men in rockets will be able to reach the moon . . . in the next 50 years" had increased from 15 to 38 percent.[77] By 1957, following the launch of *Sputnik 1*, 41 percent agreed that humans would reach the Moon within twenty-five years, with the largest number correctly predicting that it would happen in "about 10 years."[78] All of this occurred before the creation of NASA and the decision to go to the Moon.

The public was better prepared to accept not only the reality of space travel by the mid-1950s but also the vision advanced by space advocates. Space travel was consistently depicted as humans exploring the next frontier; alternative approaches received scant attention in the organs of popular communication. Given the preponderance of attention devoted to space travel as human exploration, the public had difficulty imagining any other type. This helped promote the belief, dominant within many sectors of government, that space flight could not survive within political circles unless it emphasized human flight.

Top NASA officials embraced the dominant space flight paradigm, even before NASA was formed. Congress created NASA in large part through the transformation of the National Advisory Committee for Aeronautics (NACA), a forty-year-old collection of government laboratories whose employees spent most of their time conducting research on airplanes. In early 1958 NACA was one of a number of government agencies vying for control of the new space effort.[79] Even skeptics such as NACA director Hugh L. Dryden, who was not an enthusiastic booster of human space flight, helped further the space flight dream. "The topic of our day is the new frontier, space," Dryden announced in a policy statement prepared for the annual meeting of the Institute of the Aeronautical Sciences in New York on 27 January 1958: "The escape of objects and man himself from the earth into space has long been the subject of science fiction writers and the comic strip artists. More recently, it has been a matter of interest to a growing number of serious-minded scientists. Now it has acquired a new sense of imminence and reality. Space travel has stirred the imagination of man to an extraordinary degree."[80] Dryden announced that NACA was prepared to play a leading role in the exploration of space, a position officially endorsed by the agency's advisory board eleven days earlier.[81] "In my opinion," he said, "the goal of the program should be the development of manned satellites and the travel of man to the moon and nearby planets."[82]

Shortly after NASA came into being on 1 October 1958, engineers who had only days earlier worked for NACA established the Research Steering Committee on Manned Space Flight, commonly known as the Goett committee, to prepare a set of recommendations "as to what future missions steps should be." At the first meeting in May 1959, committee members agreed that they "should not get bogged down with justifying the need for man in space in each of the steps but outrightly assume that he is needed inasmuch as the ultimate objective of space exploration is manned travel to and from other planets."[83] Max Faget, who achieved fame as the designer of the blunt, conical-shaped spacecraft that took the first American astronauts into space and back, urged the committee to select a lunar expedition as its immediate goal, "although the end objective should be manned interplanetary travel." George Low, who would play a leading role as NASA's chief of manned space flight, concurred with the lunar objective, "because this approach will be easier to sell."[84]

At the May meeting, committee members adopted the exploration program promoted by space boosters. The official NASA program of space flight adopted every one of the steps proposed by Clarke in *The Exploration of Space:* suborbital tests of spacecraft designed for human flight, orbital flights, a trip around the Moon followed by a Moon landing, reconnaissance of Mars and Venus by automated spacecraft, and human expeditions to those nearby orbs.[85] In only one respect did committee members depart from the dominant vision. Members could not agree on whether to build a large space station as a prerequisite to lunar and planetary exploration.[86] They agreed on the need for a small orbiting laboratory occupied by human beings. The much larger space station, however, might wait. In their 1959 long-range plan, NASA officials announced that at least the first human flights around the Moon could be conducted before the establishment of a "permanent near-earth space station."[87]

The paradigm of human exploration became part of NASA's organizational culture. People in NASA assumed that this was the way that space flight was done. During the 1984 debate over the proposed Earth-orbiting space station, one of NASA's top engineers admitted that the civilian space agency would still want to build the facility "even if it could be proved that functionally everything conceived of today could be done by robots." NASA would build the station, he said, because "we think it is NASA's charter to essentially prepare for the exploration of space by man in the twenty-first century."[88]

The philosophy of space travel as human expansion dominated the thoughts of policy planners in the decades that followed. In 1969, looking past the flights to the Moon, NASA officials proposed a course of action for the following twenty years. The NASA administrator who had overseen the crash program to get to the Moon, James E. Webb, had turned the agency over to his forty-seven-year-old deputy, Thomas Paine. Paine was a self-described "swashbuckler," an analogy he drew from his navy days. The son of a U.S. Navy commodore, Paine had served as a submarine officer in the Pacific during World War II. Though he left the navy for a career as a business manager and research engineer, the thought of naval traditions never left him. The Constitution permits Congress to grant letters of marque and reprisal, a device by which the government empowers private citizens to become buccaneers on behalf of the national interest. As NASA administrator, Paine urged his officers to adopt that "swashbuckling, buccaneering, privateering kind of approach" to promoting the spacefaring dream.[89]

Paine proposed a bold post-Apollo space program nearly identical to the

CONSTRUCTION OF ENVIRONMENTAL ENCLOSURE
FOR MARS COLONY

MSFC-70-PD-4000

Shortly after the creation of the National Aeronautics and Space Administration in 1958, agency officials adopted the long-range plan of lunar and planetary exploration promoted by space advocates such as Willy Ley and Wernher von Braun. "Manned exploration of the moon and nearer planets must remain as the major goals," they wrote. This artist's rendition depicts the accomplishment of a key step in the completion of the spacefaring vision—construction of an environmental enclosure for a colony on Mars. (NASA)

one set forth on the pages of *Collier's* magazine seventeen years earlier. His proposal (subsequently endorsed by the special vice-presidential Space Task Group) envisioned a very large space station, a winged space shuttle, a lunar base, a space station in polar orbit around the Moon, a nuclear-powered transportation system for deep space travel, and an expedition to Mars to be launched by 1983—all occupied or conducted by human beings. Robotic activities, though well represented, were relegated to a lesser role. Members of the advisory group set as their overall purpose "the opening of new regions of space to access by man."[90] Though his proposals were not adopted, Paine's relentless proselytizing delighted advocates of the dominant vision. It also irritated White House officials so much that they removed him from his post as NASA administrator in 1970.

Paine reappeared in 1986 as chair of the National Commission on Space, having spent the interim as a captain of industry in the aerospace sector. The commission was created by Congress and the president to chart the nation's space goals for the twenty-first century. Its report opened with the famous 1952 Chesley Bonestell diorama portraying hopes for a permanent space station, winged space shuttle, and large space telescope. Underneath it, a more recent illustration by space artist Robert McCall depicted the same facilities actually completed or approved: the current design for the *Freedom* space station, the existing NASA space shuttle, and the Hubble Space Telescope. Reflecting on the past, the commission members wrote that, "while predicting the future can be hazardous, sometimes it can be done."[91]

In the report, commission members looked fifty years into the future: "We are confident that the next century will see pioneering men and women from many nations working and living throughout the inner Solar System. Space travel will be as safe and inexpensive for our grandchildren as jet travel is for us."[92] Members of the commission lavishly illustrated their report with paintings of lunar settlements, orbiting spaceports, transportation systems, robotic spacecraft, space prospecting, and bases on Mars.

The dream continued. In 1987 NASA administrator James Fletcher established the Office of Exploration, set up to help win support for human expeditions to the Moon and Mars.[93] In his last year in office, President Ronald Reagan approved a national space policy that included a goal "to expand human presence and activity beyond Earth orbit into the solar system."[94] In July 1989 a special NASA working group submitted a conceptual plan for a lunar outpost and a human expedition to Mars.[95] In a speech later that month from the steps of the National Air and Space Museum in Washington, D.C., President George H. W. Bush endorsed those goals. In addition to endorsing the lunar base and Mars expedition, Bush went beyond, becoming the first chief executive to suggest that humans would someday leave the solar system for nearby stars: "You who are the children of the new century—raise your eyes to the heavens and join us in a great dream—an American dream—a dream without end."[96] Bush said Americans "will follow the path of Pioneer 10," referring to the U.S. spacecraft that had already flown beyond the orbits of Neptune and Pluto: "We will travel to neighboring stars, to new worlds, to discover the unknown. And it will not happen in my lifetime, and probably not during the lives of my children, but a dream to be realized by future generations must begin with this generation."[97]

The resulting Space Exploration Initiative quietly disappeared in the presence of congressional opposition and concern over its cost. Fifteen years later President George W. Bush tried again. "We will build new ships to carry man forward into the universe, to gain a new foothold on the moon and to prepare for new journeys to the worlds beyond our own," he pledged as he announced a renewed effort to achieve the fifty-year-old vision. "We do not know where this journey will end. Yet we know this: Human beings are headed into the cosmos."[98]

The idea of human space travel began as a fantastic, unattainable dream. Space boosters during the 1950s convinced the American public that the vision could be achieved. Public officials and advisory committees set forth proposals that endorsed the primary initiatives. Winning the massive amounts of financial support necessary to carry out the vision, however, was (and continues to be) very difficult. Yet an opportunity appeared. Using that opportunity to unlock the billions of dollars necessary to take the first steps in the spacefaring vision required an additional step in imagination.

3

The Cold War

Control of space means control of the world.
—Lyndon B. Johnson, 1958

Chicken Little was right.
—David Morrison, 1991

By themselves, the early efforts to promote the exploration of space were not sufficient to unleash the billions of dollars necessary to undertake the endeavor. The spirit of adventure and discovery to which much of the early promotional efforts appealed did not justify such a large commitment. Promotional efforts allowed the public to envision explorers making trips to the Moon and planets, but most politicians in the mid-1950s could not imagine appropriating the money to fund such expeditions. Members of an official U.S. science advisory committee repeated a familiar refrain when they admitted in 1958 that human expeditions to the Moon and nearby planets would occur at some time in the future, but that "it would be foolish to predict today just when this moment will arrive."[1]

Having helped convince the American public that space travel was real, boosters faced an additional challenge: they had to conjure images that would promote the will to act. For this purpose space advocates found a ready supplement in public anxiety about the Cold War. Advocates not only promoted human space flight through tales of adventure but also sought government support by scaring the American public. Public fears played a critical role in unleashing the billions of dollars necessary to begin large-scale activities in space. Enthusiasm for the grand vision drew force from cosmic nightmares as well as from pleasant dreams of human space flight.

Public policies often emerge as the result of lengthy periods of preparation punctuated by precipitating events. From the late 1940s through the mid-1950s, advocates of space exploration prepared the American public with elaborate visions of promise and fear. The precipitating events occurred on 4 October and 3 November 1957, with the launching of *Sputnik 1* and *Sputnik 2* by the Soviet Union. Both before and after these events, President Dwight D. Eisenhower and members of his administration sought to fashion a practical space effort that differed considerably from the grand vision of human exploration. The launch of the first Earth-orbiting satellites, combined with the previous promotion of space flight in the popular culture, overwhelmed the Eisenhower alternative and led subsequent political leaders to pursue more ambitious goals.

When the Cold War ended, so did much of the rationale for the ambitious program of lunar and planetary exploration that fear had motivated. Space advocates, nonetheless, did not abandon the motivational use of terror. Rather than discard their dreams, they sought out new dangers with which to promote interplanetary travel. Although not as powerful as the Cold War, the new nightmares followed the old formula, preparing public hopes and fears in anticipation of precipitating events that might eventually alter the policy agenda.

During the Cold War not everyone interested in space exploration shared the enthusiasm for the grand adventure of large space stations, lunar bases, human space travel, and outposts on nearby orbs. Not everyone accepted the prophecies of Wernher von Braun and other space pioneers. During the 1950s, a number of scientists and public officials put forth an alternative view. It was not as well presented as the adventurous vision, it never recruited a spokesperson as charismatic as von Braun, and the press tended to treat it as a dissent to the dominant vision rather than as a free-standing option. The alternative, nonetheless, enlisted one powerful advocate: Dwight D. Eisenhower, the first U.S. president to preside over the exploration of space.

Much of the attractiveness of Eisenhower's alternative arose from its low cost. During the 1950s, government officials identified a number of priorities on which to spend large sums of money. Expeditions into space were not among them. The United States undertook a $16 billion crash program to create a fleet of intercontinental and intermediate-range ballistic missiles in the hope of preventing nuclear Armageddon.[2] It initiated a $26 billion program to complete the interstate highway system.[3] It intended to spend large sums of money to modernize the nation's schools.[4] Gigantic outlays for space adventures did not sit high on the list of national priorities.

In helping to prepare the articles on the conquest of space, the panel of experts advising the editors at *Collier's* magazine estimated the cost of the first major initiative in space. An Earth-orbiting space station and the fleet of large rockets necessary to support it would consume approximately $4 billion over ten years, they said.[5] When the *Collier's* panel issued its estimate in 1952, aggregate federal spending totaled only $68 billion. Early estimates for an expedition to the Moon were even more staggering.[6] The editors at *Collier's* rightly pointed out that an ambitious human flight program would require a national effort as expensive as the World War II Manhattan Project that had led to the development of the atomic bomb.[7]

The division between advocates of the grand vision and those supporting a less-expensive alternative surfaced during the second Hayden Planetarium Symposium on Space Travel held in New York City on 13 October 1952, which took place as the second and third issues of *Collier's* devoted to space travel reached their readers. In his remarks to symposium participants, Wernher von Braun displayed the *Collier's* articles as proof that full-scale exploration of the Moon "will be well in the realm of possibility" once work on an Earth-orbiting space station was complete.[8] Von Braun devoted most of his presentation to his primary specialty, the development of large rocket ships necessary to begin moving humans and their equipment into space.

Dr. Milton W. Rosen, who immediately preceded von Braun on the podium, disagreed. Like von Braun, Rosen was a pragmatic rocket engineer. He was an active member of the American Rocket Society, the spacefaring group founded by science fiction fans twenty-two years earlier, which had changed its name in an effort to attract serious scientists and engineers. Following his college education, Rosen had accepted a job at the Naval Research Laboratory, a military research facility in southeastern Washington, D.C. Within the agglomeration of government agencies competing for a share of rocket and missile activities, the laboratory was somewhat conservative, its scientists and engineers devoted to the development of small rockets and satellite technology.

When NASA was established, those people formed the nucleus of the satellite and science facility known as the Goddard Space Flight Center in Greenbelt, Maryland. Officials in the Eisenhower administration hardly could have found an existing government facility more devoted to its alternative vision of space exploration than the Naval Research Laboratory.

This was not the last time Rosen and von Braun would clash. In 1955 defense department officials gave the Naval Research Laboratory the responsibility for launching the first Earth-orbiting satellite, an assignment von Braun's army rocket team desperately wanted. At the time of the Hayden symposium, Rosen was the scientific officer in charge of the rocket program. The Viking rocket was the first stage of the launch vehicle that the Naval Research Laboratory would use in its attempt to launch its Vanguard satellite; Rosen would become technical director for Project Vanguard. When the rocket failed in a spectacular launchpad explosion in late 1957, von Braun grabbed the satellite assignment, and using his own Juno I rocket, launched the first U.S. Earth-orbiting satellite on 31 January 1958.[9]

At the symposium, Rosen explained the difficulties involved in launching even a small rocket toward space. Thousands of separate components had to work properly in order to avoid a failure. There were even greater obstacles to surmount in order to take humans into space. Reliability of components, cosmic and solar radiation, vehicle skin temperatures, rocket motor technology, and vehicle recovery all stood in the way of von Braun's dreams. Rocket scientists "have almost exhausted our store of basic knowledge," Rosen observed. "The engineer who has drawn the ingredients from the cupboard of basic research now finds that the cupboard is bare."[10] Rosen wished that the government would devote funds to space flight but admitted that this was not currently feasible. Too great a share of the nation's resources and scientific talent were devoted to military preparedness for the Cold War.

Given all of these limitations, what sort of a space program might the United States successfully undertake? "If we cannot build a space ship today," Rosen asked, "what can we do to advance the cause of space travel?" Rosen called for fundamental research on the problems of space flight to replenish the store of basic knowledge and urged his listeners to support a program of small Earth satellites. "Before we can attempt to transport human beings in a ship that orbits around the earth," he asserted, "we must produce a practical, reliable, unmanned satellite."[11]

Attuned to the newsworthy quality of controversy, the *New York Times* and

New York Herald Tribune headlined the Rosen–von Braun debate. Both newspapers accompanied their stories with pictures of the two individuals posed alongside a twelve-foot-high scale model of von Braun's proposed three-stage rocket ship. "Two Rocket Experts Argue 'Moon' Plan," the *Times* headlined. "Army Expert Sees Platform 1,000 Mi. up in 15 Yrs," the *Herald Tribune* noted. "Navy Scientist Skeptical."[12] Both articles opened with von Braun's assertion that the United States within ten to fifteen years would construct a station one thousand miles out in space, serviced by large rockets, and both treated Rosen's comments as a dissent to von Braun's vision. The *Times* quoted Rosen's assertion that von Braun's designs "are based on a meager store of scientific knowledge and a large amount of speculation." The *Herald Tribune* featured Rosen's contention that the United States "would be throwing its money away" if it undertook "any one of the fantastic projects for a space ship that have been proposed in the last few years."[13]

Other outlets featured the dissenting view. After *Collier's* magazine claimed that humans would "Conquer Space Soon," writers at *Time* magazine responded with a cover story more skeptical in tone. "Separating facts and fancy about space travel is almost as difficult as a trip to the Moon," the writers observed. *Time* attempted to lay out the facts. A rocket ship capable of sending a small spaceship beyond the Earth's gravitation well would have to be "as big as an ocean liner." Without scientific breakthroughs in areas such as rocket propulsion, the construction of large launch vehicles and space stations would be "a reckless leap into the blind future . . . a gigantic fiasco." This had not deterred the purveyors of public imagination, however. With little attention to the technical details, the magazine writers complained, toy shops, science fiction magazines, and television programs had "already zoomed confidently off into the vast ocean of space."[14]

Scientists waiting to conduct research in space ridiculed the grandiose schemes of human space flight advocates; some attacked von Braun personally. "He is the man who lost the war for Hitler," said one critic, who claimed that von Braun had drained the best brains and material away from the German war effort and now was trying to do the same to the United States as it sought to win the Cold War.[15] The manifesto for dissent appeared as a thirteen-page pamphlet on outer space issued by President Eisenhower's science advisory committee, chaired by James R. Killian. Though trained as an engineer, Killian had developed a reputation as a person who could represent scientific points of view. He had spent his entire professional career in academia, rising from the modest position of assistant editor for a scientific journal published by the alumni association at the Massachusetts Institute of Technology to the presidency of that institution. During his years at MIT, he promoted a number of scientific initiatives, such as radar development and missile guidance systems, that found their way into military use. In the fall of 1957, following the launch of *Sputnik 1*, President Eisenhower asked Killian to come to the White House to serve as special assistant for science and technology. From that position, Killian issued the short *Introduction to Outer Space*.[16] Eisenhower found the pamphlet "so informative and interesting" that he ordered the Government Printing Office to reproduce and offer it to the American public for just fifteen cents per copy.[17]

The pamphlet set out a series of scientific questions that could be addressed by a well-constructed program of satellites and automated spacecraft. "Scientific questions come first," the statement argued.[18] Earth satellites could study cosmic rays and solar radiation, assist with weather forecasts, transmit television broadcasts, and improve the clarity of astronomical observations. Automated spacecraft could examine the origin of the Moon, search for life on Mars, and study the atmosphere of Venus.

The pamphlet placed little emphasis on putting humans in space. Because humans were such adventurous creatures, the time would come when they would want to go into space and see the results of scientific efforts for themselves, but Killian and his colleagues were reluctant to predict when that might occur. "Remotely-controlled scientific expeditions to the moon and nearby planets," they argued, "could absorb the energies of scientists for many decades."[19]

Subsequent science advisory groups repeated this message. A special 1961 presidential transition committee cautioned incoming president John F. Kennedy that "a crash program aimed at placing a man into an orbit at the earliest possible time cannot be justified solely on scientific or technical grounds." The United States led the world in space science, committee members argued, a position that an extensive human flight effort would hinder "by diverting manpower, vehicles and funds."[20] A subsequent science advisory committee issued a similar plea in 1970. The extraordinary cost of human space flight could not be justified solely on the grounds of science, technology, and practical applications, its members asserted. Instead, an excellent program of space science could be conducted for about half of the money then being expended on civil space.[21]

In spite of the care with which the dissenting view was presented, it never captured the imagination of the American public in the same way the romantic vision did. It did not become part of the popular culture, with colorful magazine stories and Hollywood films devoted to its promulgation. It did not take the form of a well-coordinated alternative to the dominant point of view, but rather appeared as an objection more noteworthy for what it opposed than the future it would represent.

This is not to say that the public accepted the grand vision of human space travel unconditionally. Public opinion in the United States is notoriously ambivalent, a quality that affects many government activities including civilian space travel.[22] By the mid-1950s, most Americans believed that ventures such as a trip to the Moon would occur soon; at the same time most believed the government should not lay out large sums of money to accomplish the task. When asked in 1960 whether the United States should spend upwards of $40 billion "to send a man to the moon," 58 percent of the respondents to a Gallup poll responded no. Fifty-two percent of the same respondents nevertheless agreed that the venture would be accomplished within ten years.[23]

This pattern persisted throughout the formative decades of the U.S. space program. Public expectations remained high while the willingness to spend money remained low. Even by the mid-1980s, when interest in space stations and expeditions to Mars rebounded, the number of people who wanted to undertake these projects exceeded by a factor of two the number of people willing to increase the space budget to pay for them.[24]

As the dreams of space pioneers encountered the dissenting realism of American scientists, realism initially prevailed. The efforts of human flight advocates to win financial support for their grand adventure had practically no effect on the Eisenhower administration. Eisenhower and his aides possessed a concept of space exploration quite different from the one that dominated U.S. popular culture in the 1950s. Drawing on the dissenting views of American scientists, Eisenhower created what became the most visible alternative to the aims of space pioneers, and for a brief time, his vision defined U.S. space policy.

Eisenhower had spent most of his life in the U.S. Army. A 1915 graduate of the U.S. Military Academy at West Point, he so impressed his superiors with his skill at military planning that they promoted him over the heads of more than three hundred senior military officers to be supreme commander of all Allied forces in Europe. Though a career soldier, Eisenhower's simple Kansas upbringing left him suspicious of empire builders in the Pentagon. He warned Americans in his farewell address that industrialists and military officials were acquiring "unwarranted influence" over the U.S. government, creating large, hyperexpensive projects for which there was no justifiable need.[25]

As his biographer Stephen Ambrose has affirmed, Eisenhower came from a generation of military leaders for whom the Japanese surprise attack on Pearl Harbor was "burned into their souls."[26] He was absolutely determined to prevent a similar move by the Soviet Union on the United States or its allies and was far more interested in using space for this purpose than as a theater for some sort of exploratory opera. He characterized the promotional efforts of people who ignored security in favor of space stations and flights to the Moon as "hysterical."[27]

From his wartime experience, Eisenhower knew that aerial reconnaissance, which revealed the movement of troops and munitions, was the key to preventing unforeseen attacks. The Soviet Union had raised the so-called Iron Curtain around itself and its European empire in part to prevent such reconnaissance. Existing reconnaissance technology, such as aircraft overflights and balloons, violated the airspace of countries behind the Iron Curtain. Soviet leaders viewed aircraft overflights as hostile incursions, a point underscored in 1960 when the Soviet Union shot down and tried U.S. spy plane pilot Francis Gary Powers.

Seeking a means to conduct effective aerial reconnaissance, Eisenhower approved the creation of the Technological Capabilities Panel, whose 1955 report identified various scientific options for defense preparedness. The panel was dominated by scientists such as James Killian and Lee DuBridge, who, like Eisenhower, favored the alternative approach to space.[28] With regard to space activities, the panel proposed that the United States develop a small scientific satellite to be placed in orbit around the Earth. In addition to its scientific value, the satellite would establish the principle that objects in orbit did not violate the territorial sovereignty of nations over which they flew. The Soviet Union, the panel reasoned, would be less likely to object to an internationally sponsored scientific satellite making the first orbital incursion than a satellite launched by the U.S. military. The scientific satellite thus would establish the important principle of free access, which in the future would allow the United States to fly military reconnaissance satellites across the Soviet Union without provoking protest or retaliation.[29]

The Soviet Union inadvertently established the freedom-of-space principle when it launched *Sputnik 1*, which crossed the continental United States shortly after it was launched. Because the Soviet's launch accomplished one of the primary goals of Eisenhower's space policy, some historians have speculated that the president was not as surprised by the Soviet achievement as the contemporary popular press portrayed.[30]

Eisenhower's vision for space relied to a great extent upon satellite technology. For Eisenhower and the members of his administration, satellites and automated craft accomplished most of what they wanted to do in space. Satellites reinforced the important freedom-of-space principle, gave the United States the aerial reconnaissance platforms it needed to detect military preparations and monitor arms agreements, and allowed scientists to conduct space-based research. They did so, moreover, at a fraction of the cost of human missions, an important consideration in the budget-conscious Eisenhower administration.

Commensurate with the emphasis on satellite technology, Eisenhower's space policy deemphasized the role of human space flight. In 1958 Eisenhower approved an effort to develop a single-seat space capsule capable of placing an astronaut in orbit around the Earth. Project Mercury, as it was known, was designed simply to test whether humans could perform any useful functions in the void.[31] Ignoring calls for more ambitious ventures, Eisenhower steadfastly refused to approve any human excursions that went beyond the single-seat Mercury capsule. In late 1960, shortly before leaving office, Eisenhower specifically disapproved NASA's request to fund the keystones of its advanced human space flight agenda—a three-person Apollo spacecraft and a powerful, liquid hydrogen–fueled rocket. In his final budget message to Congress, Eisenhower recommended that the United States undertake human space flights beyond Project Mercury only if "testing and experimentation" could establish "valid scientific reasons" for doing so, a not-so-subtle dissent to the proposition that space flight should be used to boost national prestige.[32] In an accompanying report, his Science Advisory Committee dismissed the motives for advanced human space flight as "emotional compulsions."[33]

Eisenhower's alternative space program placed a great deal more emphasis on space science than on engineering feats. Eisenhower was prepared to compete with the Soviet Union in pursuit of scientific discoveries. Yet he wanted to weigh the value of research in space against the benefits to be gained from pursuing the same discoveries from terrestrial sites. "Many of the secrets of the universe will be fathomed in laboratories on earth," his advisers wrote, and the national interest required that "our regular scientific programs go forward without loss of pace."[34] Eisenhower had no interest in entering into a race with the Soviet Union that depended upon large rocket boosters, a field in which scientific questions took a back seat to engineering capability and the Soviets held a commanding lead. Interviewed after leaving the White House in 1962, he questioned the wisdom of President Kennedy's decision to engage the Soviet Union in a race to the Moon: "Why the great hurry to get to the Moon and the planets? We have already demonstrated that in everything except the power of our booster rockets we are leading the world in scientific space exploration. From here on, I think we should proceed in an orderly, scientific way, building one accomplishment on another, rather than engaging in a mad effort to win a stunt race."[35]

Eisenhower did not want to set up a crash program to explore space. He did not want to set up a Manhattan Project for space. He was already disturbed by the degree to which the arms race had bolstered what he called the "military-industrial complex" and did not want to give the aerospace industry another project on which to expand its base.[36] His desire for a balanced budget would not allow him to respond to every Cold War contingency, and he feared that a gigantic human space flight undertaking would divert resources from more pressing needs. As he wrote after President Kennedy approved Project Apollo, "This swollen program, costing more than the development of the atomic bomb, not only is contributing to an unbalanced budget; it also has diverted a disproportionate share of our brain-power and research facilities from other equally significant problems, including education and automation."[37]

NASA executives serving during the Eisenhower administration did not want to set up separate programs for human and robotic flight. Instead, they sought to merge NASA's "manned" and "unmanned" space activities. They created a single office of space flight at NASA headquarters for both human and robotic flight. They planned to create a single space projects center for both human and machine flight at what eventually became the Goddard Space Flight Center in Maryland.[38] That center was supposed to oversee both the recently approved Mercury human space flight project and NASA's scientific satellite programs, but this intent quickly evaporated once President Kennedy approved the race to the Moon. As part of a full-scale 1961 reorganization to gear up for Kennedy's objective, NASA officials created a separate headquarters office for manned space flight.[39] Texas politicians, including Congressman Albert Thomas, head of the House appropriations subcommittee that oversaw NASA's budget, urged agency executives to build an entirely new field center near Houston to oversee human flight.[40]

Eisenhower's alternative established a space program with far more emphasis upon satellites, robotics, and science than space boosters favored. It avoided the "boom and bust" cycle of crash programs and likely would have led to a closer relationship between human and robotic space activities. Such an approach would have dampened the schism that subsequently developed between "manned" and "unmanned" activities. Although the president normally plays a leading role in defining the scope and direction of the U.S. space program, in this case images fed to the public mind undercut Eisenhower's alternative, doing so because advocates of grander schemes adroitly played upon public fears about the Cold War.

Support for Eisenhower's space program within the U.S. government collapsed with the launch of *Sputnik 1* in early October 1957 and the launch of *Sputnik 2* one month later. *Sputnik 1* was the media event of the decade. Much of the public's sense of security during the early years of the Cold War rested on the assumption that the Soviets were scientifically and technologically inferior. *Sputnik 1* shattered this assumption. The supposedly backward Soviet Union succeeded in becoming the first nation to break free from earthly bonds. More frightening, the event suggested that the United States was no longer secure from a sudden nuclear attack. Early orbits of *Sputnik 1* crossed the United States, and readers of the *New York Times* had to be reassured that the satellite was not large enough to carry nuclear bombs that could be dropped on citizens below.[41] In a frequently repeated theme, one writer for the *Times* observed that the space satellite fit the Soviet propaganda scheme "by implying that Soviet rockets can deliver heavy nuclear weapons."[42] In this respect, *Sputnik 2* was more significant than *Sputnik 1*, because the second satellite had a mass of 1,121 pounds, which *Time* magazine observed was "heavier than many types of nuclear warheads."[43]

President Eisenhower tried to downplay the significance of the event. On the day *Sputnik 1* was launched, he left the White House for a weekend of golf and rest at Gettysburg, Pennsylvania. The White House press secretary, James Hagerty, sought to pacify the press. In spite of reports to the contrary, Hagerty said, the feat had not caught the administration by surprise.[44] Defense Department officials discounted the military significance of the event. The launch provided no evidence of Soviet superiority in either missile development or military capability; the satellite could not be used to bomb Americans while they slept.[45] Eisenhower himself joined the effort to calm the American public. "I see nothing at this moment . . . that is significant . . . as far as security is concerned," he announced at a 9 October press conference devoted to the subject.[46] In a televised address on American science and national security the evening of 7 November, he again sought to reassure the nation with the statement that "the overall military strength of the free world is distinctly greater than that of the communist countries."[47]

The effort hardly accomplished its purpose. *Newsweek* called the Soviet satellite launch the "greatest technological triumph since the atomic bomb."[48] *Time* reproduced Washington senator Henry Jackson's complaint that it had been "a week of shame and danger."[49] News outlets compared the event to the splitting of the atom and the discovery of America by Columbus.[50] Outblitzed in the media, Eisenhower watched his popularity fall twenty-two points from its postelection high.[51]

In the wake of the Sputnik crises, the House Space Committee attacked Eisenhower's alternative as a "beginner" space program that failed to show "proper imagination and drive." Committee staff urged the administration to mobilize facilities throughout the nation in order to develop manned space stations, build large launch vehicles, and dispatch rockets to nearby planets.[52] Both the U.S. Air Force and U.S. Army drew up plans to put the first human into space. The director of the National Advisory Committee for Aeronautics, Hugh Dryden, announced that his agency was prepared to supervise "the travel of man to the moon and nearby planets."[53] Majority Leader Lyndon B. Johnson assembled a Senate preparedness investigating committee that held hearings from November to January and issued recommendations to strengthen the nation's missile and satellite programs.[54] Discontent with the Eisenhower space alternative reached near-hysterical proportions during this time.[55] Attempts by people in the Eisenhower administration to dispel space fears simply encouraged the belief that the president was inept and did not understand the nature of the challenge.

It is hard for people now separated from the events of the 1950s to appreciate how much the possibility of nuclear war dominated American popular culture. U.S. citizens had emerged from an exhausting world war during which their homeland was essentially secure from enemy attack to find that horrible weapons could now reach their shores. The Soviet Union learned how to fabricate atomic and thermonuclear weapons. Soviet agents had infiltrated the U.S. government and stolen atomic secrets, conservatives charged. School boards required children to practice civil defense drills and master techniques such as diving beneath their desks at the sign of the first thermonuclear flash. Whole cities practiced evacuations, and warning sirens wailed weekly in civil-defense tests. Citizens constructed bomb shelters in their basements and backyards. The activities may seem eccentric by modern standards, but they contributed significantly to public anxiety about an atomic attack during the 1950s.

For the most part, those promoting space exploration through the popular media did so because it promised adventure and discovery. Adventure and discovery, however, could not elicit the billions of dollars required to mount an aggressive exploration effort.[56] National security considerations could, particularly among those who believed space to be the "high ground" from which the Cold War would be decided. Space boosters increasingly tied their ambitions to popular fears about the nuclear age. The parsimony that motivated the Eisenhower space program seemed incomprehensible to Americans worried about the outcome of the Cold War. George Reedy, one of Lyndon Johnson's principal aides, predicted that Eisenhower's inattentiveness to this issue "would blast the Republicans out of the water."[57]

The perception of space as the high ground of the nuclear age had gained popular acceptance in the late 1940s. As part of their effort to promote space exploration, the editors at *Collier's* magazine published a story in their 7 September 1946 issue on the colonization of the Moon. In the article, American space flight advocate and science fiction writer G. Edward Pendray issued a stern warning about the military significance of the undertaking:

> Its gravitational attraction is so small that rockets only a little faster than the German V-2s could bombard the earth from the moon. With the aid of suitable guiding devices, such rockets could hit any city on the globe with devastating effect. A return attack from the earth would require rockets many times more powerful to carry the same pay load of destruction; and they would, moreover, have to be launched under much more adverse conditions for hitting a small target, such as the moon colony. So far as sovereign power is concerned, therefore, *control of the moon in the interplanetary world of the atomic future could mean military control of our whole portion of the solar system.* Its dominance could include not only the earth but also Mars and Venus, the two other possibly habitable planets.[58]

At the end of his article, Pendray suggested that the Moon "may be the fortress of the next conqueror of the earth."[59] In making this claim, Pendray adopted a method of presentation repeated by space boosters for a decade thereafter: he offered an assertion without accompanying details. His magazine statement on the military significance of the Moon is quoted here in its entirety; he offered no more. The absence of technical details allowed space advocates to maintain their assertion without proving it, which they could not have done since the assertion was in fact not true. The Moon provides a relatively poor platform for launching nuclear warheads toward the Earth, as a comparative analysis of lift weights and trajectories will reveal.

To reinforce Pendray's claim, editors at *Collier's* showed their readers how such a nuclear strike could actually occur. In a 23 October 1948 article titled "Rocket Blitz from the Moon," the story opened with an illustration of two V-2–shaped rockets rising out of lunar craters with a dome-shaped control center in the background. On the adjoining page, two large fireballs spread across an aerial view of New York City.[60] The nuclear blasts, drawn with stark realism by space artist Chesley Bonestell, were part of a larger literature portraying the effects of nuclear war on the United States. Once again *Collier's* assured readers that the Moon "could be the world's ideal military base."[61]

Promoters of space exploration repeatedly warned the American public of the military importance of space. In the 30 August 1947 issue of *Collier's*, science fiction writer Robert A. Heinlein teamed up with Captain Caleb Laning of the U.S. Navy to explain how the absence of a space corps would leave the U.S. defenseless: "Once developed, space travel can and will be the source of supreme military power over this planet—and over the entire solar system— for there is literally *no* way to strike back from ground, sea, or air, at a space ship, whereas the space ship armed with atomic weapons can wipe out anything on this globe."[62] The assertion reappeared frequently in the years that followed, commonly without elaboration. In explaining why the governments of Earth might be interested in constructing a space station, the author of a 1952 children's book on space travel asserted that "from such a station it would be possible to launch giant rockets with explosives or atom bombs against almost any spot on earth."[63]

Wernher von Braun advanced the same argument in his famous article in the 22 March 1952 issue of *Collier's*, calling for the construction of a space station: "There will also be another possible use for the space station—and a most terrifying one. It can be converted into a terribly effective atomic bomb carrier. Small winged rocket missiles with atomic war heads could be launched from the station in such a manner that they would strike their targets at supersonic speeds. . . . In view of the station's ability to pass over all inhabited regions on

Although public interest in space flight grew, political support for major space spending remained thin. To win financial support, exploration advocates appealed to public fears about the Cold War, arguing that "control of space meant control of the world." To illustrate the point, *Collier's* ran an article describing how a hostile power could use the Moon as a platform to launch nuclear missiles against the Earth. (Reproduced courtesy of Bonestell LLC)

earth, such atom-bombing techniques would offer the satellite's builders the most important tactical and strategic advantage in military history."[64]

In addition to its use as an "atomic bomb carrier," von Braun believed that his proposed space station would revolutionize military reconnaissance. He proposed that the space station be placed in a polar orbit, which would allow its occupants to act as global observers as the Earth turned beneath them every twenty-four hours. "Troop maneuvers, planes being readied on the flight deck of an aircraft carrier, or bombers forming into groups over an airfield will be clearly discernible," von Braun stated. Using special optical instruments, he said, the view from orbit would be as clear as that from an observation plane flying close to the ground. "Because of the telescopic eyes and cameras of the space station, it will be almost impossible for any nation to hide warlike preparations for any length of time."[65]

In the introduction to their series on what they called the inevitable conquest of space, the editors at *Collier's* magazine warned of the consequences of falling behind: "The U.S. must immediately embark on a long-range development program to secure for the West 'space superiority.' If we do not, somebody else will. That somebody else very probably would be the Soviet Union."

Just as V-2 rockets from Germany fell on London during World War II, space advocates warned that nuclear-tipped missiles from space could fall on New York during the Cold War. In this Chesley Bonestell rendition, one warhead has exploded near the Empire State Building, another in Queens. *Collier's* ran this painting alongside the illustration of rockets rising from the lunar surface in a vivid two-page introduction to the 1947 article "Rocket Blitz from the Moon." (Reproduced courtesy of Bonestell LLC)

A space station under the control of the free world, the editors argued, "would be the greatest hope for peace the world has ever known." No nation could make undetected preparations for war, effectively tearing down the Iron Curtains of secrecy that communist leaders had erected around their nations. In the wrong hands, a space station, the importance of which the editors likened to the development of the atomic bomb, would allow ruthless dictators to rule the world.[66]

Writers at *Collier's* insisted that "what you will read here is not science fiction."[67] In fact, the arguments were more fictional than real, a point underscored when Hollywood picked up the theme. The most critically acclaimed work of fiction to contain the message was the 1950 movie *Destination Moon*, which opens with an unsuccessful rocket launch and the news that the U.S. government has canceled support for the rocket development project. Seeking financing for a nuclear-powered rocket that can fly as far as the Moon, project leaders turn to private industry. At first industrial leaders are skeptical. Funds flow freely, however, once a retired military general explains: "We're not the only ones planning to go there. The race is on, and we better win it, because there is absolutely no way to stop an attack from outer space. The first country that can use the moon for the launching of missiles will control the earth. That, gentlemen, is the most important military fact of our century."[68]

In retrospect, few of the early warnings about the military significance of space turned out to be true. Weapons technology allowed the construction of Earth-based missiles that could carry bombs to distant parts of the planet with far more speed and accuracy than rockets based on the Moon. Large missile-carrying submarines allowed military planners to conceal the location of nuclear weapons, a substantial advantage over the predictable location of orbiting platforms. Recognizing these facts, U.S. and Soviet leaders concluded by the mid-1960s that neither side could gain a military advantage by placing nuclear weapons in space and thus in 1967 signed a treaty agreeing not to do so.[69] Likewise, technology overtook the claims about military reconnaissance uses for space stations. Scientists developed methods by which precise images could be transmitted remotely from satellites in orbit. As a consequence, automated spy satellites played a far more important reconnaissance role than astronauts peering through optical instruments.[70] It took a number of years for people in the U.S. government to realize this, but they eventually did. By the mid-1970s, the U.S. Air Force had abandoned its efforts to place sentries in space and accepted the virtues of satellite technology. Eisenhower's alternative space program, with its emphasis upon robotics, eventually came to dominate military space policy in the United States.[71]

During the 1950s, various groups tried to advance realistic statements about the military uses of space. At the time, few of their assertions took root. President Eisenhower's Science Advisory Committee, for example, added a special section on military applications to its famous 1958 brochure, correctly observing that communication and reconnaissance satellites would provide the most important military uses of space and predicting that satellites with telescopic cameras would be able to instantly transmit high-quality images back to Earth. As for the proposals to place bombs in space, committee members characterized these as "clumsy and ineffective ways" of delivering nuclear weapons: "Take one example, the satellite as a bomb carrier. A satellite cannot simply drop a bomb. An object released from a satellite doesn't fall. So there is no special advantage in being over the target. Indeed, the only way to 'drop' a bomb directly down from a satellite is to carry out aboard the satellite a rocket launching of the magnitude required for an intercontinental missile." Even if the weapon were given a small push and allowed to spiral in gradually, that would mean "launching it from a moving platform halfway around the world." Schemes to drop bombs from space had "every disadvantage" compared to launching bombs from Earth.[72]

Advocates of human space travel typically ignored arguments such as these as they continued to win converts through their claims about world supremacy. A number of factors allowed this to occur. First, because public understanding about the reality of space travel was terribly thin, based as much on science fiction as on real experience, public perceptions permitted the application of misleading images about the new frontier. Second, people who misunderstood orbital mechanics tended to draw incorrect analogies from the bombing campaigns of World War II. People who knew that bombs fell from airplanes were misled into viewing space stations and automated platforms as extensions of airborne bombers. Third, primitive technology also supported false claims. Images from first-generation spy satellites had to be retrieved through clumsy reentry schemes involving film drops, parachutes, and airplanes with skyhooks.[73] This awkward technology lent credibility to the notion that space stations would become a valuable reconnaissance tool.

The national media could have played an important role in educating the public about the realities of space. Unfortunately, they did not. They behaved irresponsibly, apparently sensing that hysterical assertions about threats from space were more newsworthy than calming assurances. Even the 1952 *Time* magazine cover story, the classic rebuttal to the *Collier's* series, did not challenge von Braun's warning about atom-tipped guided missiles spiraling downward through the atmosphere from orbiting space stations. Instead, the article characterized doubters as "timid military planners" and suggested that von Braun's schemes would sound even more practical if the public could view the military secrets that purportedly supported them.[74]

The willingness of the public to believe the worst drew much of its force from a larger concern with the way in which nuclear war might cause the world to end. This preoccupation lent credibility to the popular acceptance of space as the place from which both Armageddon and salvation might arrive. Without the cultural obsession with nuclear war and final days, the more outrageous claims about the need to "conquer" space would not have been so believable.

Since biblical times various people have advanced apocalyptic predictions, and public hysteria about impending doom has periodically swept across the face of civilization. Apocalyptic literature enjoyed wide popularity in both Jewish and Christian communities between 200 B.C. and A.D. 200. The book of Revelation, composed some fifty years after the death of Christ, provided Christian believers with a rich though symbolic description of the final days, including predictions of war, famine, plague, earthquakes, and falling stars.[75] Establishing the exact time of the apocalypse has preoccupied many prophets and religious figures. A nineteenth-century New England farmer, William Miller, recruited some fifty thousand followers by deciphering one biblical chronology. His followers set the exact date of the end at 22 October 1844 and gathered together in their usual places of worship to pray and await the Lord.[76]

Public interest in the end of the world remained strong from the years just preceding World War II through the first decade of the atomic age. The thought that civilizations were preparing for another massive war joined with eventual understanding of the weapons developed for that war encouraged members of the public to entertain various scenarios. In earlier times, religious leaders explained how the world might end; in the twentieth century, the public listened to scientists who looked to the sky.

In 1937 the Hayden Planetarium in New York presented a show that gave planetarium viewers a ringside seat on the end of the world.[77] The show drew upon works of popular science that had appeared with increasing frequency after the turn of the century.[78] All of the threats presented by planetarium curators came from outer space. *Life* magazine publicized the show with a 1937 article summarizing "Four Ways the World Might End," illustrated by Rockwell Kent. The article offered illustrations of death by fire due to the explosion of the sun and a permanent ice age occasioned by its contraction. The latter, the authors incorrectly told their readers, would occur "within a few million years." The orbit of the Moon was unstable, they further explained, which would eventually cause it to come crashing toward the Earth. A final scenario portrayed the breakup of the planet due to a celestial body passing too close to the globe. For this scenario, Kent portrayed a strange phenomenon. The gravitational attraction of a very large body would for a split second snatch human beings off the surface of the Earth and propel them skyward, creating an astronomical equivalent of the ascension.[79]

Scientific descriptions of the end of the world entered popular culture through vehicles both popular and fanciful. As part of their eight-part series on "The World We Live In," editors at *Life* magazine gave their readership an illustrated view of the beginning and end of the Earth. Counterbalancing colorful illustrations of the birth, the article featured a two-page painting by Chesley Bonestell portraying the dawn of the day on which the sun would explode.[80] Employment of spaceships to escape impending doom was a frequently used theme. Reporting on the Hayden Planetarium show in 1950, a writer for *Popular Mechanics* magazine announced that astronomers should be able to detect the impending catastrophe in sufficient time to give humans "a few thousand years to get busy colonizing another solar system." The script for the Hayden Planetarium show ended with the promise that humans might evacuate the Earth "and as a refugee on Venus find safety and peace—if there is a Venus then."[81]

After 1945, with the explosion of the atom bombs at Hiroshima and Nagasaki, scientists added nuclear war to the list of doomsday scenarios. Scientists who had earlier sought to explain far-in-the-future ends of the Earth found that they had helped create the means for an imminent conflagration. Writing for a 1947 issue of *Atlantic* magazine, Albert Einstein predicted that "little civilization would survive" after a nuclear war.[82] In 1950 he extended his prediction by announcing in a speech broadcast over NBC television that "radioactive poisoning of the atmosphere and hence annihilation of any life on earth has been brought within the range of technical possibilities." By the early 1950s, atomic war was well established among writers of popular science as one of the ways in which the world could end.[83]

Throughout World War II, skillful wartime photographers exposed the American public to the face of war. Photographs of mass destruction, much of it caused by aerial bombardment, filled American magazines and newspapers.[84] With the advent of the atomic age, it became clear that such destruction could be visited upon the United States, only thousands of times worse. The popular press portrayed nuclear Armageddon in much the same way it had reported World War II. Speculation about the effects of atomic bombardment increased following the news in September 1949 that the Soviet Union had exploded its first atom bomb.

An article in the 21 April 1953 issue of *Look* magazine was typical. The article opened with a drawing of a hydrogen bomb exploding in the sky above New York City, prepared by Chesley Bonestell. Authors explained the technical process for building a hydrogen bomb and presented a map showing how a bomb dropped over the Capitol in Washington, D.C., would knock down buildings and burn people as far away as Baltimore.[85] An earlier issue of *Collier's*, with a cover illustration by Bonestell, included an imaginative news story describing the effects of an atom bomb exploding in lower Manhattan.[86] *Collier's* later published an article describing "The Third World War," complete with another Bonestell illustration that showed Washington, D.C., in flames.[87]

None of these stories forecast the end of the world, a scenario apparently judged too depressing for periodicals that depended upon subscriptions purchased in advance. Instead, the stories promised deliverance through deterrence or safety through civil defense. More pessimistic scenarios found their way into fiction.[88] In Ray Bradbury's classic *The Martian Chronicles*, humans who have settled on Mars watch in horror as the Earth explodes in an atomic conflagration.[89] Nevil Shute's chilling novel *On the Beach*, also a top-grossing movie, describes the breakdown of society in the aftermath of an atomic war as survivors wait for radioactive fallout to engulf them.[90] In one of the better mutant movies—*Them!*—giant ants charge out of an atomic test site in New Mexico to challenge civilization.[91]

In the mid-1950s, nuclear war preoccupied the American mind. Most Americans believed that another world war would occur within their lifetimes and that hydrogen bombs would be used in it. Many believed that all of humankind would be destroyed in an all-out nuclear exchange. In a series of public opinion polls throughout the 1950s, Americans consistently identified the threat of war as "the most important problem facing the entire country today."[92]

Science revealed terrible ways in which the planet might be destroyed. All of the doomsday scenarios, both astronomical and human in origin, fell from the sky. This led naturally to the conclusion that activities taking place above the surface of the Earth would determine the future of the world. One of the most bizarre manifestations of this belief was the unidentified flying object (UFO) phenomenon. Beginning in 1947, Americans in ever-increasing numbers began to report sightings of mysterious flying objects.[93] Media interest increased following the January 1950 issue of *True* magazine, which carried a story by Donald A. Keyhoe arguing that the objects were spacecraft under the control of intelligent beings from another planet.[94] In 1952 individuals began to report actual contacts, some personal and others via mental telepathy. Radio and television programs devoted to extraterrestrial encounters thrived. Societies of believers arose, some of which attempted to spread the message that visitors had come to save humans from impending destruction.[95]

To those who believed in their extraterrestrial origin, flying saucers were like messengers from a higher civilization. Unlike angels and other religious figures who floated down from heaven in the past, flying saucers were not presumed to be the creation of the Deity. They were the product of advanced science, evidence of the length to which technology could carry civilization. Many viewed the saucers as a source of salvation from the treat of atomic war. In the original version of the classic 1951 science fiction film *The Day the Earth Stood Still*, a flying saucer lands in Washington, D.C., where its captain orders world leaders to control the nuclear arms race or face annihilation by a race of robot police. "The Universe grows smaller every day—and the threat of aggression by any group anywhere can no longer be tolerated. There must be security for all—or no one is secure. . . . It is no concern of ours how you run your own planet, but if you threaten to extend your violence, this earth of yours will be reduced to a burned-out cinder."[96]

The saucer phenomenon helped foster an increasing distrust of government in American society. A central tenet among UFO partisans was the belief that certain members of the U.S. government were engaged in a vast conspiracy to shield the truth about flying saucers from the public at large. From this perspective, public officials had dropped flying saucers behind the cloak of national secrecy that shielded other technological developments in the nuclear age. UFO partisans alleged that government officials had obtained evidence as early as 1947 supporting the premise that flying saucers were real spacecraft from another planetary system.[97] Efforts by the U.S. Air Force to explain UFO sightings, as with Project Blue Book, simply inflamed accusations that the government was hiding the truth.[98]

Reports of alien visitors did not originate in the modern era. In medieval times people spoke of encounters with demons and other creatures from the underworld. Demons were even thought to engage in sexual unions with humans, a theme later repeated in reports of extraterrestrial encounters. The otherwise reliable nature of these reports, both medieval and modern, suggests that the phenomena may represent a form of mass hallucination, triggered in both eras by apocalyptic fears.[99] Once science overpowered religious superstition, aliens from other civilizations replaced evil spirits as the visitors from beyond.

One striking piece of evidence supports this thesis. After an initial burst of interest, the rash of UFO sightings dropped off during the mid-1950s to an average of forty-six per month. This suddenly changed following the launch of *Sputnik 1*, when the number of reports rose sharply, with over six hundred sightings in the final three months of that year.[100] Barring the unlikely possibility that aliens actually stepped up observations of Earth, one is left with the plausible explanation that the launch of the Soviet satellite excited fears that caused people to scan the heavens and, as a result, detect more unknown objects in the sky.

From atom bombs to flying saucers, objects from space fell onto the public consciousness during the 1950s. Reassurances by political and scientific elites who told the public not to worry only fueled suspicions that people in the government were not disclosing all they knew. To the public at large, space technology seemed to be the source from which national salvation or doom would come. It was easy for Americans to believe that control of space would determine the future of the world.

Governmental leaders, aware of the power of public opinion in a democracy, struggled to balance their sense of reality with the perceptions of the citizenry. This proved especially challenging, because so much of the discourse about space travel was based on a vision. Politicians who adopted the language and symbols of the vision gained an advantage in the quest for public support over those who promoted a reality that few members of the public yet understood. Space travel was new and the temptation to use popular language and metaphors of imagination in political discourse pushed official policy toward those priorities the vision contained.

The images promoted by advocates of human space flight and amplified by the news media became part of the political discourse, especially after the precipitating events of the Sputnik launches. By the late 1950s, public officials in increasing numbers adopted the language of popular anxiety about space and the Cold War. On 7 January 1958 Senate majority leader Lyndon Johnson stood before the Democratic Conference (the assembly of Democratic senators) and elevated space policy to the top of the Senate agenda. His words echoed the warnings of people such as Wernher von Braun and forums such as *Collier's* magazine: "Control of space means control of the world, far more certainly, far more totally than any control that has ever or could ever be achieved by weapons, or troops of occupation. . . . Whoever gains that ultimate position gains control, total control, over the earth, for purposes of tyranny or for the service of freedom."[101] Johnson insisted that his perspective was endorsed by "the appraisal of leaders in the field of science, respected men of unquestioned competence." He did not point out that most scientists greeted his viewpoint with considerable skepticism, nor did he observe that the "control of space" argument had been advanced by people more interested in human adventure than scientific discovery.[102] Johnson gained no political advantage by emphasizing the skepticism that still transfixed the scientific community.

Instead, Johnson saw a political advantage in appealing to the public impression of space as the battleground of future conflict. He had grown up in the hill country of Texas, the son of a local politician of limited means. In spite of his election from the conservative South, Johnson remained a rural progressive and disciple of Franklin Roosevelt. Johnson knew that conservative opposition to Roosevelt's New Deal social agenda arose from the same concerns that inflamed conservative reluctance to launch a big space program. As he told members of the Democratic conference, "Our decisions, more often than not, have been made within the framework of the government's annual budget. This control has, again and again, appeared and re-appeared as the prime limitation upon our scientific advancement."[103]

Johnson's remarks were directed at not only budget-conscious members of the Eisenhower administration but also fiscal conservatives within the Democratic Party who did not want the federal government to play a large role in domestic affairs. Johnson believed that conservative Democratic senators, particularly those from the South, could be motivated to vote for a large government presence in space as a matter of national security. This, he believed, would create the precedent he needed to press for government largess in areas such as health and education.[104] Johnson also understood how vulnerable Republicans had made themselves by failing to allay public anxiety.

Other Democratic senators repeated Johnson's Cold War rhetoric. In a Veterans Day address in November 1957, Democratic senator Stuart Symington of Missouri announced that "the race for the conquest of space is today's major engagement in the technological war. We must win it, because the nation which dominates the air spaces [*sic*] will be in a position to dominate the world."[105] During his 1960 presidential campaign, Senator John F. Kennedy insisted that the United States win the race to conquer space: "Control of space will be decided in the next decade. If the Soviets control space they can control earth, as in past centuries the nation that controlled the seas dominated the continents. . . . We cannot run second in this vital race. To insure peace and freedom, we must be first."[106] Public concern about the space race helped Kennedy maintain the most important illusion of his 1960 campaign, the non-existent "missile gap" that Democrats charged the Eisenhower administration with creating.[107] Space spectaculars by the Soviet Union, accompanied by U.S. failures, helped perpetuate the illusion that the United States trailed the Soviets in the production of ballistic missiles, a charge Republicans failed to refute for fear of revealing the intelligence sources that provided the government with information on Soviet nuclear capability.

Officials within the government who stood to gain from an expanding space program issued similar warnings. Brigadier General Homer Boushey, the deputy director for Air Force Research and Development, quoted Johnson's words in a speech before the Aero Club of Washington, D.C., on 28 January 1958. Asserting the military advantages of controlling space, Boushey announced that "he who controls the moon controls the earth." Boushey believed that the Moon would provide the United States with a platform for a powerful nuclear deterrent. If the Soviets attacked the United States, he explained, they would receive "sure and massive destruction" from the Moon some forty-eight hours later. "The moon," Boushey concluded, "represents the age-old military advantage of 'high ground.'" Boushey's remarks were reproduced in the 7 February issue of *U.S. News & World Report*.[108] Other military leaders issued similar warnings.[109]

Such statements exasperated Eisenhower and his advisers, who mightily tried to calm public fears. Interviewed for a 1958 issue of *Reader's Digest*, Chief of Naval Operations Arleigh Burke tried to explain that missiles launched from the Earth offered a far more practical deterrent than orbital platforms.[110] The president's Science Advisory Committee summed up the arguments about the military significance of space: "Much has been written about space as a future theater of war, raising such suggestions as satellite bombers, military bases on the moon, and so on. For the most part, even the more sober proposals do not hold up well under close examination."[111]

These reassurances did little to alleviate the impression that the space race had turned the Eisenhower White House into what pundits labeled "the tomb of the well-known soldier."[112] National news outlets reporting the president's reassurances gave equal time to space Cassandras preaching national doom. In a feature article in *Life* magazine shortly after the launch of *Sputnik 1*, scientist George R. Price argued that "unless we depart utterly from our prevent behavior, it is reasonable to expect that by no later than 1975 the United States will be a member of the Union of Soviet Socialist Republics."[113] Even the president's science adviser, James Killian, acknowledged the difficulty of overcoming public perceptions: "Sputnik I created a crises of confidence that swept the country like a windblown forest fire. Overnight there developed a widespread fear that the country lay at the mercy of the Russian military machine and that our government and its military arm had abruptly lost the power to defend the homeland itself, much less to maintain U.S. prestige and leadership in the international arena."[114] Eisenhower was correct, but to little avail. In early 1961 President John F. Kennedy set aside Eisenhower's approach to space flight and embarked upon a massive buildup designed to make the United States first in space. In March Kennedy approved the allocation of funds necessary to accelerate development of the Saturn rocket program, and in May he established the goal of placing the first Americans on the Moon.[115] NASA's budget increased tenfold, from $524 million in the last full year of the Eisenhower administration (1960) to $5.3 billion in fiscal year 1965.

Even skeptics within NASA became advocates for the spacefaring dream. This outcome was by no means predestined. The groups that filed in to form the newly created civilian space agency represented a wide range of views, from romantic visionaries such as Wernher von Braun to practical conservatives such as Milton Rosen. Most of the employees who formed NASA were neither romantic visionaries like von Braun nor advocates of Eisenhower's conservative satellite program. They were pragmatic American engineers, people such as spacecraft designer Max Faget. Faget worked for the National Advisory Committee for Aeronautics, which contributed all of its eight thousand employees to the newly formed civil space effort. The son of a renowned public

Exploration advocates skillfully exploited public fears about Soviet achievements in space to encourage a strong U.S. response. President Dwight Eisenhower remained calm, but public concern overwhelmed his commitment to a modest space program. To demonstrate national resolve, the United States built bigger and better rockets, culminating with the rollout of the giant Saturn 5 rocket that dispatched the first humans to the surface of the Moon. (NASA)

health physician, Faget became an engineer so that he could design airplanes. After a tour of duty as a submarine officer in the U.S. Navy, Faget went to work for the NACA laboratory at Langley, Virginia. He was assigned to what Langley executives called the Pilotless Aircraft Research Division, a euphemism for work on rockets and guided missiles. Faget and his colleagues fastened models onto rockets and shot them over the Atlantic at supersonic speeds to see how the models would fly. It was the closest thing to spacecraft testing that NACA did, but NACA engineers would not call it that. NACA employees did research on airplanes, a romantic undertaking in its own right. For most of the organization's existence, as historian James Hansen points out, *space* was a dirty word, and until the space race began, NACA employees were not much interested in what many regarded as that "Buck Rogers stuff."[116]

Pragmatic NACA employees might have created a balanced vision of space flight, halfway between the romantic dreams of followers of people such as Wernher von Braun and the overly cautious Eisenhower alternative. They did not, and they failed to do so for a fairly simple reason. Like most Americans at the time, they found themselves caught up in the momentum of the Cold War, the battleground of which had become space. If Americans lost that battle, many believed, the Russians would rule the world. Pragmatic American engineers jumped in to help beat the Russians in space, implementing the romantic vision of human space flight that would dominate American civil space policy for decades to come. Even Rosen joined the bandwagon, publishing a series of articles on the importance of human endeavors. "As soon as the Sputnik went up," Faget explained, "all bets were off. It was pretty much a free for all."[117] Eisenhower's alternative disappeared under the force of Cold War necessities, and memory of it faded as the vision of human dominance filled the public mind.

As is often the case, the things people feared most did not occur. Bombs did not fall from orbiting satellites or rain down from the Moon. The Soviet Union did not conquer the world, even though it became the first nation to establish a small Earth-orbiting space station in 1971. Public preoccupation with the threat of nuclear war began to wane as early as 1963.[118] The Cold War itself ended with the collapse of the Soviet Union in 1991. As concern about the Cold War weakened, so did the connection between the control of space and national security. This created a fundamental problem for people promoting the U.S. space program. The fears that furthered their romantic vision of space exploration no longer preoccupied the public mind. The United States found itself with a space program fashioned in response to an image of the world that had disappeared in fact and in imagination. Without the Cold War, the civil space program became a cause in search of an explanation.

Social scientists for some time have studied the effects of failed beliefs.[119] True believers do not automatically abandon their cause just because reality intrudes in discomforting ways. They rarely admit they were wrong or change their behavior, especially when they keep meeting with other people who share their fantastic ideas. Believers so sustained often seek new beliefs to validate old behavior and new interpretations to explain why prophecies fail to materialize. Sometimes believers deny that their prophecies in fact miscarried. Advocates of space exploration did not abandon their cause just because public interest in the Cold War waned. They did not retreat to less ambitious space efforts. Instead, they sought new rationales for continuing endeavors, and concocting fears remained part of that strategy.

Some insisted that control of space still meant control of the Earth, a position advanced by promoters of the Strategic Defense Initiative ("Star Wars"), a space-based missile defense proposal that President Ronald Reagan endorsed in 1983. Commanders fighting the 1991 Gulf War, in which satellite technology played a decisive role, reaffirmed the importance of military assets in space. For these individuals, space remained the new high ground of modern, technological war, and they insisted that the United States retain its capability to operate freely within it.[120]

Some partisans of space travel attempted to revive the specter of superpower conflict. In 2003, the People's Republic of China completed its first orbital flight of a Chinese "taikonaut," followed in 2005 by the country's first two-person mission and in 2008 by its first flight of a three-person spacecraft with an accompanying spacewalk. U.S. officials had excluded China from the international effort to construct a large orbiting space station and, partly for reasons of national prestige, Chinese officials had decided to pursue their own human space flight capability. Commentators warned of another two-nation space race. China intended to create a "Red Moon," some claimed, by establishing a Communist military base on the lunar surface.[121] China intended to develop the capability to shoot down U.S. satellites, they said, a claim that China helped to confirm in 2007 when it tested the concept by destroying one of its own weather satellites with an antisatellite weapon. Most dramatically, the partisans claimed, the Chinese planned to race the United States into space.

The Chinese intentions as reported by Western sources strikingly conformed to the dominant vision of space travel that captured public attention in the United States during the Cold War, suggesting that the people describing the Chinese program were looking not so much through a window as into a mirror. When public officials in the United States announced their intent to return to the Moon by 2020, Western observers warned that the Chinese could be there by 2017. When officials in the United States publicized their desire to send astronauts to Mars by the middle of the twenty-first century, observers said that the Chinese wanted to arrive by 2040. Partisans of human space flight had long dreamed of sending humans beyond the Moon to gravitationally stable Lagrange points. Sources said that the Chinese were developing that capability too.[122]

Officially, the Chinese government had made no such statements. Of the two white papers issued by the Chinese government as of the dates of the Western reports, neither mentioned the Moon or Mars. One Western expert ascribed the U.S. reports of Chinese intentions to "poor translation or misunderstanding of Chinese comments." Reporters for the Western press, he said "frequently quote Chinese media sources that have no credibility within their own country."[123] The statements, nonetheless, continued to appear.

At the height of the Cold War, Russian achievements in space prompted a dramatic U.S. response. Peer competition, technological fear, and the threat of new and terrible weapons pointed at Europe and America encouraged politicians to associate space pursuits with the outcome of superpower conflict. With the exception of the Chinese proclivity for satellite destruction, which does concern people in the U.S. defense community, modern Chinese activities did not have a similar effect.[124] Like a Hollywood sequel, statements about Chinese intent appealed mostly to people who had loved the first release. China's endeavors helped to maintain support for the existing level of space activities—but so did the activities of the large number of countries, including American allies, developing their own capabilities in space. China's presence in space did not trigger an enlarged investment of the magnitude experienced during the first two decades of the Cold War.

As concerns about superpower conflict diminished and bombs did not fall from space, advocates of space exploration looked for other objects that did. Increasingly, natural objects such as comets and asteroids found their way into the politics of fear. In 1991 Clark Chapman of the Planetary Science Institute and David Morrison of NASA's Ames Research Center announced that a person's chance of being killed as the result of a large asteroid strike was "more than six times greater than your chance of dying in a plane crash." Chapman later revised this estimate (the probabilities are about the same), but the statement drew serious concern. The small number of people who die in commercial airline crashes in the United States each year produces a cumulative total over many thousands of years that does not exceed the danger posed by a single nation-killer that wipes out nearly everyone once during the same period.[125] Chapman and Morrison's statistic was contrived, but prophetic. People should not be less worried about asteroid strikes simply because they occur infrequently.

A major asteroid strike might not occur for three hundred thousand years (it might with equal probability occur tomorrow). When it does, however, it could devastate the Earth. The massive impact could create a continent-wide

When the Cold War ended, so did much of the motivation for very large undertakings in space. Seeking other nightmares with which to scare up political support, space advocates warned of the need to protect the Earth from comets and asteroids. Celestial objects had struck the Earth in the past, these individuals observed, and any advanced civilization would need the capability to deflect such threats to ensure its long-term survival. (Artwork by Don Davis; courtesy of NASA)

firestorm. Nitric acid produced from the burning of the atmosphere might acidify lakes and soil and even poison the surface of the ocean. The temperature of the Earth could alternatively plunge and rise, first as dust blocked out sunlight and second as carbon dioxide pushed into the stratosphere produced a greenhouse effect. "Agriculture and commerce would probably cease," Chapman and Morrison warned, "and most people in the world would die."[126]

Entertaining the view that doom and salvation come from above, writers of science fiction have for some time contemplated the consequences of large objects striking the Earth. In 1951 producer George Pal used the possibility of a celestial flyby to portray the end of Earth in the movie *When Worlds Collide*. In the film, which won an Oscar for its special effects, two wandering planets intersect Earth's orbit. Before the second planet collides with the Earth, a small group of humans escape to the first on a large rocket ship, where they begin civilization anew.[127] In *Lucifer's Hammer*, residents of California are bombed back into a new dark age by the effects of a massive comet smacking the Earth. Arthur C. Clarke showed how Earthlings might deflect approaching asteroids in the 1993 novel *Hammer of God*, and in 1998 Hollywood released two blockbusters—*Armageddon* and *Deep Impact*—in which humans in spaceships do exactly that. Interplanetary collision films combine the best traditions of science fiction with the fascination accorded disaster stories.[128]

Public interest in massive collisions was not enlarged by fiction alone. In 1980 Luis and Walter Alvarez made the startling suggestion that a large asteroid strike may have been responsible for the death of the dinosaurs. Cross sections of rock formations show evidence of a catastrophic event at the end of the Cretaceous era sixty-five million years ago; analysis suggests that an asteroid in the ten- to fifteen-kilometer range was responsible. This was not a freak event, moreover. Astronomers and geologists believe that really big asteroids fall every ten to thirty million years, whereas smaller objects with the potential for global transformation (at least one kilometer in size) hit Earth every three hundred thousand years.[129]

Three hundred thousand years is a long time. Remote possibilities can motivate governmental spending, nonetheless, when real events publicize the prophecies of doom. Such an event occurred in the summer of 1994. A comet perhaps ten kilometers across, named Shoemaker-Levy, split into a string of fragments that struck the planet Jupiter in mid-July. The largest fragment left visible shock waves in the Jovian cloud cover as wide as the Earth. Space advocates reminded the public that much smaller strikes could have devastating consequences on the Earth. In 1908, for example, a small asteroid about sixty meters across entered the atmosphere above Tunguska, Siberia, where it exploded and flattened trees over an area twice as large as the city of New York. In March 2009 a Tunguska-sized asteroid unexpectedly passed within forty-five thousand miles of the Earth, a planetary near miss in astronomical terms. Astronomers detected the object only two days in advance of its passing. Had the asteroid struck the Earth near a populated area like New York, the effects would have been as catastrophic.[130]

Space advocates employed the possibility of asteroid strikes to resurrect a variety of proposals for extending humans and their capabilities into space. At the least, advocates argued, the swarm of medium- to large-sized comets and asteroids, thousands in number, whose paths intersect that of the Earth should be cataloged, their orbits calculated. The asteroid belt should be explored, they added. As one advocate group observed, asteroids are the nearest extraterrestrial neighbors beyond the Moon and "logical sites to develop the techniques of human deep-space exploration." Rocket ships parked in Lagrange points could easily reach them. Many offered techniques for deflecting the orbits of comets and asteroids. Leading candidates included mass drivers, perhaps using fuel manufactured from the objects themselves, or nuclear explosions in a standoff mode. If asteroids could be deflected, others observed, they could also be moved into safe orbits around the Earth and mined for propellants, metals, or gold. Ultimately, some said, the presence of so many objects intersecting the Earth's path would force the human race to disperse itself onto other habitable spheres. Any advanced civilization, astronomer Carl Sagan warned, must become spacefaring in order to protect itself from cosmic bombardment. The eventual choice, Sagan concluded, "is spaceflight or extinction."[131]

In August 1994, shortly after Shoemaker-Levy struck Jupiter, the U.S. House of Representatives passed legislation directing NASA, in conjunction with the U.S. Department of Defense and other countries, to "identify and catalogue within 10 years the orbital characteristics of all comets and asteroids that are greater than 1 kilometer in diameter and are in an orbit around the sun that crosses the orbit of the Earth." The effort, termed Spaceguard, took its name from a similar project suggested by science fiction writer Arthur C. Clarke in his novel *Rendezvous with Rama*.[132] Most of this work is conducted using ground based telescopes, however, not spacecraft. The physics of deflection, moreover, encourage interception as far from Earth as possible, which in turn would favor low-cost missions using small, robotic spacecraft, substantially different from the vision of astronauts traveling deep into space to push asteroids around the solar system.

Warnings about Chinese astronauts and killer asteroids elicited a response feeble by comparison to the motivational power of the Cold War. The Cold War really scared people, and modern activities in space do not. The former motivated otherwise pragmatic individuals to embrace an extraordinary dream. The latter left space advocates with a mission in search of a motive.

4

Apollo
The Aura of Competence

If we can send a man to the moon, why can't we clean up
Chesapeake bay?

—Tom Horton, 1984

Shortly after taking office, President John F. Kennedy approved the crash program to put Americans on the Moon.[1] By shifting the civil space effort into high gear, he established the modern undertaking with its emphasis on large-scale engineering, big science, and human exploration. The budget of the National Aeronautics and Space Administration increased tenfold.[2] Eisenhower's approach to space, with its stress upon satellite technology and scientific achievements, disappeared.

President Kennedy's commitment was strong but not unalterable. During the 1960s, a number of efforts were made to release the United States from the self-created mandate to be first in space, particularly the goal, "before this decade is out, of landing a man on the Moon and returning him safely to earth."[3] Even Kennedy himself questioned the wisdom of the decision he had made.

The most serious effort to derail Project Apollo occurred in early 1966. Budget Director Charles Schultze informed President

Lyndon Johnson that fall that expenses in the space program were eating up the funds that Johnson needed to run his War on Poverty and prosecute the war in Vietnam. Schultze urged Johnson to cut the NASA budget by a total of $600 million, a decision that would have effectively abandoned the Moon race and deferred the lunar landing "into the 1970s."[4]

Nations engage in space exploration for a variety of reasons. They explore space for scientific discovery and understanding. They use space as a high ground for national defense. They derive commercial benefits, both directly as in the case of communication satellites and indirectly through "spinoffs" from space exploration. They do so for reasons of national prestige. President Kennedy undertook the race to the moon largely for the latter—to impress uncommitted nations that the United States had the capability and the will to best the Soviet Union technologically. As Project Apollo progressed, this rationale diminished in force. Cold War concerns did not preoccupy Americans in the mid-1960s to the same degree that they had a decade earlier, especially as the United States pulled ahead in the space race.[5]

Diminishing interest in the use of space exploration for the purpose of advancing national prestige in an international setting undercut the principal justification for Project Apollo. In its place arose a new rationale, also tied to national prestige, but in this case directed at domestic consumption. The space program became a means for demonstrating national competence. If the United States could land Americans on the Moon, the nation could do anything else to which its citizens set their minds. Project Apollo and the human flight program became a means to bolster American self-confidence. No president could defer the national space effort without injuring the national aura of confidence. President Johnson refused to follow his budget director's advice and none of the presidents that followed dared to suspend the underlying vision.

A generation of Americans grew up during the 1960s watching NASA astronauts fly into space, beginning with fifteen-minute suborbital trajectories and culminating in an eight-day trip to the Moon. Television coverage of real space adventures was long and intense; there were moments of danger and seasons of accomplishment. In the public mind, the civilian space agency established a reputation as a government organization that could take on difficult tasks and get them done. Project Apollo was widely viewed as triumph of effective government as well as a demonstration of technology.[6]

NASA's ability to demonstrate competence helped boost public confidence in the capability of government in general. Although it was not the sole contributor, the new space effort joined other forces to produce a record level of public confidence in government. Responding to an opinion survey in 1964, 76 percent of Americans polled expressed confidence in the ability of government "to do what is right" most or all of the time.[7] An all-time high in the history of polling, this level of public trust allowed the government to move ahead with other large initiatives during the 1960s, including the War on Poverty and civil rights reform, and helped maintain faith in the value of government-sponsored science and technology.

In the beginning, the aura of competence surrounding the early space flight program owed as much to perception as reality. Without question, NASA's early space flight record could have supported accusations of incompetence as easily as images of success. During its formative years, NASA completed all of its human space flight missions without enduring a major accident or a nationally televised fatality. This was a great accomplishment at the time, given the well-known tendency of rockets to explode or fly off course.[8] Even so, the new agency frequently suffered technical difficulties and launch delays. There was plenty of ammunition for an assault on NASA's capabilities had the media wanted to launch one.

Rocket boosters often misperformed, and even if a launch went well, its mission might flounder in space. During the first year of operation (1958), all four of NASA's major launch attempts misfired. The situation was little better in 1961, when nine of twenty-four major launch attempts failed.[9] Fortunately, none had people on board. Beginning in 1961, NASA launched the first in its series of Ranger spacecraft, a robotic mission designed to produce the first close-up pictures of the Moon. NASA employees made six consecutive attempts to fly Ranger spacecraft between 1961 and 1964; all six failed. Congress launched an investigation, but the incidents hardly tarnished NASA's reputation. NASA recovered and carried out the mission with the remaining three spacecraft.[10] During the second attempt to push an American astronaut to the edge of space in 1961, the Mercury space capsule sank into the Atlantic Ocean before navy helicopter pilots could retrieve it, nearly drowning astronaut Gus Grissom.[11] Some suspected that Grissom had panicked and prematurely blown the capsule's side hatch into the water, but the press portrayed "little Gus" as a national hero.[12] John Glenn's 1962 orbital flight was repeatedly delayed by bad weather, including a nationally televised 27 January scrub just twenty minutes before liftoff. The press moaned but did not openly complain.[13] When the Vanguard rocket had exploded on its launch stand during the first attempt to orbit a U.S. satellite in 1957, the press had roasted the Naval Research Laboratory, the directing organization.[14] NASA failures escaped similar treatment.

Although neither the public nor the media seemed chagrined by NASA mishaps, the people advising John F. Kennedy were. Kennedy's aides frequently discussed the political ramifications of potential space flight disasters as they organized themselves to run the government. The issue surfaced during the 1960 presidential campaign and persisted through the spring of 1961 as officials in the Kennedy administration pondered whether to commit the United States to a space race with the Soviet Union. Aides considered how they might distance the new president from Project Mercury, the agency's high-profile human space flight initiative, should an accident prevail. A special transition team openly urged Kennedy to disassociate himself from the project or risk blame should a failure occur.[15]

Kennedy's advisers had good reason to urge caution about Project Mercury. Blunders plagued virtually every aspect of the project during this period. In the summer of 1960, while Kennedy was campaigning for the presidency, the first test of the Mercury-Atlas system ended in calamity when the rocket exploded mysteriously one minute after liftoff. Although no astronaut was on board, the accident delayed the flight program by six months. On election day 1960, the capsule's escape system misfired. Two weeks later, the Mercury-Redstone combination lifted four inches off the launch pad and shut down. In March 1961 the escape system failed again, and in April a Mercury-Atlas test flight had to be destroyed by the range safety officer when the autopilot sent the rocket astray.[16] The aerospace trade journal *Missiles and Rockets* warned its readers during the summer of 1960 that Project Mercury "could easily end in flaming tragedy."[17] A space flight disaster with an astronaut on board would have undercut public confidence in an already besieged administration and derailed discussions leading up to the lunar commitment. Understandably, Kennedy looked grim as he watched Alan Shepard's nationally televised fifteen-minute suborbital flight from the White House on 5 May 1961.[18]

Shepard's successful mission contributed significantly to a general sense of euphoria and pleasure with NASA's space effort.[19] After a string of malfunctions, the program seemed to be off the ground at last. Step by step, NASA rocket engineers identified weak points in their technical program and corrected them. Five more Mercury astronauts followed Shepard into space, with John Glenn becoming the first American to orbit the Earth on 20 February 1962.

In the beginning, the civil space program was marked by exploding rockets and crashing spacecraft. During the first test of the Mercury-Atlas system, an automated test with no astronaut on board, the rocket exploded one minute after liftoff. Flight engineers reassembled the Mercury capsule from debris collected from the ocean floor. Advisers warned president-elect John F. Kennedy that if he did not disassociate himself from the space project he would risk blame for its failure. (NASA)

All of the flights with astronauts on board were nationally televised, and thousands of people made the pilgrimage to the beaches at Cape Canaveral to watch the rockets blast into space. One hundred thirty-five million Americans watched John Glenn's orbital flight on national television.[20] The flight stopped foot traffic in New York's Grand Central Station, as thousands of commuters paused to watch the lift-off on a sixteen-foot television screen.[21] It was wonderful theater, contrasting sharply with the shroud of secrecy surrounding the Soviet space program.

With such intense national interest, the press did not need to highlight problems in order to maintain viewer attention. Television commentators frequently downplayed dangers that would have captivated public attention had they been properly explained. Instead, press coverage reflected the calm "can-do" attitude that NASA sought to portray. During John Glenn's orbital flight, instruments suggested that the heat shield on his *Friendship 7* spacecraft had come loose. In a desperate measure, flight controllers told Glenn to re-enter the atmosphere with his retrorockets still strapped on. Only after Glenn had returned safely to Earth did the press softly explain what the consequences might have been. Had the signal been accurate (it was not), Glenn would have faced a fiery death as his unprotected space capsule reentered the atmosphere.[22] Time after time, a government agency whose space program had begun with exploding rockets put its reputation on the line and carried out another mission, each more complex or daring than the last. That, at least, was the appearance NASA transmitted during the 1960s. Errors and malfunctions were forgotten as a feeling of competence grew around the U.S. space program.

Performance alone did not maintain the sense of competence that came to envelop the American space effort during the 1960s. Improving performance certainly helped the image of success, but it was not sufficient to explain the reverence in which the space flight program was held. NASA did not operate an error-free space effort during this time, and, in fact, the actual record of human and instrumented flight could have supported an alternative interpretation had opinion leaders wanted to go that way. Why then did the impression of competence prevail?

One of the most important factors that helped create the aura of competence was the personalization of the American space effort through the astronaut corps. By promoting the astronaut corps, agency advocates and media leaders were able to reduce complex technical issues to personal values such as bravery and patriotism. The first astronauts were remade into the embodiment of American values in such a way that few people wanted them to fail. Accusations of failure would have called into question the special values that defined American culture.

NASA officials presented the seven Mercury astronauts to the public at a press conference in Washington, D.C., on 9 April 1959, just six months after

On 9 April 1959 NASA officials presented the seven Mercury astronauts to the Washington press corps. The apparent willingness of these individuals to risk their lives on shaky spacecraft and flaming rockets enthralled the American public. The astronauts became instant heroes and the most visible representatives of the expectations placed on the newly created civil space venture. (NASA)

the creation of the space agency. To the surprise and ultimate consternation of some NASA leaders, the astronauts immediately became national celebrities and the leading symbols of the fledgling space endeavor.[23] Both NASA officials and the press contrived to present the seven astronauts, whose public images were as carefully controlled as those of movie idols or rock music stars, as embodiments of the leading virtues of the American experience in the 1950s.[24]

The press was fascinated by the apparent willingness of the astronauts to risk their lives for the good of a national cause. Tom Wolfe captured the method of this imagery some twenty years later in *The Right Stuff*. The astronauts were not brave in a stupid, unknowing way. Any fool could sit passively in a rocket ship and throw his or her life away: "No, the idea here . . . seemed to be that a man should have the ability to go up in a hurling piece of machinery and put his hide on the line and then have the moxie, the reflexes, the experience, the coolness, to pull it back in the last yawning moment—and then to go up again *the next day*."[25]

Remembering that first press conference, Wolfe cut to the essential question the reporters had circled around. Were the astronauts afraid they were going to die? "They had volunteered to sit on top of rockets—which *always blew up!* They were brave lads who had volunteered for a suicide mission! . . . And all the questions about wives and children and faith and God and motivation and the Flag . . . they were really questions about widows and orphans . . . and how a warrior talks himself into going on a mission in which he is bound to die."[26] In introducing the Mercury Seven to their readers, the editors of *Life* magazine ran a two-page photograph that showed the astronauts at a long table smiling broadly with their hands raised toward the sky. The editors explained the hazards, "yet when asked for a show of hands by those who thought they would come back alive, the answer came unhesitatingly, unanimous." John Glenn, who would become the first American to orbit Earth, lifted both hands above his head.[27]

The skill and bravery of the astronauts touched emotions deeply seated in the American experience at that time. Young and courageous, each sat alone in the single-seat Mercury capsule, like the "lone eagle" Charles Lindbergh crossing the Atlantic Ocean thirty-two years earlier. Facing personal danger, they fit the myth of frontier law enforcers, whose grit filled the substance of Hollywood matinees and television screens.[28] As military test pilots, the astronauts recalled the sacrifices required to produce the Allied victory in World War II. Their personal exploits even evoked the substance of one of America's most popular sporting events: in exploring the appeal of the astronauts, Wolfe drew on his interviews with Junior Johnson, a legendary transporter of moonshine whiskey whose degree of physical courage propelled him to the top ranks of professional stock car drivers.[29]

Test pilots and race car drivers were thought to be a hard-living, hard-drinking lot.[30] In private, a number of the astronauts behaved this way, a fact revealed by Wolfe after the Mercury program ended. At the time of their presentation, however, the astronauts were hardly cast in this light. To a public clamoring for personal details, the Mercury Seven were presented by the press as the personification of the clean-cut, all-American boys whose mythical lives popularized family-oriented television programs during the 1950s and 1960s.[31] They were portrayed as brave, God-fearing, patriotic individuals with loving wives and children. Addressing a joint session of Congress after his orbits around the world, astronaut John Glenn announced to the wildly cheering crowd, "I still get a real hard-to-define feeling down inside when the flag goes by" and got away with it.[32]

One year after their presentation to the American public, the Mercury Seven signed a contract by which they gave *Life* magazine the exclusive right to their personal, firsthand stories.[33] Astronaut stories ran in twenty-eight issues of the weekly magazine between 1959 and 1963.[34] Although Life reporters followed the astronauts at home and on the road and witnessed occasional indiscretions, such tidbits did not find their way into the mainstream press.[35] One of the reporters for *Life* later explained: "There was no explicit editorial direction, but the deal Life made with NASA and the seven individuals created a strong bias toward the 'Boy Scout' image, because all pieces under the astronauts' bylines had to be approved by them as individuals, as a group, and by Shorty Powers [NASA's public affairs officer for the astronauts] and whomever happened to be in charge at the moment in Washington." The astronauts, the reporter continued, were the "main architects" of the image[36] and were using their status as national heroes to enhance their influence in a flight program dominated by rocket scientists and engineers. NASA as well was anxious to perpetuate the mythology of the astronauts, although government officials were nervous about unrestricted access.[37] NASA used its astronauts to promote the space program, parading them through the White House and across Capitol Hill. Reporters cooperated because it made great copy and permitted them to tag along like technology groupies on a great American tour.

The press knew the astronauts endured the same personal difficulties as any other cross section of well-educated, middle-class Americans. In fact, the astronauts probably endured more, given the pressures of the flight schedule and the temptations afforded celebrities. Said one of the writers for *Life:* "I knew, of course, about some very shaky marriages, some womanizing, some drinking and never reported it. The guys wouldn't have let me, and neither would NASA. It was common knowledge that several marriages hung together only because the men were afraid NASA would disapprove of divorce and take them off flights."[38] *Life* introduced its readers to the astronauts' "brave wives and bright children." Wrote the staff: "If the U.S. was getting a bargain in its calm, brave astronauts—and it was—it could also take pride in the wives they had waiting back home."[39] When NASA made its first flight assignments, *Life* ran a two-page picture of three astronauts gathered with their wives and children on the beach at Cape Canaveral watching a test flight of the Mercury-Redstone system that would carry the first two astronauts into space. Drawing on the conjunction of the common and the exceptional, the staff wrote: "In shorts and summer hats, carrying cameras and field glasses, the group looked like sightseers whose next stop might be Cypress Gardens or Marineland. But the men were not vacationers. They were Astronauts from Project Mercury, the prime candidates for a violent, historic event."[40] In 1962 *Life* ran a poignant story about astronaut Scott Carpenter, returning with his father and four children to the Colorado mountains that he had loved to explore as boy. The family climbed rocks and entered caves, a personal sanctuary from the difficulties of the world.[41] It was a moving portrayal of a moment most fathers hope to share with their progeny. The story gave no indication that Scott Carpenter's marriage with his wife Rene was headed for divorce.[42]

The astronauts appeared at a time when the advocates of space exploration desperately needed to inspire public trust in their ability to carry out national space goals. Rockets might explode, but the astronauts shined. They seemed to embody the personal qualities in which Americans of that era wanted to believe: bravery, youth, honesty, love of God and country, and family devotion. How could anyone distrust a government agency represented by such people? The trust the public placed in the astronauts spread through NASA and to the government as a whole. As one of the *Life* reporters summarized: "Life treated the men and their families with kid gloves. So did most of the rest of the press. These guys were heroes, most of them were very smooth, canny operators with all of the press. They felt that they had to live up to a public

President Kennedy looked noticeably concerned as he watched the first flight of an American astronaut on a suborbital trajectory into space. He need not have worried. The Redstone rocket performed flawlessly, and the U.S. Navy recovered astronaut Alan Shepard and the Mercury space capsule only eleven minutes after its water landing. Thus began a twenty-five-year string of U.S. space missions in which NASA always brought its astronauts back alive. (John F. Kennedy Library)

image of good clean all-American guys, and NASA knocked itself out to pre-serve that image."[43]

Of equal importance in building the sense of competence was the glorifi-cation of science and technology. The early space flight program commanded public attention at a time of great public respect for what became known as Big Science. For much of the twentieth century, progress had seemed to sprout

from invention and scientific discovery. The ability of governments to organize very large scientific undertakings like the Manhattan Project had helped
the Allied nations win the Second World War. Big Science and technology were
widely viewed as the means by which the United States would win the Cold
War. In a phrase popularized by David Halberstam, the "best and the brightest" applied the energy of intelligence toward the solution of national problems.[44] The space program epitomized this trend and reminded Americans of
the achievements that could be realized through the peaceful uses of science
and technology, an image not overlooked by its promoters.

The history of the Seattle World's Fair well illustrates the growing public
interest in science and technology. In the mid-1950s, civic leaders in Seattle
sought participation for what they called the Century 21 Exposition. At the
time, Seattle was a remote city noted for its bad weather and airframe industry.
Shortly after the launch of *Sputnik 1* and *Sputnik 2* in 1957, civic leaders, who
were struggling to identify a theme that would attract financial support for
the fair, linked arms with officials in science agencies who wanted "to awaken
the U.S. public to the significance of the general scientific effort and the importance of supporting it."[45] The latter wanted to rectify the slight to scientific
developments that had taken place in the U.S. pavilion at the 1958 Brussels
World's Fair. That exhibit had featured hamburgers and wide-screen movies,
whereas the Soviet Union had displayed Earth satellites and spacefaring dogs.[46]
Together with Washington senator Warren G. Magnuson, scientists and civic
leaders won government funding for a museum-sized pavilion that sought
to communicate "the innate joy of science," including exploration of the universe.[47] "No one knew whether a single huge presentation devoted solely to
science would interest the public," its designers wrote.[48] They did not need to
worry. By the time the Seattle World's Fair opened in April 1962, the U.S. space
program was in full throttle and the public was clamoring for information
about science and technology.

The exhibit treated science as a process of discovery. Although a number
of exhibits within the pavilion's five large buildings dealt with space science,
including a large planetarium, the science center deliberately avoided the display of space technology.[49] In apparent compensation, NASA constructed a
separate building around the corner from the science pavilion. In it, the agency
presented exhibits on rocketry, satellites, technology, tracking and communication, and human space flight.[50] The highlight of the NASA exhibit was John
Glenn's *Friendship 7* space capsule, returned to Earth and eventually sent to

Seattle as part of its world tour.[51] Space exploration was also featured in the Washington state pavilion and the Ford Motor Company's simulated journey into space.

The following February Glenn's capsule was installed in Washington, D.C., in a small tin hanger on the south side of the national Mall. It joined a modest but enormously popular exhibit of aircraft and space memorabilia in what was known as the National Air Museum. That exhibit in turn formed the core of the National Air and Space Museum, the construction of which Congress authorized in 1966.[52] In Huntsville, Alabama, home of NASA's Marshall Space Flight Center, local leaders won support for what became the U.S. Space and Rocket Center and U.S. Space Camp.[53]

During the early 1960s, the public possessed a seemingly insatiable appetite for information about space science and technology. Sensing this interest, television networks and print journalists devoted extensive resources to the process of informing the public about the details of space flight, hiring science reporters and displaying their work. Newspapers provided elaborate accounts of rocket technology, orbital dynamics, life support, guidance and control, and reentry mechanics. They explained communication blackouts, space medicine, rendezvous and docking, and a host of other details regarding space exploration.[54] Words in print followed the tradition established by writers of popular science two decades earlier, whose books led readers step by step through the details of space flight without much interpretation or philosophizing.[55] Even Madison Avenue joined the bandwagon, using public interest in space technology to market a wide variety of products such as Tang, a powdered orange drink supplied to the astronauts for their voyages into space.[56]

Communication experts wondered whether space flight could be presented effectively in the new medium of television. In the beginning, extended space flight did not seem to lend itself to visual presentation. Although the launches were spectacular, the voyages offered little visual excitement. Television cameras did not go into space for real-time coverage until 1968.[57] Except for liftoff and splashdown, the astronauts and spacecraft were out of sight. Moreover, launch preparations were monstrously slow, snippets of action hardly punctuating hours of inactivity. Media critics wondered how television producers could fill the great void of airtime in a medium that required constant action and moving pictures. There was plenty of airtime to fill. When John Glenn became the first American to orbit the Earth, the three television networks canceled the regular Tuesday schedule and devoted their entire daytime programming to the flight.[58] Adding the scrubbed launch on 27 January and the parades and ceremonies that followed, total television coverage of the story averaged twenty-nine hours per network.[59]

How did the networks fill all of that time? They rarely sought to explain the NASA bureaucracy, except to compliment the NASA-industry team. They hardly ever spoke about the politics of national space policy, even though controversies remained. Very little time was devoted to space propaganda or to selling future missions. Instead, when not presenting personal interest stories about the astronauts and their families or public reactions to the flights, the media treated the American audience to a huge, decade-long lesson on space science and technology. Given the state of television technology at the time, this was done with great creativity. Carpenters build mock-ups so that newscasters could explain spacecraft maneuvers, reporters acted out scientific experiments in makeshift laboratories, networks commissioned animation sequences to demonstrate rocket staging and reentry, and newscasters took cameras to industrial plants in which space capsules were built.[60]

Television easily adapted to space flight and, in fact, enjoyed a particular advantage over print journalism. Newspapers and magazines had to describe space technology with diagrams and words, whereas newscasters could show viewers the machinery with their hands. Newscasters such as Jules Bergman became media celebrities by explaining space technology using mock-ups or other props. During the flight of Scott Carpenter, the second American to orbit the Earth, Bergman placed himself in one of the biomedical harnesses that the astronauts wore to monitor their vital signs. The harness revealed that Bergman suffered as much stress during the twelve hours of television coverage as Carpenter did in orbit.[61] One media critic observed that "the moment the rocket goes out of sight is precisely when television can let its imagination soar."[62]

Much of the coverage of early space ventures placed newscasters and journalists in the position of professors explaining the wonders of science and technology to attentive students. Humorist Art Buchwald spoofed the tendency of newscasters to lay science lessons over periods of inactivity with his parody of the press covering an early morning launch from Cape Canaveral:

"The sun has just come up, David, and it's quite a sight to see."

"Could you describe it to us?"

"Well, from where I'm standing, it's round and looks like a great big fiery ball.

Scientists have informed me it's 85 million miles from the earth and it's very hot. As you can see, it's rising from the east. . . . I have been told that without the sun the earth might not sustain life."[63]

Science and technology can be frightening. Doomsday scenarios of atomic warfare terrified Americans during the 1950s, as did images of malfunctioning nuclear reactors in subsequent decades. The national space program, however, provided an example of science under control. It helped to maintain the faith that science meant progress and discovery.

Media emphasis upon science and technology served to displace other, more pejorative, images of government. It displaced the language of exposé, cost overruns, and management errors. Those accusations fell on the space program, especially in later years, but in the early days such matters were drowned out by the chorus of confidence. Science lessons displaced the language of partisan wrangling and policy disputes. Some people complained about the funds lavished on space, but those opinions were background noise set against the wider symphony of general praise.[64] Rarely did the media trounce on the space program as an example of the horrors of Big Government. Project Apollo was one of the largest undertakings ever set upon by the federal bureaucracy, but the engineers and scientists who ran NASA were cast as citizen soldiers in the Cold War, not empire-building desk squatters with lifetime sinecures. To a generation of Americans in the early 1960s, the space program was a national joy ride. It was a demonstration of personal character, of the power of science and technology under control, and of the ability of Americans to complete a task simply because it was hard.

Drawing a slogan from the circus, the *Herald Tribune* called the space program not "'The Greatest Show on Earth,' but . . . The Greatest Show, period."[65] Even critics gushed over the early space program. Editorial writers at the *New York Times*, which had scoffed at space flight when Robert Goddard first proposed a rocket shot to the Moon, praised President Kennedy's decision to conduct a space program "second to none" and called the flight of John Glenn "one of our finest hours."[66] Editorial writers at the *Washington Post*, often sour on the value of human space exploration, compared Glenn's flight to Columbus's discovery of America. Responding to the Gemini 3 flight a few years later, *Post* editorial writers exclaimed that "this is wholesome competition in a race in which the ultimate winner is mankind itself."[67]

Much of the worthiness of the national space effort arose from the difficulty of the undertaking. Human flights to the Moon and robotic ventures to the planets seemed incredibly perplexing to a public barely accustomed to rocketry. They even appeared difficult to NASA engineers.[68] The space race thus provided a national self-examination, a trial of the ability of Americans and their government to surmount great obstacles, just as the mobilization for World War II had tested the American system two decades earlier.

The political language supporting the early space program emphasized this perspective. Politicians understood the degree to which external impressions about the challenges of space fostered confidence in government, both at home and internationally, and their understanding can be traced through their own words. Speaking at Rice University in September 1962, President Kennedy provided the most complete statement of this perspective. "We choose to go to the moon in this decade and do the other things," he said, "not because they are easy, but because they are hard, because that goal will serve to organize and measure the best of our energies and skills."[69] A May 1961 memorandum to Kennedy from Defense Secretary Robert McNamara and NASA administrator James Webb had argued that achievements in space "symbolize the technological power and organizing capacity of a nation."[70] Kennedy used similar language in proposing the Moon race on 25 May 1961. Defending his choice, Kennedy told Congress that no other objective in space "will be so difficult or expensive to accomplish."[71]

As a consequence of its symbolic value, Kennedy realized much of the advantage of the lunar mission simply by setting the goal. The decision to go to the Moon signaled the willingness of the U.S. government to contest the Soviet Union at all levels of military and technological competition. Kennedy and other members of his administration understood the symbolic importance of the decision and rarely sought to defend it on scientific or military grounds, where they knew the United States to be well ahead.[72] In many ways, the actual mobilization for the voyage was anticlimactic to the decision to go, which helps explain why Kennedy became reluctant to complete the voyage once that advantage had been gained.

Realizing almost immediately that Project Apollo would empty the federal treasury, Kennedy quietly sought a way to curtail the endeavor. He could not back down alone without calling into question governmental capability and Cold War commitments. Instead, he began to explore the possibility of reshaping the program from one of competition into one that fostered international cooperation.[73] In June 1961 Kennedy made a direct proposal to Soviet premier

Nikita Khrushchev at the summit meeting in Vienna for a joint expedition to the Moon.[74] He repeated the call in 1963 in a speech before the United Nations: "Why, therefore, should man's first flight to the moon be a matter of national competition? Why should the United States and the Soviet Union, in preparing for such expeditions, become involved in immense duplications of research, construction, and expenditure? Surely we should explore whether the scientists and astronauts of our two countries—indeed of all the world—cannot work together in the conquest of space." Kennedy closed his speech by urging, "Let us do the big things together."

Kennedy's cooperative vision was steadfastly resisted by the Soviet Union. In public, Soviet leaders were noncommittal, dismissing the proposal as premature, but in private, they viewed the offer as a ploy to gain advantage in the Cold War. Cooperation in space would open the Soviet society to U.S. scrutiny and steal attention from one of the few arenas of Soviet success. At the Vienna summit, Khrushchev linked space cooperation to other Cold War issues, insisting that Kennedy first withdraw his military forces from bases along the Soviet border.[75]

Kennedy's cooperation initiative had a second, less-intended consequence. It undercut domestic political support for the race to the Moon, thereby providing an opportunity for critics to propose large reductions in the budget for Project Apollo. Buoyed by the thought that the Moon race was no longer the Cold War priority it had once been, opponents moved to cut funds for what one called "a manned junket to the moon." The House of Representatives began the assault in 1963 by removing $600 million from Kennedy's $5.7 billion NASA budget request. The administration objected on the grounds that the cuts would interfere with the deadline for reaching the Moon. Members of Congress questioned the White House commitment to a deadline that had apparently lost its priority. In the Senate, Arkansas senator J. William Fulbright moved to cut 10 percent more from the NASA appropriation. The president's allies prevailed on that vote, but assaults on Apollo's budget followed almost yearly thereafter.[76]

By 1965 a campaign against civil space expenditures was fully under way. Space expenditures, running at more than $5 billion per year, dwarfed other priorities, such as the $1.8 billion War on Poverty and the $2 billion set aside to improve elementary and secondary education.[77] When asked in early 1966 to identify ways to cut the growth in federal spending, Budget Director Charles Schultze, a pragmatic economist with a strong commitment to social reform, recommended that President Lyndon Johnson defer the lunar landing until the 1970s and abandon post-Apollo space efforts, a recommendation that produced the second largest sum of savings in Schultze's litany of cuts. Failing to win Johnson's approval in 1966, Schultze tried again the following year. He warned that NASA might fail to meet the lunar goal even if Johnson restored the funds: "It would be better to abandon this goal now in the name of competing national priorities, than to give it up unwillingly a year from now because of technical problems."[78]

Johnson agonized over the prospective cuts. He did not want to abandon the commitment to put Americans on the Moon by the end of the decade. At the same time, he knew that he had to endorse substantial cuts in federal spending in order to win congressional support for the 10 percent tax surcharge needed to finance the war in Vietnam. Johnson spared Project Apollo but decimated NASA's spending plans for exploration efforts beyond. Responding to the protests of NASA administrator James Webb, Johnson replied that he personally did not want "to take one dime from my budget for space appropriations" but agreed to do so in order to satisfy the budget cutters supporting his tax bill.[79] Exhausted by the budget battles, Webb resigned the following year.

In resisting efforts to undermine the lunar goal, Johnson acknowledged the degree to which space flights contributed to domestic confidence. "Somehow the problems which yesterday seemed large and ominous and insoluble, today appear much less foreboding," he announced after the completion of the *Gemini 5* flight in August 1965. Americans, he proclaimed, did not need to fear problems on Earth when they had accomplished so much in space.[80] In his last month as president, Johnson welcomed the crew of *Apollo 8* to the White House. Using words that recalled the difficulty of the endeavor, Johnson said, "If there is an ultimate truth to be learned from this historic flight, it may be this: There are few social or scientific or political problems which cannot be solved by men, if they truly want to solve them together."[81]

Through the 1960s, the space program provided successive examples of a government program that worked well. This in turn inspired one of the greatest feats of imagination in American history. NASA's accomplishments allowed members of a society founded on the mistrust of government to believe that their political institutions could organize an incredibly complex endeavor and accomplish it successfully. As Lyndon Johnson suspected it would, this level of trust allowed the federal government to proceed with other ambitious initiatives. Some of them, such as the Vietnam War, failed. The space program, however, continued to shine.

Excited by the sense of national confidence, exploration advocates contemplated their post-Apollo initiatives. They dreamed of space stations, space shuttles, orbital observatories, lunar bases, robotic tours, and expeditions to Mars. Such desires required a level of government funding equal in value to that provided during the height of the Apollo moon effort.[82] To begin these initiatives, NASA officials and their allies had to ensure that the hardware and technical work force assembled for the flights to the Moon did not vanish once the landings began. In 1965, as spending for space exploration peaked at $5.5 billion, NASA officials established the Saturn-Apollo Applications Office at NASA Headquarters and began to press for new missions that would maintain the production of Saturn V and Saturn IB rockets and related spacecraft. The office sought approval for additional scientific and technological missions "in earth orbit, lunar orbit, and on the lunar surface . . . through use of Saturn/ Apollo hardware." The main proposals to emerge from this near-term endeavor were an orbital workshop, later named *Skylab*, and a large telescope to be operated by astronauts working in space.[83]

The demonstrations of competence that transformed the space program into an object of national pride and provided a basis for optimism about future plans coexisted alongside a growing level of public crankiness toward technology. Up until that time, technology had remained a force that in the mind of the public seemed to produce more good than evil. Broad-scale attacks on technology such as the "ban the bomb" movement had remained on the fringe of American political life.[84] Other antitechnology protests possessed a mystical quality that placed them well outside the mainstream of American public opinion. The October 1967 March on the Pentagon illustrated the madcap quality of the latter. The march was an early event in the protests against the war in Vietnam. As a gesture of defiance against American military technology, protesters planned to encircle the Pentagon with a sufficient number of demonstrators to perform a rite of exorcism that would tear the building from its foundations and cause it to levitate above the ground. The government, which had no apparent objection to the levitation, would not permit the demonstrators to form the human chain around the structure that was essential to the rite on the grounds that the chain would block employee access. Arrests inevitably followed.[85] To the participants, the exorcism was no more absurd than the public's faith in the saving power of technology.

As the sixties progressed, attacks on technology became more respectable, moving away from the protest fringe. In her 1962 book *Silent Spring*, Rachel Carson questioned the value of the chemical revolution that had transformed agriculture and industry, warning Americans that they were poisoning their environment.[86] In 1968, in *The Population Bomb*, Paul Ehrlich raised the specter of environmental collapse in a world plagued by uncontrolled population growth and dwindling resources, an image that became more plausible as pessimism about technology spread.[87] Works of fiction amplified the theme. Joseph Heller and Kurt Vonnegut, in two 1960s classics, portrayed the application of military technology as acts of insanity, a popular counterculture theme.[88]

Throughout the 1950s, Big Science had been presented to the American public as the foundation of a secure nation. Big Science in the form of the Manhattan Project had helped to end the Second World War. Big Science protected America from a Soviet nuclear attack, and Big Science in the form of Project Apollo would help win the Cold War. Thousands of middle-class Americans went to work for Big Science institutions like the aerospace industry, fulfilling public causes while providing themselves a degree of job security unprecedented in previous times. Although Project Apollo involved far more engineering and management than science, it defined Big Science. The term came to represent an approach to national priorities characterized by large-scale initiatives funded by government appropriations and staffed by expert personnel, a definition that easily fit the race to the Moon.[89]

No event did more to transform American attitudes toward Big Science and its technologies than the constant bombardment of images from Vietnam. An extremely powerful military-industrial complex made possible by scientific and technical elites visually failed, day after day on the broadcast news—a conclusion presented as unavoidable after the January 1968 Tet Offensive. With all the technology it could muster, the U.S. military could not contain a poorly equipped guerrilla band assembled by a Third World country with a subsistence economy.

A sharp downturn in aerospace employment accompanied the failures in Vietnam. Skilled workers who had come to rely upon government support for their economic security found themselves the objects of layoffs, cutbacks, and scorn. The sense of disillusionment arising from these developments undercut trust in government, even among its strongest supporters, and disillusionment made possible that which was unthinkable one decade earlier: the abandonment of new technologies.[90]

Defeat of the Supersonic Transport (SST) marked an important point in the emergence of this trend. Congress had supported the SST throughout the 1960s, ever since President Kennedy proposed that the government assist American industry with this "logical next development" in aviation technology.[91] But media attacks and citizen protests mushroomed as the sixties progressed. In 1967 William A. Shurcliff, a Harvard physicist, formed a citizen's organization with the relatively innocuous purpose of protesting the assault on tranquillity generated by sonic booms from a fleet of supersonic aircraft. By 1970 this modest objection had been joined by premonitions of ecological disaster. Exhaust from the high-flying aircraft, opponents claimed, would cause global warming and damage the protective ozone layer in the upper atmosphere.[92] Economic arguments also played a role, but the final defeat of the SST was due largely to the fact that it became a symbol of misguided progress among a public increasingly suspicious of new technologies. The congressional votes in March 1971 to discontinue government funding for the SST were a slap in the face to those who believed that new technologies always bred progress.[93]

At least initially, the U.S. space program seemed immune to such attacks. The civil space program provided the one best example of technology both benign and effective. The only major event tarnishing its reputation during this period was the fire at launch complex 34 on 27 January 1967 that claimed the lives of astronauts Virgil Grissom, Edward White, and Roger Chaffee during a test of an Apollo spacecraft.[94] Rather than reduce support for the civilian space effort, however, the tragedy actually served to strengthen public resolve, as opinion polls revealed.[95]

Public support for NASA space flights peaked toward the end of 1968, elevated by the famous Christmas voyage in which astronauts Frank Borman, James Lovell, and William Anders became the first humans to leave the gravitational well of Earth and travel to the Moon. On Christmas Eve, from their orbit around the Moon, the astronauts sent back close-up television pictures of the lunar surface. The pictures conformed remarkably to the images anticipated by space artist Chesley Bonestell some twenty years earlier. As the lunar surface passed below them, the crew read the story of Creation from the first ten verses of the book of Genesis. "Merry Christmas and God bless all of you," they closed, "all of you on the good earth."[96]

Coverage of the flight was uniformly favorable. "Not since Christopher Columbus's first voyage to the 'new world' have men embarked upon a journey comparable to that begun by Apollo 8," gushed editorial writers for the normally dour *New York Times*. The voyage "vividly juxtaposed science as the instrument broadening man's reach into the universe with science as the source of weapons that may destroy humanity." Editorial pages portrayed the *Apollo 8* capsule as a Christmas star.[97] The voyage of *Apollo 8* provided a moment of high hope and serenity in what had been an otherwise ugly year. The American people had endured riots, the assassinations of Robert Kennedy and Martin Luther King Jr., the Tet Offensive in Vietnam, and a turbulent contest for the presidency. Said editors for the *Los Angeles Times*, "The Apollo 8 flight, therefore, comes as a welcome talisman of future good fortune—a kind of reassurance that we are still a nation capable of great enterprises."[98]

During the 1960s space travel escaped nearly all the apocalyptic portrayals that afflicted other images of technology. Misused military technology provided the basis for the classic 1964 satire on official stupidity and nuclear war, *Dr. Strangelove*, as well as the surprise ending for the popular 1968 film *Planet of the Apes*, in which Charlton Heston discovers that the simian planet on which he has landed his spacecraft is really a post-apocalyptic Earth. Environmental destruction motivated the 1971 release *Silent Running*, in which the last hope of species rescued from a sterile Earth resides with keepers of large, terrarium-like structures in space. Stupid government officials order keepers to destroy the greenhouses and all they protect.[99]

The optimism accorded space technology was carried to spiritual levels with the 1968 release of Stanley Kubrick's *2001: A Space Odyssey*. The film stunned audiences with the remarkably quality of its special effects, setting new standards for the cinematic portrayal of space travel. The screenplay, by Arthur C. Clarke, suggests that space technology, both Earth-based and extraterrestrial, has the capacity to radically transform human evolution in positive ways. After a battle with the onboard computer (the HAL 9000 provides the one example of technology gone awry), astronaut Dave Bowman travels through a strange monolith to an unexplained destination in a corner of the universe. There he is reborn as a superbeing, a star child who is transported back to the solar system to gaze down at planet Earth.[100]

Throughout the 1960s, space travel supported the tradition of technological optimism. Social commentators continued to associate it with new beginnings and social justice, themes emphasized in the most popular television shows to deal with space travel during that decade. In *Lost in Space*, which began its three-year run in 1965, the Robinson family is selected from more than two million volunteers to leave the critically overpopulated Earth. Their sabotaged spacecraft flies off course, allowing them to meet a variety of alien characters in what becomes an extraterrestrial morality play. A similar impulse guided the highly influential *Star Trek* series, which began its initial three-year run in1966. The crew of the starship *Enterprise* sets out to explore worlds where no humans have gone before. Creator Gene Roddenberry used *Star Trek* to address the social issues of the times: superpower conflict, fascism, civil rights, and interracial sexuality. These two series eschewed the more terrifying aspects of space exploration examined in alien monster movies such as *The Thing* (1951) or the bizarre plot twists exploited in the *Twilight Zone* (1959-64).[101]

One of the few widely viewed examples of a dangerous space technology during this period appeared with Michael Crichton's *Andromeda Strain*, published as a novel in 1969 and released as a film the following year. In the story, a *Scoop VI* satellite searching for biological weaponry brings a virulent extraterrestrial organism back to Earth, where the life-form threatens the general population after killing all but two residents of a small southwestern town. The story excited fears that space exploration would bring some unknown retribution on the Earth, a possibility that prompted NASA officials to quarantine the first astronauts to return from the Moon.[102] Occasionally, as with the 1979 release of *Alien*, producers and writers terrified the public with the misuse of space technology. In *Alien*, the crew of the space freighter *Nostromo* is sent to retrieve twenty tons of mineral ore and, without the crew's consent, a vicious alien that proceeds to kill every human on board except Sigourney Weaver.[103] In general, however, space travel in popular culture continued to be associated with rebirth and hope, themes repeated in later releases such as *Close Encounters of the Third Kind* (1977), *E.T.: The Extra-Terrestrial* (1982), and *Cocoon* (1985).[104]

NASA's own flight program helped maintain the image of space technology as benign. Astronauts from Earth landed on the surface of the Moon on 20 July 1969. The landing was far more difficult than the public knew. Steve Bales, a twenty-six-year-old flight controller at NASA's Mission Control in Houston, told astronauts Neil Armstrong and Buzz Aldrin to ignore an alarm from the spacecraft computer during the final descent. To compound that problem, Armstrong and Aldrin left themselves less than thirty seconds of fuel as they searched for an adequate place to land. Had the fuel tanks run dry or the alarm

Even as other big government initiatives failed, the U.S. space program remained one of the few examples of a large government undertaking that worked. From the summer of 1969 to the end of 1972, successive teams of American astronauts explored the surface of the Moon. As the moon landings ended, public officials concerned about the high cost of human space flight nonetheless would not allow the endeavor to end. (NASA)

been real, the onboard computers would have fired the ascent engines in an attempt to separate the upper half of the landing module and the astronauts from their moving platform, a dangerous maneuver.[105]

NASA officials prevailed again when one of the oxygen tanks supporting the flight of Apollo 13 exploded halfway to the Moon, depriving the crew of the breathing air and electricity it needed to survive. NASA engineers pieced together an emergency plan for the astronauts who squeezed into their powered-down lunar lander and rode it like an interplanetary lifeboat back to Earth. As the rescue surged from crises to crises, NASA made the return trip look almost routine.[106]

Repeatedly during this period, the press praised the capabilities of the men and women who managed the nation's space effort. In spite of the obvious negligence that necessitated it, the rescue of the Apollo 13 mission was widely hailed as NASA's "finest hour," an impression that grew stronger in subsequent years. By the time director Ron Howard made the big-budget *Apollo 13* film more than twenty years later, NASA's astronauts and flight controllers had become legends in American popular culture.[107]

NASA's reputation for competence lasted until 1986, well after other icons had fallen from public favor. It ended with the first live television broadcast of a space flight fatality, the 28 January loss of the seven astronauts on the space shuttle *Challenger*. With the accident, NASA's image moved from that of an agency that could do no wrong to a bureaucracy that could do little right. Forty-seven percent of those responding to a Media General/Associated Press public opinion poll indicated that their confidence in NASA had been shaken by the event, and only one-third of them said after two years that it been restored.[108] A succession of errors amplified the impression of an agency in decline—the myopic Hubble Space Telescope, an antenna problem with the *Galileo* Jupiter spacecraft, the mysterious loss of the billion-dollar *Mars Observer* expedition, and more exploding rockets, initially with only instruments on board. Seventeen years after the *Challenger* accident, in what astronaut Sally Ride characterized as an "echo" of the original error, NASA lost the space shuttle *Columbia* and its seven-person crew. Big Science lost its luster. Congress canceled the Superconducting Super Collider and nearly abolished the International Space Station. NASA's continuing travails with the remaining space shuttles clearly revealed that the transport vehicle had failed to meet its original cost, schedule, and reliability goals.[109]

In actuality, NASA conducted a less accident-prone flight program during the period when it lost the *Challenger* than it had during the 1960s when the sense of competence arose. NASA endured far more launch and mission failures during the 1960s, when the space program was less routine and more experimental.[110] The resulting image of the Apollo years raised the performance bar, however, and as history shows, facts do not always play the dominant role in the creation of images about the U.S. space program.

In the early 1970s, after the first landing on the Moon, President Richard Nixon examined various options for the future of the U.S. space program. He confronted the most important decision affecting the U.S. space effort since President Kennedy had challenged Americans to race to the Moon. Drawing on the accumulation of public confidence, space advocates pushed hard for an ambitious set of post-Apollo ventures that would take robots to the outer regions of the solar system and humans to Mars. Even though public confidence in the goodness of space travel remained high, public support for extensive government investment remained low. By 1973, 59 percent of American's responding to a Gallup public opinion poll said they wanted the government to cut spending on space, compared to only 7 percent who wanted to spend more.[111] At the same time, public confidence in the overall space effort remained high. The public did not want to spend money on space exploration, but they were proud of the nation's achievements.

Various factions within the White House fought over the future of the national space effort. Many opposed any large new human flight initiative. Economic inflation and the resulting need to cut government spending had forced White House officials to set a $3 billion cap on NASA appropriations, far less than the budget levels attained during the Apollo years. Much of the interest in space-based competition with the Soviet Union had disappeared.[112]

On the other hand, the U.S. space program remained one of the few shining examples of a big government undertaking that worked. Further cuts, warned Caspar Weinberger, then deputy director of the president's Office of Management and Budget, would send a signal that "our best years are behind us." In an August 1971 memorandum to the president, Weinberger mused that "America should be able to afford something besides increased welfare, programs to repair our cities, or Appalachian relief and the like."[113]

Nixon read the memo and contemplated his options. He did not want to lavish funds on the national space effort and felt no obligation to continue the legacy of President Kennedy, his opponent in the 1960 presidential election. Concurrently, he did not want to be the one to shut down such a source of national pride, especially so soon after the defeat of the SST. (Nixon was also motivated by his desire to win votes in aerospace states like California and Texas for his 1972 reelection campaign.) He picked up his pen and scribbled in the margin of Weinberger's memo "I agree with Cap."[114] Nixon continued the human space flight program through one new initiative, an Earth-to-orbit space transportation system that preoccupied NASA's human space flight energies for the next quarter century.[115]

Despite occasional bouts of pessimism, the presidents who followed Nixon continued to support the doctrine of human flight. They did not confer as many resources on the national space effort as President Kennedy had, but neither did they feel obliged to end the human space flight dream. The Cold War, the primary justification for the first large steps toward the planets and Moon, disappeared, but the flight program lived on. A principal reason was the sense of national achievement that the effort conferred. Project Apollo continued to occupy special status as a bright shining example of what a great nation could accomplish with sufficient understanding and commitment. When Al Gore proposed a national effort to fight global warming some forty years later, he employed a well-worn example.[116] If humans could fly to the Moon, he said, they could surely do anything else to which they set their will.

5

Mysteries of Life

If we find the answer to why it is that we and the universe exist, it would be the ultimate triumph—then we would know the mind of God.

—Stephen Hawking, 1993

Space exploration received government support during its infancy largely as a result of nationalistic concerns. The desire to rekindle national pride, to grasp military advantages from the cosmos, and to build confidence in the U.S. system of government created governmental majorities for large space expenditures. Space exploration promised to do all these things. Without these promises, it is doubtful that the civil space effort would have progressed much beyond the modest research program advanced by the Eisenhower administration.

To its most devoted advocates, however, space exploration promises far more: to continue the quest begun centuries earlier to supplant religious dogma with science as a means of understanding the universe. Space exploration addresses the great mysteries of life. How did the universe begin? Where did the solar system come from? Are humans alone, or is the universe teeming with life? How will it all end? Since Galileo Galilei employed a pair of converted spectacles to observe the moons of

Jupiter, advocates of this new way of understanding have argued that natural observation will provide answers to questions such as these.

For many in the exploration business, this quest has a spiritual quality, promising answers to cosmological questions that have intrigued humans throughout history and have inspired great myths and religions.[1] By probing the mechanics of the universe, humans can find answers to questions that have encouraged spiritual introspection since thinking began. It will allow humans "to know the mind of God."[2]

To a great extent, the images that portray space exploration constitute a form of cultural anticipation. Not only do such images frame questions; they also anticipate answers. Scientists may urge caution, but popular culture advances conclusions. The answers people expect to find in the cosmos already exist in their own minds, created by works of fiction and cultural traditions. The universe itself places a heavy burden on these expectations. Initial reports from the void, sent back by automated satellites and planetary probes, reveal a solar system that confounds traditional beliefs: it does not conform to the images that dominate popular culture. This has not defeated the advocates of space exploration, however. Rather than revise their views, they simply look harder.

No expectation has had more influence on support for space exploration than the belief that humans are not alone. Those who promote exploration possess an almost universal faith in the doctrine that life exists on other spheres. Even the most skeptical scientists have difficulty concluding that God or whatever natural force created life did so on a single planet in the entire universe.

For much of modern history, exploration advocates advanced the view that life, either primitive or advanced, could be found within the solar system, most probably on Venus and Mars. This belief drew sustenance from the stimulating but factually incorrect theory of solar system evolution widely known as the nebular hypothesis, advanced by the French mathematician and astronomer Pierre Simon de Laplace. Laplace postulated that the planets arose from a vast rotating nebula extending into space beyond the orbit of the farthest planet. The mechanics of rotation forced the nebula to separate into rings, which in turn coalesced into planets, beginning with the spheres most distant from the sun and ending with the planet Mercury.[3]

The proposition that natural forces created the planets in stages, accompanied by growing acceptance of the theory of biological evolution, suggested to the popular mind that the planets constituted a sort of evolutionary time machine. On Mars, people could view Earth as it would become millions of years hence, once its oceans dried up and its atmosphere grew thin. Any life that arose on Mars would be ancient and any civilizations far older than those on Earth. The planet Venus, by contrast, was viewed as an embryonic planet still shrouded in primeval haze. As late as 1934 a writer for *Nature* magazine reported that "the opinion is quite generally held by astronomers that Venus is not as far advanced in evolution as the Earth, while Mars is a much older world."[4]

Improvements in telescopic imagery during the nineteenth century allowed astronomers, both professional and amateur, to make detailed observations of Earth's neighbors in space. A rash of reports accompanying these improvements purported to present evidence of life on nearby spheres. Some were perfect hoaxes, such as the wholly fictional report by the *New York Sun* in 1835 that astronomers had sighted animals on the Moon. Others were more serious. Twelve years earlier, Munich astronomer Franz von Paula Gruithuisen announced that he had discovered a walled city on the lunar surface. What Gruithuisen observed was a strange formation of mountain chains near the crater Schroter, amplified by an active imagination. The announcement, according to space historian Willy Ley, "created a stir which can be traced through the whole literature of that time."[5]

Notions about an extraterrestrial presence on the Moon persisted even into the twentieth century. Frustrated by the absence of visible evidence, promoters of popular culture suggested that fortifications or other structures might be placed where they could not be seen. In their three-part television series on space exploration, animators working for Walt Disney made such a suggestion in portraying an imaginary trip around the Moon. The four-person crew makes a startling discovery as it crosses the terminator on the back side of Earth's nearest neighbor, the side previously hidden from human view. The crew reports: "Captain, I'm getting a high geiger count at 33 degrees. . . . Contour mapper shows a very unusual formation at about 15 degrees 7 latitude and meridian 210. . . . Get some flares in that area quick."[6] As flares light up the lunar surface, the crew looks down on the ruins of a lunar colony of unknown origin on the back side of the Moon. Arthur C. Clarke and Stanley Kubrick continued this tradition in the 1968 classic *2001: A Space Odyssey*, a film in which lunar explorers discover a monolith of extraterrestrial origin buried beneath the surface of the Moon by a spacefaring civilization millions of years ago.[7]

Detailed observations of the Moon, even in the nineteenth century, revealed a stark and lonely sphere with little evidence of life either local or extraterrestrial. Rather than discourage speculation, these findings simply moved speculation to new realms. Astronomers watching the planet Venus observed a faint luminosity in the atmosphere of its night sky. This led to suggestions that its inhabitants might be holding giant festivals, or burning large stretches of jungle to produce new farmland, or even attempting to communicate with Earth. The physical similarities between Venus and Earth led the popular science writer Camille Flammarion to conclude that its inhabitants should be "but little different from those which people our planet. As to imagining it desert or sterile, this is a hypothesis which could not arise in the brain of any naturalist."[8]

The notion of Mars as an ancient planet held special appeal and helped create the Martian myth, which dominated popular culture throughout the twentieth century. The myth drew support from reports that Mars harbored conditions suitable for the development of life (an atmosphere and liquid water) that had been present at least as long as they had been producing life on Earth. This hope gave rise to a number of popular accounts promoting the notion that most certainly vegetation and possibly advanced life-forms could be found on Mars. The widely disseminated findings of Giovanni Schiaparelli and Percival Lowell, who saw markings on the Martian surface that looked like canals, encouraged such views.

Schiaparelli, who observed the markings in detail during the close approach of 1877, called them *canali*, or "channels." Schiaparelli was a well-trained astronomer, with more than twenty years of experience in the field. As director of the Milan Observatory, he had already established his reputation by observing asteroids, comets, and meteor showers. He reserved his judgment on the meaning of the Mars markings in spite of the fact that the *canali* and surrounding areas seemed to change with the expansion and contraction of the Martian north pole. It was not necessary to attribute the markings to the work of intelligent beings, he wrote in 1893, as they could as easily be natural features on the surface of the planet, like sea channels on the surface of Earth.[9]

In English, however, *canali* became canal, a translation with tremendous popular significance. The greatest works of engineering on Earth during the nineteenth and early twentieth centuries were canals, including the Suez Canal, completed in 1869, and the Panama Canal, begun in 1904. The ability to organize the construction of canals was, in the nineteenth century, the signature

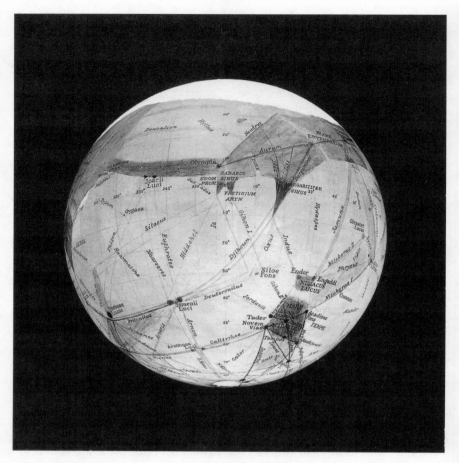

Efforts to understand the nature of the cosmos invited strange explanations. Percival Lowell created one of the most enduring with his turn-of-the-century drawings of canals on Mars. Lowell insisted that the objects he viewed were the work of an ancient civilization struggling to maintain its existence on a dying world. (Lowell Observatory Archives)

of an advanced civilization. To the informed public, Schiaparelli's findings required little interpretation. If canals crisscrossed the surface of Mars, the planet must harbor a highly intelligent civilization.

This vision excited the imagination of Percival Lowell. Born into a famous Boston Yankee family in 1855, Lowell had a fortune sufficient to pursue whatever fancy intrigued him without the inconvenience of formal training in the

subject. After graduating from Harvard in 1876, he devoted himself to literature and travel, producing a series of books on the Orient. As an amateur astronomer, Lowell was aware of the findings on Mars. He returned to the United
States with the intent of building his own observatory and making his own
investigations during the planet's close approach in 1894. Friends at Harvard
suggested a site on a high mesa near Flagstaff, Arizona. With technical support from Harvard astronomers, supplemented by his family's ample resources,
Lowell completed the observatory by the spring of 1894. He gazed with amazement at the polar ice caps and located by his own account nearly two hundred
canals.[10] Between 1895 and 1908, Lowell published three books that helped
create the popular image of Mars.[11]

Lowell concluded that the canals he mapped were the product of an ancient civilization struggling to survive on a dying world, a world far older than
Earth. "His continents are all smoothed down; his oceans have all dried up,"
he reported in 1895.[12] Beings on Mars confronted a growing scarcity of the
fluid essential to existence, he asserted, the only available water being that
which came from the semiannual melting of the caps of snow at the Martian
poles. Lowell detected a pattern to the canals—straight lines crisscrossing the
surface of the planet meeting in what he perceived to be oases of vegetation
and life. The pattern led Lowell to conclude that the canals must carry water
between the poles and the more habitable portions of the planet. The regularity of the lines and the knowledge that water does not run uphill led Lowell
to the unalterable conclusion that intelligent beings must be pumping water
back and forth.

Because canals crossed the entire planet, Lowell also deduced that Martians had abolished war: only a necessarily intelligent and nonbellicose race
could "act as a unit throughout its globe." The pressures of survival on a dying
world would further sharpen mental capabilities. "To find, therefore, upon
Mars highly intelligent life is what the planet's state would lead one to expect."[13] Sadly, Lowell concluded, by the time Earthlings developed the technology required to reach Mars, the civilization might have died away as the
last of its water disappeared.[14]

Scientists cautioned the public against embracing Lowell's conclusions. In
a book-length review of *Mars and Its Canals*, the famous naturalist Alfred
Russel Wallace correctly observed that the markings Lowell saw might be due
to craters and cracks on the planet's surface. Russell faulted Lowell for not
considering scientific findings to the effect that the Martian atmosphere was

exceptionally thin and appraising the effect of a thin atmosphere upon on the retention of heat. The mean surface temperature on Mars, Wallace correctly predicted, would be well below the freezing point of water and thus hardly capable of supporting a canal-building civilization.[15]

Cranky scientists did little to dim popular interest in life on the red planet, as Martian tales swept the public fancy. In 1897 British novelist H. G. Wells began to serialize *The War of the Worlds*. If the Martians of Lowell's imagination did not make war on themselves, they were certainly capable of making it on other planets. "To carry warfare sunward is indeed their only escape from the destruction that generation after generation creeps upon them," Wells explained.[16] As a student at the Normal School of Science in London, Wells had studied under the famous biologist T. H. Huxley, an outspoken supporter of Darwin's theory of natural selection. Wells wanted to examine how Darwin's theory might operate on a totally separate world.[17] In seeking to explore the views of Charles Darwin, however, Wells promoted Percival Lowell. The remarkably well-enduring tale, appearing at precisely the same time as Lowell's books, exposed thousands of readers to the notion of Mars as the abode of a superior but dying civilization.

The story's special appeal, as Wells himself recognized, lay in its power to suppose that creatures on a nearby planet might evolve to the point that they could destroy human civilization on Earth.[18] The idea that life on Earth might be primitive both technologically and biologically in comparison to other planets assaulted the views of religious authorities who opposed Darwin's theory. A superior race of Martians bent on conquering the Earth posed an inalterable challenge to the belief that God planned to make Earth the seat of intelligent life and English-speaking people rulers of the world. If Earthlings could be snuffed out by an extraterrestrial invader, as easily as humans had destroyed species or indigenous cultures on Earth, then one could hardly view *Homo sapiens* as the ultimate objective of God's plan. In his third book on Mars, Lowell admitted that the existence of Martians would force humans to revise their cosmology: "Their presence certainly ousts us from any unique or self-centered position in the solar system, but so with the world did the Copernican system [alter] the Ptolemaic, and the world survived this deposing change."[19]

Over and over again, fictional Martians provided humans with a counterpoint to earthly life-forms.[20] Beginning with the *Princess of Mars* in 1912, Edgar Rice Burroughs introduced a generation of readers to the exotic flora and fauna of this nearby planet and one of most beautiful and scantily clad heroines in modern literature. Ten more books followed as Burroughs reinforced the prevailing notion of a dying planet on which survivors of a once-mighty civilization struggled against the perils of a thinning atmosphere and warring tribes.[21] On 30 October 1938 Orson Welles made radio history with his newscast-style presentation of *War of the Worlds*, proving once again the appeal of the Martian invasion tale.[22] In 1950 Hollywood released *Rocketship X-M*, a competitor to *Destination Moon*, in which space explorers encounter a dying Mars inhabited by a race of blind, mutated survivors. As the rocket ship approaches the vicinity of Earth (insufficient fuel dooms its safe return), the remaining members of its crew warn the people below of their startling discovery: an all-out atomic war blew an advanced Martian civilization back into the Stone Age.[23] In 1953 Martians attacked Earthlings again as the movie version of *War of the Worlds* played to large audiences, and a remake appeared in 2005.[24]

By the time Ray Bradbury published his *Martian Chronicles*, Martians had disappeared, at least in their physical form. The Martians of Burroughs's imagination had evolved to the point where they disappear into each other's minds. Bradbury's Martians had moved to another dimension, a theme presented in many works of fiction that explore the consequences of evolution.[25] Settlers from Earth arriving on Mars encounter the ruins of an ancient civilization and an invisible race that uses telepathic powers to repel early explorers.[26]

In his third and final program on space exploration, broadcast in 1957, Walt Disney reinforced prevailing conceptions about Mars. Much of the program, titled "Mars and Beyond," dealt with the search for life in the solar system. Disney's animators prepared drawings of exotic plants and animal life on the Martian surface. Out of respect to the Martian myth, and its hope for intelligent life, the program concluded with pictures of flying saucers skimming through the planet's atmosphere: "When earthman finally walks upon the sands of Mars, what will confront him in this mysterious new world? Will any of his conceptions of strange and exotic Martian life prove to be true? Will he find the remains of a long-dead civilization? Or will the more conservative opinions of present-day science be borne out with the discovery of a cold and barren planet, where only a low form of vegetable life struggles to survive?" Only by going there would humans know. The first interplanetary expeditions, the series writers concluded, would give humans the opportunity to understand "the miracle of life as it exists in all its countless forms throughout an infinite creation."[27]

For much of the twentieth century, the Martian myth dominated the popular culture of space, among both the public at large and serious scientists.

The possibility of canals on Mars spurred speculation about the life-forms that might live there. Following the release of *War of the Worlds*, H. G. Wells published a magazine article in which he described how Martians might appear. His account anticipated the pervasive popular image of extraterrestrials as humanoid in form with thin frames, large heads, and distinctive eyes. (Ordway Collection/U.S. Space and Rocket Center)

Bruce Murray, a professor of planetary science who helped NASA conduct the Viking mission to Mars, observed in 1971 that Mars had grabbed hold of human emotions in such a way as to distort both popular opinion and scientific desire: "We want Mars to be like the Earth. There is a very deep-seated desire to find another place where we can make another start, that somehow could be habitable."[28]

Belief in the plurality of worlds drew strength not only from works describing conditions in outer space but also from the whole history of exploration on Earth. The hope that life could be found elsewhere in the universe found support in the remembrance of earthly expeditions. For centuries terrestrial explorers had amazed the public with tales of strange creatures, leading the public to associate ventures into the unknown with new life-forms. People promoting space exploration drew upon this expectation. It is doubtful that the general public would have paid as much attention to space exploration had its early proponents insisted on a dead and uninhabited universe.

Much of the anticipation of new life-forms flowed from what Daniel Boorstin has characterized as the effort to catalog the whole creation.[29] Since antiquity, natural scientists have sought to provide the public with descriptions of the flora and fauna that inhabit the Earth. In A.D. 77, Pliny completed his *Natural History*, an encyclopedia that included within its numerous books descriptions of known animals, insects, and plants, including legendary creatures and popular folklore. One of the greatest literary achievements of antiquity, the work remained an authoritative source for fifteen hundred years, reminding educated readers that foreign lands teemed with exotic beasts. Generously illustrated bestiaries, natural histories that described real and imaginary animals, appeared in medieval times. To the delight of readers, such books described the habits of exotic beasts, often inaccurately. Elephants had no desire to reproduce, a twelfth-century Latin bestiary reported, and other books repeated stories about an Arctic goose whose young did not hatch from eggs but grew up in shells hanging from pieces of wood.[30]

In earlier times, most people rarely ventured more than a slight distance from their homes. As a result, readers had to take on faith the portraits of exotic creatures observed by travelers to distance lands. Occasional zoos and traveling circuses provided some empirical evidence,[31] but most people lacked practical guides and could hardly distinguish between a flying fish (a real creature) and mythical creations such as the griffin, a cross between a lion and an eagle.[32] Ancient and medieval catalogs overflowed with such beasts. The fourteenth and fifteenth centuries, when, as one observer noted, "no travellers' tales seem too gross for belief," produced some of the most impressive.[33] One of the most popular sources was *Mandeville's Travels*. Its author supposedly set out from England in 1322 and traveled to the Near East, India, and China. He described species of wondrous variety, including animals with human faces, lambs that grew out of vegetables, and a race whose faces appeared on their torsos.[34]

As biologists discredited animal myths from previous reports, explorers revealed new species.[35] Journeys into the Americas produced reports of the amazing species to be found there, including the grizzly bear and pronghorn antelope.[36] Expeditions often employed naturalists for the purpose of cataloging the strange life-forms to be found in foreign lands. One of the most influential was the documentation produced by the naturalist aboard the HMS *Beagle*. Conducting an expeditionary voyage between 1831 and 1836, the crew of the *Beagle* traveled along the coasts of South America and across the South Pacific. The naturalist was Charles Darwin, twenty-two years old when the *Beagle* set sail. Darwin reported the results of the expedition in a series of books published between 1839 and 1846. He described species as wondrous as those to be found in medieval mythology: giant tortoises on the Galápagos Islands and herds of lizards that swam out to sea.[37] On the coast of South America at Tierra del Fuego, he described a race of humans so savage that he remarked that they might as well have been creatures from another world.[38]

The discovery of other spheres in space, accompanied by the rush of reports portraying exotic creatures across the Earth, strengthened the belief in a plurality of worlds. The vast variation of earthly species, living and extinct, was well established by the end of the nineteenth century, as was the existence of extraterrestrial spheres on which other life-forms might arise. Respectable scientists and other educated people treated the possibility of extraterrestrial life as a proposition worthy of serious investigation.[39] If biology had produced complex creatures on the Earth, the process seemed capable of producing beings on other worlds. Attempts to portray Earth as the sole source of life, or even as the asylum of intelligent beings, carried the philosophic burden associated with the Earth-centric view of the universe. The concept of humans as lonely travelers on the only habitable planet in the cosmos never captured the imagination of space advocates in the same way that the concept of extraterrestrial life did. For others concerned with the involvement of the Deity, the existence of life on other planets now seemed to be part of the creator's plan.

Belief in Earth as the sole source of living creatures required believers to adopt views of the cosmos that did not maintain a large following. Aristotle's argument for a single world, for example, depended upon a boneheaded theory of physics. In Aristotle's view, heavy elements in the universe coalesced toward the center to form the Earth, whereas lighter elements such as fire rose toward the heavens to form stars.[40] Astronomical observations spoiled that argument. Saint Thomas Aquinas based his belief in a single world on the premise that God could not create an imperfect one.[41] An evolutionary process that proceeded on the basis of chance would produce many errors. One perfect world created intentionally, Aquinas reasoned, reflected God's magnificence. Darwin's theory of natural selection pretty well extinguished that point of view. To the extent that God was involved in the evolution of species at all, he had apparently created a wide variety, leaving many to die. The process of natural selection sorted out the successes from the errors. So much for the idea of a single perfect world.

That left the single-world theory sitting on a one-legged stool: the doctrine of catastrophe. Advocates of catastrophe suggest that rare but powerful events affect the evolution of life. On most planets, those events snuff life out. On Earth, catastrophes worked the other way. They promoted life. The doctrine eventually gained credibility, but at the dawn of the space age it suffered from an association with bad science and efforts to impose religious beliefs upon modern astronomy.

According to one mid-twentieth-century theory of catastrophic occurrence, a freak event created the planets of the solar system when another star passed by so close that it ripped material from the Sun.[42] If true, then most stars would not bear planets, a supposition not supported by later investigation. According to another theory popularized by the psychiatrist Immanuel Velikovsky, human life evolved in response to catastrophic upheavals recorded in ancient texts as the miracles. Between 1950 and 1955, Velikovsky published three books that set out to reconcile astronomy with biblical accounts of antiquity. Velikovsky reported that the planets of the solar system had periodically interfered with each other. According to his analysis of ancient texts, the planet Venus (then a comet) brushed Earth around 1500 B.C., parting the Red Sea and reversing the rotation of the Earth. The comet returned some fifty years later, according to this interpretation, when it again interrupted Earth's spin and accounted for such terrestrial phenomena as the collapse of the great walls of Jericho during Joshua's siege.[43]

The pseudoscientific quality of mid-twentieth-century catastrophic theories, as well as their persistent allegiance to biblical texts, repelled serious thinkers. Since the time of Copernicus, scientists have fought against the weight of revealed truth and religious dogma. A primary battleground in that fight has been the centrality of human beings. Centrality theories hold that God created humans on Earth in order to populate the planet with creatures in his own image and, according to some, set the universe to rotate around the planet.[44] Through natural observation, astronomers removed the Earth from the center of the universe, placing it instead on a remote spiral arm of an inconspicuous galaxy. Biologists, in the meantime, advanced their theories of evolution, suggesting that life advanced randomly according to biological principles present everywhere. If the processes kindled life on Earth, they could do the same elsewhere.

Armed with these beliefs, intelligent beings from Earth speculated on the shape of creatures from other worlds. A variety of individuals both serious and fanciful put their imaginations to work. Bernard de Fontenelle, a French writer of popular science and history, composed a widely read seventeenth-century tract. *Entretiens sur la pluralité des mondes* (Conversations on the Plurality of Worlds) appeared in 1686 and was a huge popular success. Its bestiary of planets described luminous birds that lit the night sky on Mars and sluggish creatures on Saturn that moved slowly to protect themselves from that planet's extreme cold.[45] Following the publication of *War of the Worlds*, H. G. Wells wrote an article for *Cosmopolitan* magazine in which he showed how Martian creatures might adapt themselves to local conditions, captivating readers with descriptions of a local ruling class nine to ten feet tall, feather-covered, with tentacles in place of hands.[46] In 1951 Kenneth Heuer described the types of beings that other planets in the solar system might produce, "a conclusion to which the mind is almost necessarily led." Living pillars with a sodium-based chemistry graced Jupiter, and on Venus the first explorers might encounter thinking and talking trees.[47] In 1966 astronomer Carl Sagan speculated on the life-forms that could arise on other worlds, suggesting that "organisms in the form of ballasted gas bags" might swim like plankton-eating whales through the thick Jovian atmosphere.[48] These interplanetary bestiaries departed little from medieval tomes, with fact indistinguishable from fancy. A 150–pound bat could glide through the air in a world that retained a dense atmosphere, and animals with six legs could hop across a low-gravity world.[49]

Writers of science fiction contributed even stranger life-forms to such bestiaries of space.[50] In the 1951 version of *The Thing*, scientists discover a carrot-shaped monster frozen in a spacecraft in the Arctic ice shield.[51] In *Alien*, a 1979 release, Sigourney Weaver battles a predatory creature utterly hostile to human life.[52] George Lucas produced the ultimate science fiction bestiary for his movie *Star Wars*. Drawing on the tradition of the frontier tavern in the Hollywood western, Lucas created the Mos Eisley Cantina, an intergalactic bar occupied by a grubby assortment of alien low-life.[53] Four of the five top-grossing motion pictures of all time as of 1991 dealt with life-forms from other planets, led by the children's tale, *E.T.: The Extra-Terrestrial*.[54]

Such descriptions inspired theological speculation. For much of history, people have wanted to believe in the Earth as the center of life. "Though the belief that our world was the material center of the Universe has long been dead," astronomer Henry Norris Russell observed in 1943, "the supposition that it was . . . unique in being the abode of creatures who could study the Universe has lingered long."[55] The supposition gave meaning to life by elevating the role of *Homo sapiens* in the cosmos. It supported the view that humans were the result of a deliberate process designed to create beings who could contemplate their existence and imagine God. Now science was suggesting that evolution occurred elsewhere. If just a single speck of vegetation was found on a planet such as Mars, British astronomer H. Spencer Jones observed in 1940, it would follow that "life does not occur as the result of a special act of creation or because of some unique accident, but that is the result of the occurrence of definite processes." On any planet capable of supporting it, life will gain a foothold. "Given the suitable conditions," he concluded, "these processes will inevitably lead to the development of life."[56]

The new cosmology did not banish God from the universe. Rather, it encouraged insights into the Creator's intent.[57] If it was God's plan to create life, then he must have done it elsewhere. "God exists," insisted one early twentieth-century astronomer, summing up the implications of life on Mars, "and He did not create habitable spheres with no object." Habitable spheres were created for the purpose of being inhabited.[58] Writing some fifty years later, Harvard astronomer Harlow Shapley suggested that the idea of a plurality of worlds "inspires respect and deep reverence." If theologians found it difficult to accept the idea that the God of humanity was also the God of gravitation and hydrogen atoms, at least theologians might be willing to consider the reasonableness of extending to other thinking beings "the same intellectual or spiritual rating" they give to us.[59] So powerful was the expectation of extraterrestrial life that even religious figures acknowledged its creation as God's will.[60]

The search for living beings elsewhere in the universe excited spiritual debate, if for no other reason than the possibility that the God of humans might
be busy overseeing other planets.[61] By advocating the plurality of life, scientists addressed one of the great mysteries of life. They could no more avoid the
religious implications of this message than the followers of Charles Darwin
could avoid the need to debate the church.

The actual search for life on other spheres began with the planet Mars. For
decades the general public had been bombarded with images of life-forms on
that nearby planet. Although serious scientists discounted the possibility of
finding intelligent life, they held high hopes for the discovery of lower forms.
For years following the Schiaparelli-Lowell discoveries, the popular press assured the public that some form of life would be found by the first explorers
to reach Mars. Fuzzy photographs taken through Earth-based telescopes showed
the seasonal advance and retreat of Martian polar caps and captured images of
dark areas that seemed to vary in shape and color from season to season. Displaying a sequence of images in a 1944 issue of *Life* magazine, writers for that
popular outlet announced that "it is logical to conclude that the vast regions
on Mars that change from green to brown in seasonal cycles are covered by
vegetation."[62] Confirmation of even dry mosses or lichen, the first observable
evidence that the life process had begun elsewhere, would be a stunning discovery.[63] Mars provided the first practical test of the plurality-of-worlds thesis.
The results, to say the least, were disappointing.

On 4 November 1964 NASA launched a 575-pound robotic traveler toward
the planet Mars. The speculations of nearly a century rode on the *Mariner 4*
spacecraft as it glided along a curving trajectory toward the planet of so many
dreams. The spacecraft arrived on 14 July 1965. Scientists at NASA's Jet Propulsion Laboratory cheered as sensors on *Mariner 4* located the planet and began
taking the first of twenty-two pictures. Digit by digit, the spacecraft's radio
transmitted the coded data back to Earth. The third picture revealed a surface
feature, a circle of sorts some twelve miles across. By the seventh photograph,
the features had become clear. Mars was covered with craters. The close-up
pictures showed a surface pocked with ancient craters. There were no oceans,
no artificial canals, and no oases of vegetation. Mars looked lifeless. It looked,
as the horrified scientists were forced to acknowledge, like the Moon.[64]

"Mars is dead," announced *U.S. News and World Report*. "There are no cities,
oceans, mountains or even continents visible on Mars."[65] Inspecting the photographs, President Lyndon Johnson observed that "life as we know it with

The real Mars dashed expectations. In the summer of 1965, the *Mariner 4* spacecraft produced the first close-up pictures of the red planet. This is the first picture in which viewers could distinguish surface features. There were no canals, no signs of life. Mars was pocked with craters. To the disappointment of exploration advocates everywhere, a series of robotic expeditions beginning with *Mariner 4* revealed no evidence of surface lifeforms. (NASA)

its humanity is more unique than many have thought."[66] The extreme lunar-like features shocked even the scientific teams overseeing the mission. Many had hoped to see evidence of weathering that might suggest the presence of free water so essential to life on Earth. As *Mariner 4* sped past the planet, it beamed radio signals back to Earth through the Martian atmosphere, measuring its density. The results were just as depressing as the photographs. The atmosphere proved remarkably thin, hardly capable of protecting embryonic surface life from the constant bombardment of cosmic rays and ultraviolet radiation.[67]

Scientists tried to stem the general disappointment. Astronomer Carl Sagan compared the photographs from *Mariner 4*, taken from six thousand miles above the Martian surface, with similar satellite photographs of Earth that also showed no signs of life. Astronomer Clyde Tombaugh announced that photographs of a large crater on Mars contained markings coinciding with the position of a short canal mapped more than a half century earlier by Percival Lowell.[68]

NASA tried again in 1969, as *Mariner 6* and *Mariner 7* zipped by Mars. "All is not lost for the astronomical romantics," the *Washington Post* reported. "The famed canals of Mars showed up clearly."[69] The announcement prompted Arizona senator Barry Goldwater at a special congressional hearing to request the pictures that showed the long-sought-after canals. NASA officials sheepishly explained that no such pictures existed. What the pictures in fact revealed were discolored areas on the Martian surface, light and dark splotches that resembled broad canals "under poor seeing conditions."[70] In spite of the discouraging results, NASA executives hoped that scientific interest in Martian exploration would not wane. One Mars expert hired to interpret the photographs for NBC television, however, was not so sure. "If no one told you, you wouldn't know if this was Mars or the moon," he complained.[71]

The search for life got a small boost when *Mariner 9* swung into orbit around Mars in late 1971. The planet was shrouded under a global dust storm when the spacecraft arrived, and scientists had to wait several months before the atmosphere cleared enough to permit detailed photography. Appearing first above the dust clouds were four tall volcanoes, the largest more than twice the elevation of Mount Everest. As the dust finally settled, scientists watched in amazement as the spacecraft transmitted images of canyons and chasms that looked like nothing less than dry riverbeds.[72] If water had carved these features, life might lurk in protected pockets of the planet's soil.

To further the search for life on Mars, NASA dispatched two Viking spacecraft, which arrived in their Martian orbits in the summer of 1976. Each carried a 272-pound lander (plus fuel) that descended to the surface of the planet. The landers carried a variety of instruments, including a slow-working camera and a biological package to test the soil for living microorganisms. Excited scientists let their imaginations soar. "I keep having this recurring fantasy," Carl Sagan reported. "We'll wake up some morning and see on the photographs footprints all around Viking that were made during the night, but we'll never get to see the creature that made them because it is nocturnal." Sagan wanted a night light put on the Viking landers. He also joked about putting out bait.[73]

Once again, investigation deflated expectations. The cameras took pictures of a cold, dry, rock-strewn surface. The biology package produced chemical reactions like those one would expect in the presence of living organisms, but efforts to detect organic molecules in the Martian soil failed.[74] Mars was "deader than Elvis," wrote *Washington Post* staff writer Kathy Sawyer.[75] Concluded one commentator, "The exploration of the planet Mars in the 1960s and 1970s, culminating in the landings of two instrumented Viking spacecraft in 1976, ended a long dream of Western civilization. Since these explorations obtained a negative answer to their most interesting question, that of Martian life, they are doubtless regarded by many as a failure."[76]

Mars continued to tantalize the attentive public with suggestions of possible life, perhaps clinging to pockets of existence in subterranean sanctuaries. In 2008 NASA's Mars *Phoenix* lander uncovered water ice, which vaporized as soon as the lander's robotic arm exposed the ice to the thin Martian air. The presence of liquid water would provide favorable conditions for the processes necessary to create life. Earlier, in 1997, the Mars lander *Pathfinder* and its interplanetary roving vehicle *Sojourner*, explored an area marked by extensive flooding. At some time in the planet's geological history, the area had experienced conditions more like those on the Earth than on the Moon. In 2009 telescopes in Hawaii detected plumes of methane gas on Mars. On Earth, living beings produce methane gas in great quantities—it is a signature of life. Methane also can be produced through the interaction of water, heat, and subsurface minerals, suggesting the possibility of water pockets below the cold and sterile surface of Mars.[77]

The desire to satisfy the expectation of extraterrestrial life within the solar system received substantial attention in the summer of 1996, when news of potential fossil forms exploded on the national scene. Public awareness of extraterrestrial life had been heightened that summer by the alien invasion movie *Independence Day*, a modern-day Martian invasion story. The extraterrestrials (not from Mars) arrive in an enormous spaceship with the single purpose of blasting humans off the face of the Earth. The movie resembled a cartoon, with impossible scenarios, but proved so entertaining that it set box office records that summer season.[78]

For two years before the summer of 1996, scientists at NASA's Johnson Space Center along with a university-industry team had been looking for evidence of extraterrestrial life in a rather unusual way. According to the best evidence, a large impact had blasted rocks from the surface of Mars into space sixteen million years ago. Some of the rocks eventually intersected Earth's orbit and fell through the atmosphere to the ground. Scientists dated one of the rocks, found in the Allan Hills region of Antarctica, to the early phases of the solar system, four and a half billion years ago, when the Martian climate was probably more wet and warm than it is today.

Therein ensued a fascinating detective story, in which scientists on Earth probed one of the rocks for evidence of various organic compounds and fossil-like structures that might have been produced by ancient microorganisms. The scientists planned to announce their findings in the 16 August issue of *Science* magazine, accompanied by a modest press conference at the Johnson Space Center. In its 5 August issue, however, the trade weekly *Space News* broke the discovery story with a small, three-paragraph report on the second page of the publication. Analysis of the meteorite, the weekly explained, "points to indications of past biological activity on Mars."[79]

The news hit government policy circles with the intensity of a Hollywood movie premier. Had NASA discovered life on Mars? "If the results are verified," announced astronomer Carl Sagan, "it is a turning point in human history." President Bill Clinton called it "one of the most stunning insights into our universe that science has ever uncovered." Clinton held a special press conference on the White House lawn to pledge that "the American space program will put its full intellectual power and technological prowess behind the search for further evidence of life on Mars" and announced that NASA's next Mars mission would land in less than one year on—what else—America's Independence Day. Space advocates cheered themselves with the hope that the findings would regenerate public support for the long-standing goal of a human expedition to Mars. As evidence that the story had gotten out of hand, NASA administrator Daniel Goldin issued a special news release: "I want everyone to understand that we are not talking about 'little green men.' . . . There is no evidence or suggestion that any higher life form ever existed on Mars."[80] All the plans for a low-key press conference at the Johnson Space Center evaporated as Goldin and his associates rushed to prepare a nationally televised briefing with the principal investigators and other experts.

The overall vision of space travel predisposes well-informed people to believe in extraterrestrial life. Among the public at large, Mars has enjoyed a favored position in this regard. It has been viewed as a nearby place, Earth-like in its development, that could harbor possible life-forms. With respect to the apparent fossil, scientists urged caution. The Mars of earlier imagination, with its artificially constructed waterways and intelligent civilizations, is most certainly a lie. The tubelike structures on the rock could have crystallized from mineral vapors, a process that requires temperatures too warm to permit life.[81] Evidence of life may still be found on Mars, perhaps in fossil form or in subterranean sanctuaries. But Mars as Lowell, Burroughs, and Wells saw it, with its ancient and technological advanced life-forms, does not exist.

Where else might such life be found? Spacecraft sent to investigate Venus, in size Earth's sister world, revealed a hellish sphere, hardly the primeval place of carboniferous swamps pictured by early visionaries. With its thick atmosphere, mainly carbon dioxide, Venus represents a greenhouse effect run amuck. The surface temperature is nearly nine hundred degrees Fahrenheit, and the clouds contain sulfuric acid.

Might life exist beyond Mars, in the frozen outer reaches of the solar system? The surface of Europa, one of Jupiter's many moons, is covered with ice. The ice floats on top of an ocean of slush or water. In 1996 and 1997 NASA's *Galileo* spacecraft captured images of Europa's surface that looked strikingly like the ice floes that cover Earth's polar regions. The images, moreover, revealed a red-brown substance along breaks in the ice floes that suggested an organic chemistry. The tidal forces produced by the mass of Jupiter could be sufficient to produce the internal warmth that, with water, is necessary to harbor life. Scientists want to dispatch robotic spacecraft to Europa capable of penetrating the ice and search for aquatic life-forms.[82]

The absence of complex surface life on other spheres within the solar system did not rule out intelligent life beyond. If life develops through a natural process, as most scientists believe, then beings might be found on suitable objects turning in orbits around other medium-sized yellow stars. In 1960 astronomer Frank Drake prepared a famous equation that calculated the probability that people on Earth might be able to communicate with such civilizations. The number of "communicative civilizations" in the Milky Way, he suggested, was equal to the product of the following quantities:

1. The number of stars in the galaxy
2. The fraction of stars with planetary systems
3. The number of planets per system that are suitable for life
4. The fraction of suitable planets where life might actually develop
5. The fraction of living planets that produce intelligent life forms
6. The fraction of life forms that are willing and able to communicate with other civilizations
7. The average life expectancy of a technological civilization expressed as a fraction of life of the planet[83]

All of the numbers in the equation are estimates. Some of the numbers are terribly small. The probabilities favoring the development of complex life forms on a suitable planet might be one in one hundred, and the chances that one of them creates a technological civilization might be equally scanty. Still, the number from which the calculation begins is astonishingly large. Astronomers believe that the Milky Way accommodates at least two hundred billion stars.[84] Even small probabilities multiplied by large numbers produce impressive results. Millions of civilizations with national space programs could have developed within the Milky Way alone. The galaxy could be teeming with intelligent civilizations at this very time.[85]

The potential existence of long-lived, technologically advanced civilizations is a fascinating possibility, one that has motivated authors of science fiction for some time. Between 1951 and 1953, Isaac Asimov issued the foundation trilogy, three sweeping books assembled around a plot line originally developed for a set of short stories written for *Astounding Science-Fiction*.[86] One of the most influential science fiction writers of the twentieth century, Asimov began publishing fantastic stories while a student at Columbia University. He supported himself as a professor of biochemistry until the income from the foundation series and other works allowed him to write full time. The foundation trilogy describes a galactic empire in which a spacefaring civilization has spread itself over twenty-five-million inhabitable planets. After twelve thousand years of relative peace, the civilization is collapsing. Efforts based on the science of psychology to shorten the period of impending barbarity guide the plot.

Galactic empires, human or otherwise, have provided the backdrop for some of the most popular science fiction stories of all time. In the original *Star Trek* television series, forces of the Federation do battle with the Klingon Empire, a weakly disguised effort to replicate the superpower conflicts of the

If advanced life-forms could not be found nearby, perhaps they inhabited more distant worlds. If such civilizations exist, perhaps they broadcast radio waves. The movie *Contact*, based on Carl Sagan's novel, featured the Very Large Array radio astronomy observatory in New Mexico as a listening post where scientists receive the first broadcast from extraterrestrial beings. (National Radio Astronomy Observatory / Associated Universities, Inc. / National Science Foundation)

1960s in an galactic setting.[87] In the famous *Star Wars* cinema series, Princess Leia Organa represents the resistance in a battle against the evil military leader of the Galactic Empire, Darth Vader.

In Asimov's foundation trilogy, humans travel between distant corners of the galaxy with as much ease as Americans of the twentieth century flew between cities on Earth.[88] What writers of fiction imagined, however, rocket scientists could not quickly produce. Not content to wait for the development of spaceships of the interstellar kind, scientists sought to communicate with extraterrestrial civilizations in more immediate ways. In 1960, in conjunction with the formulation of his famous equation, Frank Drake used the eighty-five-foot-wide dish of the National Radio Astronomy Observatory to listen for cosmic calling cards. Drake hypothesized that any advanced technological civilization would emit radio and television waves; some might intentionally dispatch radio signals announcing their presence to other civilizations. "At this very minute, with almost absolute certainty," Drake exclaimed, "radio waves sent forth by other intelligent civilizations are falling on the Earth."[89] After reviewing the possibilities, he tuned the West Virginia radio observatory to 1,420 megahertz and pointed it at two nearby stars.

To his astonishment, Drake received a very strong signal of intelligent origin from the apparent vicinity of Epsilon Eridani, a star some ten light years from Earth. For a moment, it appeared that Drake had found the proverbial needle in the cosmic haystack. As the first person to discover extraterrestrial life, Drake would have been as famous as Albert Einstein. Who else might be transmitting signals at that wavelength? Upon further investigation, Drake discovered that he had inadvertently tuned in to an Earth-based signal emitted as part of a secret military experiment.

This did not discourage individuals seeking to contact extraterrestrial civilizations. In the early 1970s scientists advising NASA on its *Pioneer 10 and 11* spacecraft urged the inclusion of a small plaque on each containing a symbolic message from Earth. The two spacecraft were designed to fly by Jupiter and, in the case of *Pioneer 11*, Saturn, encounters whose imparted momentum would carry the machines out of the solar system. NASA agreed to attach an engraved gold-anodized aluminum plate to the antenna support struts on each.

The plates caused an immediate controversy. In addition to locating the planet from which they had come, the plaques included figures of a man and a woman as they would appear unclothed. The vision of male genitalia and female breasts being carried on a government-sponsored space mission to an alien civilization outraged some Americans. "Isn't it bad enough that we must tolerate the bombardment of pornography through the media of film and smut magazines?" one letter writer to the *Los Angeles Times* declared. Now "our own space agency officials have found it necessary to spread this filth even beyond our own solar system." Responding a few days later, another letter writer suggested that NASA officials might "visually bleep out the reproductive organs of the man and the woman. Next to them should have been a picture of a stork carrying a little bundle. Then if we really want our celestial neighbors to know how far we have progressed intellectually, we should have included pictures of Santa Claus, the Easter Bunny, and the Tooth Fairy."[90]

Scientists and their government allies sought public funding for the effort to locate signals produced by extraterrestrial civilizations. During the early 1970s NASA proposed Project Cyclops, a powerful ground-based listening sys-

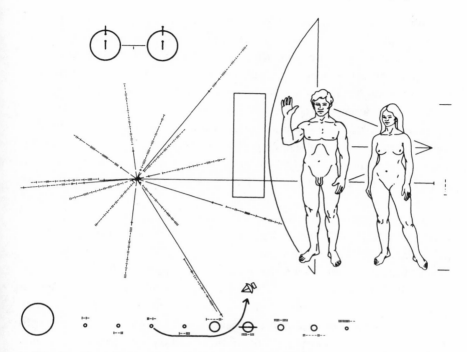

Anticipating the existence of extraterrestrial civilizations, Earth scientists dispatched radio signals and other messages to distant worlds. In 1972 NASA launched *Pioneer 10* and *Pioneer 11* on trajectories that took them past Jupiter and out of the solar system. Each spacecraft carried a gold-anodized aluminum plate designed to show any creatures that might intercept it millions of years hence where the machine came from, when it was launched, and who sent it on its way. (NASA)

tem. Some fifteen hundred antennas, each one hundred meters in diameter, would have been linked by computers into one enormous listening device.[91] Estimated to cost $20 billion, the project was never approved. In 1974 Drake and Sagan reversed the normal communication process. Aiming the Arecibo Radio Telescope in Puerto Rico at the globular star cluster M-13 in the constellation Hercules, they beamed out an electronic greeting card. By Sagan's estimate, the chances of hitting a technological civilization in the more than one-hundred-thousand-star cluster was fifty-fifty. At the speed of light, however, the message would not arrive for twenty-four-thousand years.[92]

In 1975 NASA administrator James Fletcher delivered an address to the National Academy of Engineering in Washington, D.C., in which he asserted that the universe was teeming with life. Our own galaxy, he said, "must be full of voices, calling from star to star in a myriad of tongues." He criticized the desire of pragmatists to structure the space program in such a way as to produce only direct and immediate benefits to humans on Earth. Through their preoccupation with what Fletcher called the "now syndrome," people were losing sight of the revolutionary breakthroughs that more adventurous undertakings might provide. "It is hard to imagine anything more important than making contact with another intelligent race," he observed. "It could be the most significant achievement of this millennium, perhaps the key to our survival as a species."[93] Fletcher's words were rooted in his own religious experience. A lay minister in the Church of Jesus Christ of Latter-Day Saints, Fletcher subscribed to the theological doctrine that God had created a plurality of worlds populated with intelligent beings. His position as NASA administrator allowed him to search for empirical verification of those beliefs.[94]

NASA officials continued to promote the search for extraterrestrial intelligence (SETI) throughout the 1970s and 1980s.[95] "Life as we know it here on Earth appears to be the result of universal laws of chemistry and physics," they observed in defense of their effort. "Elsewhere among the 400 billion stars in the Milky Way galaxy or in one of the hundred billion other galaxies, the same processes may have produced beings who stare at the heavens and wonder about other occupants of their universe."[96] In spite of official enthusiasm for the effort, the project received a skeptical reception in the U.S. Congress. Politicians viewed it as an extravagance the country could ill afford. One likened it to "buying fur coats for your cows while your children were freezing."[97] In 1978 Wisconsin senator William Proxmire granted SETI his Golden Fleece Award, which he periodically bestowed on projects he felt misused the public treasury.[98] Members of Congress complained about efforts to locate "little green men" and sought to eliminate SETI's funding.[99]

In the face of mounting criticism, NASA officials refocused the initiative and gave it a new name. The Microwave Observing Project would survey the heavens for microwave radio signals, the most likely means of communication between intelligent civilizations. NASA officials estimated that the project would cost a paltry $135 million over ten years, much less than the cost of a single space shuttle flight. Counting on its success, supporters called it "the biggest bargain in history."[100] Congress was unimpressed. In 1993, after a brief startup, lawmakers terminated the project. Senator Richard Bryan, the Nevada Democrat who led the cancellation effort, argued that the chances of finding intelligent life were too remote to justify the expenditure. "The SETI program has found nothing," he observed. "All the decades of SETI research have found no confirmable signs of extraterrestrial life."[101]

Even serious scientists grew skeptical. If the universe was teeming with various life-forms, where were they? A lush, wet planet like Earth would hardly go unnoticed in a universe full of spacefaring civilizations. A single civilization, moving out from its home planet at the rate of one one-hundredth the speed of light could colonize the whole galaxy in less than ten million years, a short period of time in astronomical terms.[102] Given the age of the universe and the incredible number of stars, aliens and their radio broadcasts should have arrived already.

The skepticism that silence incurs has been characterized as the Fermi paradox. Inspired by a lunchtime conversation, the physicist Enrico Fermi calculated the frequency with which a single planet should expect visitation in a universe teeming with advanced civilizations. Representatives from other civilizations should have visited Earth before human history began. Calculations indicated frequent visitation; actual experience suggested none.[103] Inconsistency invited explanation. Perhaps Earth had been visited—a proposition favored by people associated with the unidentified flying object (UFO) movement and encouraged by films such as the 1997 release *Men in Black*. In that movie and its sequels, agents employed by a top-secret independent agency monitor and disguise the activities of alien beings allowed to live secretly on Earth.[104] Another theory, advanced frequently during the height of the Cold War, suggests that technological civilizations might be remarkably short-lived. Species that simultaneous discover space travel and atomic power might quickly extinguish themselves, taking a fatal last step on the evolutionary scale. Such a possibility, Carl Sagan pointed out, would leave "no one for us to talk with but ourselves."[105] Or alien species might evolve into forms so advanced as to be unrecognizable to human beings. The latter possibility, nonetheless, would not preclude encounters of an indirect kind, a prospect advanced with the 1997 release of Sagan's *Contact* story in movie form. At the conclusion of the movie, the character played by Jodie Foster has an extraterrestrial encounter so dreamlike that no one can tell whether it really occurred.[106]

Beneath the more exotic explanations lay a most disturbing one. Perhaps the plurality-of-worlds thesis is wrong; perhaps humans really are alone. Could it be possible that within the population of solar systems accessible by spacecraft or radio communication, only the Earth has produced advanced life-forms? In

2000 astronomers Peter D. Ward and Donald Brownlee published *Rare Earth*, a powerful statement discounting the plurality of worlds.

Earth-like planets in metal-rich solar systems orbiting single yellow-orange dwarf stars are leading candidates for the development of life, especially those that revolve in a "goldilocks zone" (not too hot, not too cold) where water assumes its liquid form. Ward and Brownlee advance this concept many steps further. The Earth benefits from the placement of Jupiter, whose gravitational pull regulates the frequency with which killer asteroids intersect the orbit of Earth. The Earth benefits from the absence of gamma ray bursts, possibly a consequence of its location in an outlying section of the galaxy not prone to such astronomical catastrophes. A gamma ray burst could strip away the ozone layer of a life-bearing planet and expose its inhabitants to lethal levels of radiation. The Earth benefits from the presence of plate tectonics, a process that acts like a global thermostat by reprocessing greenhouse gases. Life on Earth benefits dramatically from a single large moon that produces exceptional tides and helps to stabilize the tilt of the Earth and the length of its days. Astronomers believe that a freak encounter between a half-grown Earth and an object of planetary proportions ripped away the material that formed the Moon.

Ward and Brownlee present nearly two dozen conditions that may be required to lead a planet through the stages necessary to produce complex and intelligent life-forms. They hypothesize that these conditions are extremely rare. "Life in the form of microbes or their equivalents," the two suggest, may be quite common. The chemistry necessary to produce simple life-forms seems to be widely distributed throughout the universe and living creatures can develop in hostile places well removed from the benefits of sunlight and clean air. For such creatures to develop into intelligent beings, or "even the simplest of animal life," Ward and Brownlee suggest, may require a combination of events so uncommon that it has occurred in our sector of the cosmos only once.[107] The slightest changes in the characteristics of the Earth would leave it a much less desirable place to live.[108] "If we were typical," said one observer trying to discredit the plurality-of-worlds doctrine, "we should not exist."[109] Such a doctrine, once unthinkable in circles devoted to the separation of science from the Earth-centered view of the universe, has grown in strength as humans learn more through space exploration.

The possibility that humans may never encounter advanced life-forms with technologies and cultures like their own undercuts one of the principle elements of the spacefaring vision. Yet in a strange way it creates a new expec-

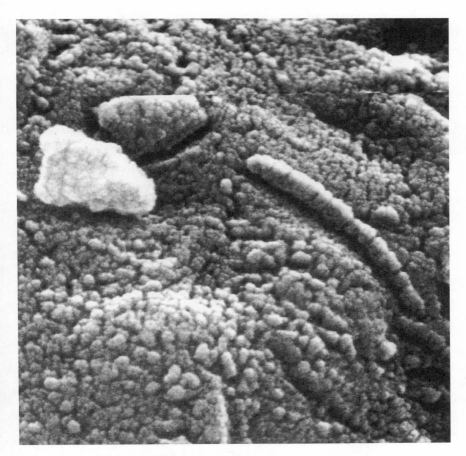

In the summer of 1996, NASA scientists announced that they had found fossil evidence of primitive life in a meteorite blasted from the planet Mars and sent on a trajectory that carried it to Earth. In the most famous (and least conclusive) bit of evidence, a high-resolution scanning electron microscope revealed a strange, tubelike structure on the rock from Mars. (NASA)

tation, one that may motivate the pursuit of space flight in centuries to come. The expectation is expressed through the anthropic principle, which space historian Steven Dick characterizes as "a secular search for the meaning of life based on physical principles rather than theological dogma."[110]

Members of an intelligent species living alone in a huge universe might look at the cosmos and conclude that the universe was created for their benefit. This conforms closely to the theological interpretation of creation. The anthropic principle turns this perspective on its head. Only in a universe with a specific combination of conditions will an intelligent species arise. Those creatures can then look at the universe and ask: "Why is the universe the way we see it?" The answer is quite straightforward. If the universe had evolved in a different way, the creatures would not be present to ask the question. Neither they nor their curiosity would arise.[111]

The various forms of the anthropic principle (from weak to strong) suggest a central purpose for space flight. If this particular universe exists so that humans can understand it, their responsibility is to understand and to preserve the unique capabilities that allow this to occur. Humans may never meet another intelligent species capable of space flight or other advanced technologies. In that case, the sole responsibility for preserving intelligent life and extending it into the universe rests with human beings. The anthropic principle does not return to a theological vision of an Earth-centered universe, but it does ascribe far more important purpose to the human endeavor than the Darwinian vision of many imperfect worlds harboring alien species with characteristics produced by mere chance.

Philosophic questions such as these tend to elevate public interest in space exploration. To learn the true nature of the universe, scientists venture to the outer limits of credibility, looking for black holes, evidence of hyperspace, planets around other solar systems, and echoes of the Big Bang. Some undertakings survive; others disappear. Congress terminated the High Resolution Microwave Survey. Even its innocuous sounding name could not disguise its intent—to search for radio broadcasts from alien creatures. The initiative suffered through its association with extraterrestrial abduction stories and flying-saucer cults. "The opponents of the program have frequently poked fun at it," observed Maryland senator Barbara Mikulski in a futile effort to save the project, "and one can understand why. . . . Have we all not seen those pictures? 'Extraterrestrial alien with Bush at Camp David,'" she said as she parodied her opponents.[112]

Searching for echoes from the Big Bang is hardly less exotic than searching for radio signals from alien creatures. The Big Bang project, however, survived, in part because of the manner in which it addressed philosophic and religious preconceptions. The $230 million *Cosmic Background Explorer* (*COBE*) sought to examine the way in which the universe began, a matter of great theological concern.

Scientists believe that the entire universe was created some ten to fifteen billion years ago from a single point of unity, smaller than the dot at the end of this sentence. During its initial stages of expansion, the material making up the present universe was intensely hot. Any hot object emits radiation and continues to do so as it cools. That is why a soldier with special goggles can see tanks and jeeps at night even after the vehicles have been parked and shut down. They continue to emit background radiation. As strange as it may sound, the background radiation from the period of the Big Bang is still around, and its frequency can be calculated with mathematical precision. Scientists can detect the radiation using antennas on Earth, but to obtain measurements with sufficient accuracy to test the Big Bang theory, they need to send their instruments into space.[113]

NASA launched the two-and-one-half-ton *COBE* satellite into Earth orbit on 18 November 1989.[114] Scientists were ecstatic as the results poured in. "They have found the Holy Grail of cosmology," exclaimed physicist Michael Turner.[115] In effect, *COBE* took a snapshot of creation a few hundred thousand years after the Big Bang. The space probe detected background radiation at precisely the predicted frequency. Moreover, it resolved one of the most significant problems with the Big Bang theory. The galaxies could not have formed from a uniformly expanding universe, because gravity would act equally on all available matter. From the ground, with imperfect instruments, the background radiation seemed just that—uniform in every direction. With a clearer view, *COBE* detected ripples or blotches in the background radiation, exactly what was needed for matter to condense in irregular ways.[116] "It is the discovery of the century, if not all time," said Cambridge professor Stephen Hawking.[117] "It's like looking at God," declared astrophysicist George Smoot.[118]

The *COBE* program and its findings were a success because they fit so well with theological images already in people's minds. All cultures possess creation stories.[119] The story found in Genesis tells of a creation beginning with a great burst of light. God then separates the light from the darkness. In the first phase of creation after the Big Bang, as scientists describe it, the universe was all light—an opaque fog so thick that light scattered as soon as it formed. *COBE* took a picture of the universe at the point at which the universe became transparent, when the darkness divided from the light. In Genesis, "God said, 'Let there be light.' . . . God then separated the light from the darkness. . . . Thus evening came, and morning followed—the first day."[120] Said veteran space journalist Kathy Sawyer: "The announcement [of *COBE*'s results] had such impact because the scientists' themes are in apparent harmony with the biblical version of creation recorded centuries ago by scribes who had no inkling of relativity, particle physics or other elements of modern cosmology."[121]

In 1895 H. G. Wells wrote *The Time Machine*, in which he sent a character from Victorian England spinning through the future almost to the end of the world.[122] Time travel is a favorite theme in science fiction but not as fantastic as writers might think.[123] In promoting the Hubble Space Telescope, NASA officials pointed out that it, like all other telescopes, acted as a cosmological time machine: "Because light travels across the universe at a finite speed (186,000 miles per second), the deeper astronomers look into space, the farther back in time they look."[124] With the unprecedented clarity that an orbital platform offers, the space telescope could peer back close to the beginning of time. "From its vantage point outside the Earth's murky atmosphere, the HST will be able to probe a distance of fourteen billion light-years—offering views of galaxies so distant that they will appear as they were when the universe was formed."[125]

A space telescope can look back in time; it cannot transport objects through it. For years the possibility of moving backward and forward in time has been the province of science fiction writers and scientific crackpots. Time travel raises a seemingly insurmountable paradox, well represented in the motion picture *Back to the Future*. In that film, Michael J. Fox travels back in time and meets his mother as a young girl. To his horror, Fox's mother falls in love with him and spurns his father. Unless he can reunite his parents, Fox will not be born.[126] In physical terms, he will create another universe in which he does not exist.

This is not as far-fetched as it sounds. The leading theoretical candidate for the unification of all known laws of nature allows for time travel and alternative universes. That is the theory of hyperspace or, in its most advanced form, superstring theory. "Everything we see around us," says popular science writer Michio Kaku, "from the trees and mountains to the stars themselves, are nothing but vibrations in hyperspace."[127] The theory suggests that the universe may operate through more than the four familiar dimensions that people commonly perceive (height, width, depth, and time). The laws of nature may fit together into one unified theory when more dimensions are added. The fact that we cannot perceive these dimensions does not alter the possibility that they exist.

The added dimensions allow for the presence of "wormholes," tunnels that create shortcuts through space and time.[128] Even more impressive is the possibility that wormholes may provide pathways to different universes. Physicists have suggested that ours is only one of a vast number of parallel universes connected to one another by passageways. Inner-galactic travel through wormholes is theoretically possible.[129] Fantastic children's stories in which young people step through looking glasses or wardrobes into other worlds may turn out to be the inspiration for twenty-first-century physics.[130]

Already the Hubble Space Telescope has photographed the signature of a black hole.[131] The vortex of material around a black hole leaves a distinctive signature, visible from a telescope above the atmosphere of Earth. The laws of mathematics suggest that matter entering a black hole may reemerge in an alternative universe.[132] Watching matter being yanked into a potential black hole, said the director of the Space Telescope Institute in Baltimore, "was like having a telephone line to God."[133]

Costly, complicated projects with armies of government-supported employees are required to test theories such as these. In 1993, after appropriating hundreds of millions of dollars to start the project, Congress voted to terminate the superconducting supercollider. This "window on creation" was designed to accelerate two beams of protons and send them speeding in opposite directions within a circular tube fifty miles in diameter. As the beams approached the speed of light, they would collide, releasing subatomic particles. Among other possibilities, the supercollider might have revealed the secrets of superstrings.

The science was simply too esoteric for most politicians. Struggling to defend the project, congressional supporters issued excuses about photon-beam cancer therapy and trains capable of traveling at three hundred miles per hour. Supporters were unable to articulate a persuasive vision of direct benefits for the simple reason that the consequences of the project were fantastic. "The costs are immediate, real, uncontrolled, and escalating," observed Congressman Sherwood Boehlert of New York. "The benefits are distant, theoretical, and limited."[134]

Another group of scientists pressed for funds to investigate nearby solar systems. For some time, astronomers have dreamed of constructing instruments so precise that they could capture images of planets around other stars.[135] In the final decade of the twentieth century, astronomers began to produce a rush of evidence confirming the presence of extrasolar planets. The objects, it seemed, were as common as dust. The presence of the planets was confirmed by indirect means, such as their gravitational effect on the central star.

Proponents of space travel hoped that renewed interest in the possibility of Martian life might resurrect support for a human expedition to that nearby sphere. In this illustration, the first visitors from Earth explore Noctis Labyrinthus, part of the vast Valles Marineris network of Martian canyons. Others favored searches in the outer solar system and instruments that could detect evidence of life-bearing planets circling nearby stars. (NASA)

To advance this work, NASA officials proposed a twenty-five-year undertaking using space and ground-based instruments to locate and eventually look directly at extrasolar planets. If life thrives in such places, it could be detected in predictable ways. Life-forms similar to those found on Earth will transform their planets in predictable ways. The presence of large quantities of free oxygen in a planetary atmosphere, for example, is a signature of life.[136] Satellites are being developed that could conduct spectrographic studies of extrasolar planets and their atmospheres. "In the not too distant future," NASA administrator Daniel Goldin promised, "we will have the technology needed to image any planets that orbit nearby stars." This advanced technology utilizes precisely flown flotillas of space-based instruments and a process for nullifying the glare of the central star called interferometry. Such an undertaking would allow scientists to answer critical questions: how rare is the Earth, and do any nearby planets possess conditions supportive of life? The discovery of another planet on which humans could live "would change everything," Goldin said. "No human endeavor or human thought would be untouched by this discovery."[137]

The program to conduct this research encountered obstacles. In 2007 public officials indefinitely postponed the Terrestrial Planet Finder, one of the key steps in the undertaking. NASA needed the funds to pursue more traditional visions of space exploration, to return to the Moon and continue flying the winged space shuttle for a few more years. Planetary Society executive director Louis Friedman called the decision "an investment in the past."[138]

During the 1950s advocates of space exploration worked hard to promote their dreams. They convinced the public that space travel was something desirable and real, not just the fantasy of a small group of believers. Drawing on cultural traditions and public expectations, they transformed fantastic ideas into a vision that produced moon trips, planetary investigations, and space telescopes. Their most far-reaching efforts led to discoveries that quickly outdistanced the vision that made the efforts possible. After just a half century of investigation, space science was pushing against fantastic ideas once more.

6

The Extraterrestrial Frontier

Space: the final frontier ...

— Captain James Kirk

At Epcot, the large theme park that opened near Orlando, Florida, in 1982, the Walt Disney Company constructed a colonial-style building called the American Adventure. Nearby exhibits honored, among other cultural traditions, Japanese art, Moroccan architecture, French cooking, and German beer. The American pavilion celebrated U.S. history. In the central rotunda, designers placed a mural depicting the evolution of NASA's space shuttle. Next to it they set a quotation from historian and philosopher Ayn Rand praising the generations of Americans "who took first steps down new roads armed with nothing but their own vision."[1]

For many partisans of space flight, the U.S. cosmic adventure promised to maintain America's special characteristic. Space travel promised far more to them than the opportunity to unlock the mysteries of the universe. The undertaking provided an opportunity to continue the practices that had spurred invention and innovation and made America great. Those practices arose from the willingness of Americans to take the "first steps down new roads," from the great migrations that brought people to the continent to the settlements and industries they founded. If space exploration is just about science, then the vision of space travel with its persistent emphasis on human flight is headed down the wrong path. To its most devoted advocates, space exploration promises more. To them, human space travel has value in its own right. It presents an opportunity for Americans to move their culture of invention and innovation outward, beyond the Earth, in ways that will make it last forever. Among the core of true believers, devotion to this point of view is extremely strong.

This vision manifests itself in the idea of space as a final frontier. For nearly all of its history, American has been a frontier nation. Waves of people have spread across the continent and learned how to live on unfamiliar lands. After 1900, when the availability of open land diminished, Americans continued the frontier tradition through expeditions to the poles, further immigration, the development of technology, and the exploration of space. Within the culture at large, the frontier experience is thought to have shaped American culture in distinct ways, encouraging ingenuity, invention, innovation, equality, democracy, and material progress. Without a continuing frontier, from this point of view, these characteristics will disappear. Americans will cease to be unique. They will become like people in more static countries and their special culture will cease to be a force advancing the future of humankind.

The vision of humans in large numbers moving out across an extraterrestrial version of a frontier trail is an ambitious one. Most Americans experienced the space program vicariously, as armchair explorers viewing television reports or images on book pages or computer screens. In the first half century of flight, only a few hundred people actually ventured into space. As the space program matures, its advocates claim, this will change. Americans in large numbers will join the venture, reactivating the vast migrations that shaped the country in the past. Humans will move into space in multitudinous throngs, repeating the process of exploration, invention, and settlement.

The vision of space as the final frontier is controversial. Realization dampened promise as humans failed to move into space in the anticipated numbers. Visionaries anticipated that space stations would house scores of people; the International Space Station holds less than a dozen. Advocates of the frontier analogy visualized lunar bases with hundreds of people and space colonies attracting millions. At least through the early phases, space has turned out to

Space advocates promised that extraterrestrial exploration would renew the frontier spirit they said had made America great. Just as nineteenth-century artists like Albert Bierstadt portrayed pioneers crossing western trails, space artists prepared pictures of Americans traversing the space frontier. In this 1993 painting by Pat Rawlings, explorers in spacesuits and a pressurized rover advance through the Ganges Chasma canyon on Mars. (NASA)

be more like the seabed than the frontier—a nice place to visit but not a great place to live.

The frontier analogy contains a strong element of utopianism, attractive to people who believe in it but curious to those who do not. Utopian societies rarely turn out to be as special as their visionaries proclaim. Professional historians question the role of frontiers in fostering the special characteristics so frequently ascribed to them. To historians, the idea of the frontier contains more myth than substance. The American public may embrace the virtues of frontiers, but to historians the popular image is a post-frontier phenomenon, perpetuated by vehicles like the Hollywood western.[2] Promoters of space as the final frontier offered their special vision just as revisionist historians mounted a full-scale campaign to debunk the notion, undercutting the analogy, at least within intellectual circles, where ideas about public policy mature. Promoters not only had to overcome skepticism about the technical feasibility of space colonization but also had to deal with suspicions about the historical accuracy of their ideas.

Still, frontier analogies continued to play an important role in the promotion of space exploration. One of the most popular analogies draws on the public's memory of sea captains, who in centuries past crossed vast bodies of water to reach distant lands. The only event comparable to the first landing on the Moon, editorial writers at both the *Washington Post* and *Washington Daily News* agreed, was "Columbus' discovery of the Western Hemisphere."[3] To commemorate the five hundredth anniversary of the first voyage of Christopher Columbus to the New World, a NASA-sponsored organization prepared a comic book for young children explaining the similarities between the challenges Columbus faced and those encountered by modern spacefarers. "Just as Christopher dreamed about opening a new trade route to the Far East, we can dream about a clean and beautiful Earth, about other space routes to Mars and colonization of our neighbor, the Moon."[4] Members of the 1986 National Commission on Space chose as the title for their report *Pioneering the Space Frontier*. Charged by the president and Congress to set out civilian space goals for the twenty-first century, members opened with the Columbus analogy: "Five centuries after Columbus opened access to 'The New World' we can initiate the settlement of worlds beyond our planet of birth. The promise of virgin lands and the opportunity to live in freedom brought our ancestors to the shores of North America. Now space technology has freed humankind to move outward from Earth as a species destined to expand to other worlds."[5]

In explaining his plans for gravity-assisted spacecraft that could cross the expanses of space between Earth and Mars, Buzz Aldrin drew on lessons from fifteenth-century maritime explorers. After his astronaut career, Aldrin worked hard to promote interplanetary exploration. Aldrin reminded Americans that European mariners of the fifteenth century had used the tropic winds that blow westward across the equator to reach the New World: "The new routes did not follow direct courses but instead looped along curving paths that sometimes appeared to carry the mariners away from their objective." The trade winds provided a pathway between the two continents, "making possible the great age of discovery."

Having established this analogy, Aldrin proposed that reusable spacecraft, called cyclers, be set into permanent orbits between Earth and Mars. The cyclers would use the force of gravity to whip by each planet, accelerating to the required velocity for the ensuing voyage. Smaller spacecraft would intercept the cyclers to move people and supplies home. Earlier routes of discovery provided the analogy that Aldrin needed to explain his plan: gravity provided a free and inexhaustible source of motion for interplanetary travel, just as trade winds had done for maritime explorers. "Like a ship sailing the trade winds," cyclers would follow a broad elliptical path rather than a more direct route between the two planets.[6]

Memories of the debate between Queen Isabella and her advisers over the Columbus voyage repeatedly appear in the consideration of modern space policy. "The many uses of space technology will make our investment in space as big a bargain as that voyage of Columbus," said governor of California Ronald Reagan. At the time, Reagan reminded his audience, support for the Columbus expeditions "was denounced as a foolish extravagance."[7] It is true that the Talavera commission set up by King Ferdinand and Queen Isabella of Spain had in 1490 issued a recommendation against financing the voyage. Although the monarchs did not follow the advice of the commission, the legend that the queen borrowed money for the expedition by pledging her crown jewels as collateral is untrue. That piece of folklore nevertheless played a prominent role in modern space policy debates. When space advocates first presented plans for a lunar voyage to President Dwight Eisenhower, they employed the Columbus analogy to justify the undertaking. Eisenhower reportedly countered that he was "not about to hock his jewels" to support a lunar expedition. Regardless of whether Eisenhower actually said any such thing, the story became an emblem of his modest space effort.[8] Queen Isabella came up again at a meeting of the Cabinet Council on Commerce and Trade in December 1983 as President Reagan and his advisers considered the proposal to start work on an Earth-orbiting space station. Budget director David Stockman, an opponent of the station, announced that the administration would never control the deficit if it continued to support such questionable projects. Attorney General William French Smith countered that Queen Isabella must have heard the same story from her advisers. The simple reference to the Columbus myth was enough to disarm Stockman's objection and make everyone, including the president, laugh.[9]

In their attempts to justify modern space endeavors, advocates have consistently appropriated the image of new frontiers. Americans have a "continuing urge to chart new paths and to explore the unknown," NASA administrator James Beggs said in announcing NASA's effort to win political support for a large space station. "That instinct drove Lewis and Clark to press across the uncharted continent. It guided admirals Peary and Byrd to the icy wastes of the poles. It drove Lindbergh alone nonstop across the Atlantic and sustained twelve Americans as they walked on the moon." The compulsion to probe unknown frontiers spurred the creativity of Americans. "If we ever lose this urge to know the unknown," Beggs argued, "we would no longer be a great nation."[10]

During the early part of the twentieth century, explorers like Robert Byrd, Roald Amundsen, and Robert Scott probed the icy regions of the South and North Poles. To many, it represented the last great era of discovery on the surface of the Earth. "The lonely explorers like Ronald [*sic*] Amundsen, Robert Scott and others who endured the rigors of Antarctica in the early decades of this century are analogous to our space pioneers today," Beggs argued in a 1984 speech. Scott had died in his effort to beat Amundsen to the South Pole, an event that transfixed the attention of armchair explorers around the world. "This is as close as you and I are going to get to setting foot on another planet," explained a member of the U.S. team conducting research on the Antarctic ice.[11]

Analogies extend from sea captains and explorers to the settlement of the American West. Speaking at the July 1982 landing of the space shuttle *Columbia*, President Ronald Reagan announced that the conclusion of the flight test program was "the historical equivalent to the driving of the golden spike which completed the first transcontinental railroad." During the easy-money years following the decision to go to the Moon, NASA actually funded a project

According to space advocates, frontiers foster innovation and discovery. When the urge to explore is curtailed, advocates warn, civilization decays. A key step in the opening of the space frontier would be the establishment of bases on the Moon, described in works of imagination both fantastic and real. (Lockheed Martin Missiles and Space Photo Archive)

that paid social scientists to determine whether the railroad analogy could be used to explain the U.S. space effort.[12] Defending his proposal to send a human expedition to Mars, President George H. W. Bush announced that "throughout our history, America has been a nation of discoverers." It would be hard to imagine Thomas Jefferson "sending a robot out alone to describe the wonders of the American Rockies and the Pacific coast."[13] Robert Zubrin, one of the most vocal advocates for the settlement of Mars, argued that only a planetary colony would provide the stimulus necessary to maintain a technologically advanced civilization. "Apply what palliatives you will, without a frontier to grow in, not only American society, but the entire global civilization based upon values of humanism, science, and progress will ultimately die."[14]

Where do these ideas come from? In large measure, they arise from Turnerism, the most influential doctrine affecting the teaching of U.S. history during the first half of the twentieth century.[15] Widely disseminated to the generation of Americans who managed the early space program, Turnerism remained popular with the educated public even as academic historians tried to debunk the doctrine. Frederick Jackson Turner was a young history professor at the University of Wisconsin when he delivered his 1893 paper on "The Significance of the Frontier in American History." He traced many of the distinctive characteristics of American society to the influence of free land across an open frontier. Inquisitiveness, inventiveness, and individualism were American traits forged on the frontier, Turner argued. Once created, these traits persisted even after the actual conditions of frontier life had disappeared. Turner traced the rise of American democracy and extended suffrage to social conditions on the frontier. Frontier life, according to Turner, bred a love of liberty that found its expression in the political doctrine of self-rule, and migration of many people to the frontier provided a powerful engine for the cross-fertilization of ideas and cultures that promoted America's sense of national identity. Turner's thesis inspired an outpouring of books on what became known as the idea of American exceptionalism.[16]

As Turner's paper observed, the American frontier closed in 1890. Quoting a brief official statement from the superintendent of the census, Turner noted that the distinctive, ever-moving line of settlement that had characterized America since its founding ceased to exist as of that year. Gone with it was the source of "this perennial rebirth, this fluidity of American life, this expansion westward with its new opportunities."[17] Other historians attacked Turner's thesis, insisting that American ingenuity and democracy could be traced to experiences other than the frontier.[18] But to advocates of space exploration, the Turner doctrine, however dimly understood, became the basis for a new adventure.

The more academic historians sought to discredit the myth of the frontier, the more space advocates exploited it.[19] Advocates of space exploration offered the extraterrestrial frontier as a place to energize the human spirit. All of the talk about technology spinoffs and scientific instruments faded in comparison to this aim. To the proponents of space exploration, modern civilizations need frontiers in order to maintain human innovation. Anticipating the colonization of Mars, the editors of *Life* magazine predicted that "a frontier ethic that celebrates courage, independence, imagination and vitality will merge with a technological bias that from necessity mothers inventions."[20]

In the minds of human space flight advocates, new challenges revitalize cultures grown stale. The exploration and eventual settlement of space offers such a challenge, and its advocates take comfort in the thought that similar animations of human spirit followed previous epochs of discovery. Even an outlet as skeptical as the *Washington Post* embraced this point of view. Commenting on John Glenn's first orbital flight in 1962, editorial writers at the *Post* likened the venture to the inspiration occasioned by an earlier age of discovery: "There is something in the very air of this space age that is not unlike the climate of another great age of discovery which took place in the fifteenth century." Europe at the end of the 1400s was gripped in a period of depression and anxiety, the *Post* writers observed. Quoting from the historian Samuel Eliot Morrison, editorial writers announced that fifteenth-century Europeans felt "exceedingly gloomy about the future." Their influence was shrinking, efforts to recover the Holy Sepulcher at Jerusalem had failed, Christianity was losing ground to Islam, and the Ottoman Turks had overrun most of Greece, Albania, and Serbia. "Then came an event that to Fifteenth Century Europe must have been quite as astonishing and breath-taking as the voyage of the Friendship VII. Into Lisbon harbor, came the Nina, sailing before a wintry gale to bring news of the discovery of the new world. That news changed the spirit of Europe. In Morrison's words: 'New ideas flared up throughout Italy, France, Germany and the northern nations; faith in God revives and the human spirit is renewed.'" Revolutions in science, philosophy, and religion followed, which the *Post*'s editorial writers optimistically ascribed to those early voyages of discovery. "So must these ventures into our space environment revive and renew the human spirit," the *Post* promised. The message so impressed the *Post*'s editorial writers that they repeated it when the Apollo 11 astronauts landed on the Moon.[21]

Maintaining a spirit of discovery is not easy. Learning science and technology is hard. Periodically, societies turn away from science and construct alternative cultures. "History offers many trenchant examples of what happens when the urge to explore and the development of new technology are forcibly curtailed," said NASA administrator James Beggs, commenting on the American withdrawal. Delivering a lecture before the Royal Aeronautical Society in 1984, Beggs once again reminded his audience of the experience of the fifteenth century, this time turning attention not to the accomplishments of Christopher Columbus but to the experience of the Chinese. Seventy-five years before the Columbus trip, Ming emperor Yung-lo authorized a series of voyages to contact Western people and participate in their affairs. Before the ships could reach Europe, conservative Chinese leaders prohibited private contacts with foreigners and forbid the launching of private voyages. Europeans moved out; Chinese turned in. By the technological and economic standards that came to dominate the world, the decision stunted Chinese civilization for centuries to come. By abandoning our exploration program, another NASA administrator added later, "we risk making the same mistake the Chinese emperors made more than 700 years ago."[22]

This admittedly ethnocentric view of the world gives little credit to the accomplishments of alternative cultures. In the minds of space advocates, however, it is how the world works. Frontiers imply conquest. History favors ethnocentrism. "The process of pushing back frontiers on earth begins with exploration and discovery, which are followed by permanent settlement and economic development," Beggs bluntly observed. Confrontations between technologically advanced civilizations and inward-looking ones inevitably work to the detriment of the latter. When societies collide, the exploring culture invariably wins. That is why the histories of inward-looking peoples are so frequently written in the language of their conquerors.[23]

Advocates of space exploration embrace a frontier philosophy that to some seems sternly paternalistic. Dominate or perish, they say. For many, it is a matter of national survival. When John F. Kennedy accelerated the space race with his decision to go to the Moon, he did so because he wanted to preserve the American way of life. "Only if the United States occupies a position of pre-eminence can we help decide whether this new ocean will be a sea of peace or a new terrifying theater of war," Kennedy argued in defending his space policy at Rice University in 1962. "No nation which expects to be the leader of other nations can expect to stay behind in this race for space."[24] The exploration of space would go ahead, he assured his audience, whether or not the United States led it.

For others, moving into this new frontier is a necessity for survival. A species cannot remain on a single planet for any extended period of time, Carl Sagan observed. On a single planet, it will certainly perish—its demise assured by astronomical events such as asteroid strikes or homegrown catastrophes: "Every surviving civilization is obliged to become spacefaring—not because of exploratory or romantic zeal, but for the most practical reason imaginable: staying alive." NASA administrator James Fletcher offered a similar argument in 1975. Any people who turn their back on the future will lose control of their destiny, he said. "Like Darwin, we have set sail upon an ocean: the cosmic sea of the Universe. There can be no turning back. To do so could well prove to be a guarantee of extinction." The great rocket pioneers Robert Goddard and Konstantin Tsiolkovsky both agreed that the navigation of interplanetary space was essential for the continuation of the human race.[25]

To all of its advocates, space frontiers promise to keep the spirit of innovation alive. "Frontiers summon the creativity, imagination, and inventiveness of the human mind," said Walter Hickel, governor of Alaska and a frequent space booster. "Civilization needs big projects, the kind that ignite the mind and inspire the soul." Imagination inspires new ideas in science and technology, James Beggs maintained. It nourishes art and literature and promotes the notions of freedom and self-fulfillment that people in democratic societies hold dear. "Small wonder," Beggs said, "that those nay-sayers and disbelievers who have ignored imagination and its potential to shape our destiny leave only a few, faint footprints on the sands of history."[26]

Promoters of space frontiers place a great deal of faith in the inspirational effects of exploration. "Looking back to the early navigators," NASA administrator Thomas Paine said in 1969, "the thing that impresses you is not the culture that they carried to continents like North and South America, Africa, Australia, and the Far East, but the effect of the culture that they brought back to Europe."[27] They brought back a global perspective that transformed the exploring nations and dominated civilization for the next five hundred years. They did not simply prove that the world was round (in any case, few educated people in Columbus's time subscribed to the flat Earth doctrine). Rather, explorers brought back a view of the world that encouraged the development of new technologies, such as sailing ships and navies that could master the seas. This New World view encouraged global commerce and global migration. It inspired scientific discoveries that would have been impaired in more repressive, inward-looking societies. In the geography of the seafaring world, Europe sat at the center of the map.

In another burst of Turnerism, Paine insisted that frontiers were responsible for the rise of democratic governments around the world. As Europeans settled new continents such as North America, they experimented with new forms of government needed to "conquer and organize a new continent." Historians agree that the absence of European institutions provided a fertile environment for the development of liberal democracies such as those that arose in Canada and the United States. Paine suggested that democratic ideas inevitably filtered back to Europe. The development of democratic governments in North America, he said, "set a new standard for governments around the world" and inspired the adoption of democratic reforms elsewhere.[28] The notion that frontier conquest promotes democratic government is not a new idea, but neither is it without controversy. Many factors encouraged the development of democracy in Europe, of which the experience in far-away America was only one.

"As we sail the new ocean of space," Paine insisted, "we are carrying out the same kind of exploration that the early navigators did when they set forth from Western Europe in their first ocean-going vessels." Fresh social transformations would surely follow as humans ventured away from the Earth.[29] Advocates of space exploration have had little difficulty imagining that these things will occur. However fuzzy or inaccurate their knowledge of history, they have not hesitated to apply it to the space frontier.

As Americans prepared to land on the Moon, enthusiasts made plans to settle space. NASA officials commissioned a number of studies in preparation for the establishment of bases on the Moon.[30] Engineers hoped to develop a twenty-five-thousand-pound module that could be launched using the Saturn V rocket and land softly on the Moon.[31] One proposal contemplated the establishment of four lunar bases, two at Grimaldi Crater, one on the far side of the Moon, and a fourth at the lunar south pole.[32] One of the more imaginative proposals appeared in the 1968 movie *2001: A Space Odyssey*, in which screenwriter Arthur Clarke describes Clavius Base, located in the second largest crater on the visible side of the Moon. In an emergency, the colony could be entirely self-supporting. Clarke anticipated that elements such as hydrogen, oxygen, and nitrogen could be produced from local rocks crushed, heated, and chemically treated on the Moon. Food was produced in an underground biosphere that also served to purify air. A variety of transportation vehicles, most moving on flex wheels or balloon tires, carried crews to various parts of the lunar surface. Eighteen hundred men and women lived and worked at the fictional base, which, in the optimism of the day, had been established by the U.S. Corps of Astronautical Engineers in 1994.[33]

The anticipation of large numbers of people on extraterrestrial bodies like the Moon excited a complementary strain of thought, one closely associated with the doctrine of frontiers. To inspire fresh ideas, Robert Zubrin argued, a frontier needs to be sufficiently remote "to allow for the free development of a new society." For Zubrin, that could "only be on Mars." Zubrin bypassed interest in lunar bases in favor of Martian colonies. The Moon was too close. "If people are to have the dignity that comes with making their own world, they must be free of the old."[34]

In his famous collection of stories assembled within *The Martian Chronicles*, science fiction author Ray Bradbury describes colonists fleeing a disintegrating Earth and establishing a new society on Mars. The colonists call themselves Martians. "Earthian logic, common sense, good government, peace, and responsibility," says one of the colonists looking back toward an Earth destroyed by atomic war. "It's not there anymore."[35]

The idea of starting over in a better place is as old as human hope. Sir Thomas More placed his fictional Utopia on an island in the New World. John Winthrop and his community of English Puritans came to Massachusetts. The Shakers, or members of the United Society of Believers in Christ's Second Appearing, settled in upstate New York, among other places. The Harmony Society, a German religious group, founded towns in Pennsylvania and Indiana. Persecuted in New York, Joseph Smith and Brigham Young led their followers toward a new city of Zion in Missouri, Illinois, and finally a western state they called Deseret. After the closing of the American frontier, British diplomat Hugh Conway set paradise in Shangri-La, a fictional community in the mountains of Tibet.[36]

Traditional utopian communities in America were largely religious in origin. Believers sought out-of-the-way places in which to establish more perfect societies based on biblical teachings. As the country grew more urban and industrialized, utopian thinking shifted toward technology. Although religious and social utopias still persist, more modern forms anticipating perfection through expertise have joined them. Supporters of the technocracy movement, transhumanism, and the idea of a technological singularity foresee situations in which humans use science and technology to create more perfect worlds inside the old one. From their perspective, knowledge from science will abolish poverty, suffering, and even death. In the early twentieth century, the technocracy movement sought to improve society by empowering scientists and other experts to make governmental decisions. The singularity is a predicted point in time, popularized by Ray Kurzweil, in which pace of technological change becomes so great that machines learn how to improve and think for themselves. Advocates of transhumanism envision a point in time at which humans live long enough to live forever.[37]

The space pioneering movement achieves its distinctiveness by combining the desire for new places with the idea that technology will make their settlement possible. As such, it contains the fervor of traditional religious utopias without the necessity of having to promote religious beliefs. Says Zubrin again: "Though the Red Planet may appear at first glance to be frozen desert, it harbors resources in abundance that can enable the creation of an advanced technological civilization. Mars is remote and can be settled. The fact that Mars can be settled and altered defines it as the New World." Zubrin embraces the classical utopian idea that righteous human beings cannot exist within a oppressive society; he offers technology as the means to create a new one. "Everywhere you look," he writes, "the writing is on the wall." Zubrin points to the concentration of wealth, the bureaucratization of life, the impotence of government, the spread of irrationality, the aversion to risk, economic stagnation and the loss of inventiveness. To a classical utopian, the solution is clear. Move on. Do so soon. "Can a free, egalitarian, innovating society survive in the absence of room to grow? Perhaps the question was premature in Turner's time, but not now. . . . Without a frontier from which to breathe new life, the spirit that gave rise to the progressive humanistic culture that America has represented for the past two centuries is fading."[38]

Shortly after the landing on the Moon, a physics professor at Princeton University electrified the devoted corps of space frontier advocates with his proposal for moving out. In 1974 Professor Gerard O'Neill published what author Michael Michaud called "one of the most photocopied science articles in history."[39] O'Neill had challenged his students to consider whether a planetary surface like the Earth was the best place for an expanding technological civilization. Their answer, in typically utopian fashion, was no. O'Neill's solution was imaginative.

Writing in *Physics Today*, O'Neill described how humans could move off the Earth into a multitude of artificially constructed colonies located at gravitationally stable points in the emptiness of space. The most efficient design, he argued, would be cylinders about four miles in diameter and sixteen miles in length. People residing on the inner edge of the rotating cylinder would live in a technologically perfected Earth-like environment, with lakes, mountains, trees, suburbs, artificial gravity, and a blue sky spotted with clouds three thousand feet "above" the inner rim. Animals and plants endangered on Earth could thrive on these cosmic arks, but insect pests would be left behind, eliminating the need for pesticides. Light from the Sun would be directed into each cylinder from large movable aluminum-foil mirrors, which would create night and day and seasons like those on the home planet. Ample electricity would be provided by steam-turbine generators, powered by the Sun and providing a clean source of energy for transportation and personal use. "With an abundance of food and clean electrical energy, controlled climates and temperate weather, living conditions in the colonies should be much more pleasant than in most places on Earth," O'Neill prophesied.[40]

The first colony could be completed just after the turn of the century, O'Neill argued, in about twenty-eight years. With the manufacturing technology in place, the number of colonies could expand exponentially. A fully developed colony, he declared, could easily support a population of ten million people, plus desirable flora and fauna.[41] Continuing those calculations, O'Neill estimated that emigration to the colonies could reverse the population rise on Earth by 2050. In another thirty years, Earth's population could be reduced "to whatever stable value is desired"—perhaps 1.2 billion people.[42] Colonists would mine the Moon for materials to build the first colonies, then excavate the asteroids. After exhausting the asteroid belt, they could tear up the moons of the outer planets. The raw materials available in the solar system, O'Neill offered in a fit of enthusiasm, could support a twenty-thousand-fold increase in the human race while reducing population pressures on Earth.[43] It seemed too good to be true, and probably was.

A number of obstacles stood in the way. To construct the first space colony, the sponsoring nation would have to move some five hundred thousand metric tons of metal, soil, rock, and water to the construction point in space. And someone had to move the first colonists, a major challenge given the astronomical cost of transporting humans and their accompanying supplies from the Earth to extraterrestrial destinations.[44]

O'Neill attacked these challenges with technological optimism. He urged colonists to acquire the bulk of the materials needed for the first space colonies from the Moon. To launch them toward the construction site, he proposed a type of recirculating conveyor belt called a mass-driver. Magnetic impulses produced by electric energy would accelerate a twenty-pound bucket of lunar material to the velocity necessary to hurl its contents toward the appropriate spot in space.[45] As for the problem of transporting people and a few essential materials from Earth, he accepted the widely held notion that reusable launch vehicles would reduce transportation costs by a factor of ten. O'Neill's supporters argued that the first space colony, a rather Spartan version, could be

Eventually humans would move into space in large numbers—some to the Moon and Mars and others to artificially constructed colonies suspended in gravitationally stable points of empty space. In 1974 physics professor Gerard O'Neill put forth a utopian proposal for a multitude of terrarium-like space colonies with living conditions superior to those found on a crowded Earth. (NASA)

constructed for about $33 billion in 1972 dollars—roughly equivalent to the amount spent to send American astronauts to the Moon. Internal NASA studies set the price closer to $200 billion.[46]

O'Neill's space colony proposal was not unique, but the degree to which it captured the public imagination was. The idea of large artificial colonies in space had been advanced previously by an assortment of writers, from the famous to the obscure. At the beginning of the twentieth century, Russian space pioneer Konstantin Tsiolkovsky had envisioned dwellings in space that could house millions of people. British scientist J. D. Bernal advanced a similar concept in 1929, and Arthur C. Clarke helped popularize the idea in his 1954 children's novel *Islands in the Sky*. Dandridge M. Cole presented plans for space colonies formed out of hollowed-out asteroids in 1964, and Krafft Ehricke, a member of the von Braun rocket team, issued his call for an "extraterrestrial imperative" in 1971.[47]

Unlike earlier proposals, which attracted a narrow audience, O'Neill's vision splashed upon the public scene. It attracted interest from the mainstream of American politics to the cultural fringes of radical thought. His concept was embraced by visionaries who wanted to pioneer space, environmentalists concerned about overpopulation and dwindling resources, futurists who saw technology as the solution to human problems in an industrial civilization, and a variety of space-age groupies in occasional need of psychiatric help. The California-based counterculture group responsible for producing the environmentally correct *Whole Earth Catalog* promoted the idea. Congress held hearings, and NASA gave financial support to a variety of supporting studies.[48]

Excited by the prospect of pioneering the high frontier, Americans throughout the 1970s in ever-increasing numbers filed into spacefaring clubs. O'Neill's vision spawned the L-5 Society, named after one of the gravitational stable regions created by the Earth and Moon at which objects such as a space colony could remain indefinitely. The main purpose of the L-5 Society, formed in 1975, was "to arouse public enthusiasm for space colonization."[49] The society attracted adherents whose exuberance about space colonization irritated people laboring on practical U.S. space activities. In response, industry and government leaders in 1975 formed the more conservative National Space Institute, at its head the aging space warrior Wernher von Braun, which sought to mobilize grass-roots support for NASA's more conventional exploration plans. In 1987 the two organizations forgot their differences and merged into the National Space Society. The society envisions "people living and working in thriving communities beyond the Earth, and the use of the vast resources of space for the dramatic betterment of humanity."[50]

Strange and wondrous groups continued to form. Distraught at the cancellation of their *Star Trek* television series, science fiction fans organized local clubs and federations as a means of keeping their enthusiasm for galactic fantasies alive. Through a massive letter-writing campaign, "Trekies" convinced the government to name the first space shuttle test model after the starship *Enterprise*.[51] Following the broadcast of the *Cosmos* television series, 120,000 individuals joined Carl Sagan and Bruce Murray in forming the Planetary Society. The society collected signatures of notable and ordinary Americans for its Mars Declaration, a statement advocating the exploration of Mars as an important step "toward the long-term objective of establishing humanity as a multi-planet species."[52]

By 1980, by one estimate, space enthusiasts had formed nearly forty major interest groups devoted to the cause of exploration and colonization. Local chapters, astronomical societies, and science fiction fan clubs pushed the total number of organizations close to five hundred.[53] They included groups promoting capitalism in space, groups advancing the role of women in space, groups set up to privately fund space activities, and groups prepared to train space pilots and pioneers. In 1988 Rick Tumlinson and some O'Neill acolytes established the Space Frontier Foundation. Taking its name from the impulse that drove adherents to dream of starting new places, foundation members dedicated themselves to "opening the space frontier to human settlement." Ten years later, Robert Zubrin formed the Mars Society "to further the goal of the exploration and settlement of the Red Planet." There was even a political action committee for space.[54]

Many books and reports appeared advancing various scenarios for the accomplishment these goals. The most lavishly illustrated, if not widely read, was the 1986 report of the National Commission on Space, chaired by former NASA administrator Thomas Paine. Members of the commission recommended that the government establish an outpost on the Moon by 2006 and a human outpost on Mars by 2015. "Many of the people who will live and work at that Mars Base have already been born," the report's authors noted.[55]

Members of the commission proposed an elaborate infrastructure in space. There would be an Earth-orbiting space station, a lunar-orbiting space station, and a station around Mars. A special spaceport at one of the gravitationally stable points near the Moon would prepare humans for the journey to Mars. There would be transfer vehicles designed to take humans between the stations. The commission proposed lunar landers and Mars landers and cycling spacecraft and special spaceships that could with a burst of speed catch the cyclers as they flew by.

Paine's report contained wondrous illustrations of the new frontier: men and women tending fruit trees and vegetables in a lunar biosphere, a spacecraft landing at a twenty-first-century Martian settlement, astronauts in space suits servicing a transfer vehicle at a gravitationally stable spaceport, and the same transfer vehicles using Earth's upper atmosphere for aerobraking maneuvers. In another illustration, a specially designed robot worked to mine propellants from Phobos, a moon of Mars. Preliminary studies indicated the presence of water, carbon, and nitrogen on the tiny moon. "If so," commission members suggested, "there is an orbiting fuel depot just 6,000 miles above the red planet to top off the hydrogen and oxygen tanks of visiting spacecraft."[56] The commission hoped that the first occupants would obtain much of the material needed to set up their lunar and Martian outposts from local resources. Oxygen in the form of metal silicates might be extracted from the lunar soil, and Paine hoped that explorers would locate water ice and other volatile compounds in permanently shadowed craters near the lunar poles. On Mars, the commission members observed, all the necessary oxygen, hydrogen, nitrogen, fertilizer, and methane needed to start a permanent settlement could be extracted from that planet's puny atmosphere. In one of the last papers he wrote before his death in 1992, Paine predicted that the processes needed to extract local materials would stimulate a new generation of industrial robots "with a hundred times the productivity of terrestrial factories."[57]

Extraterrestrial colonies would allow the human race to leave "the precious and fragile planet where it was born," and extend life "to the far reaches of the inner Solar System," commission members maintained. Without expansion into space, humans remaining on a more crowded Earth would be forced to compete increasingly for limited resources, Rick Tumlinson argued nine years later. Nearly anything that a person on such an Earth wanted to do would be "something someone else cannot. . . . Equilibrium will be the goal of the state and individual freedoms will become ever more expendable." Space colonization would break this cycle, Tumlinson insisted. It would allow humanity to prosper and grow.[58]

Engineers and enthusiasts debated the practical details of achieving the pioneering dream. Should humans work to establish a lunar colony or proceed directly to Mars? "Moon firsters" argued that the lunar colony would provide much-needed experience with the frontier technologies necessary to move farther from home.[59] Within ten years of the first landing, an industry task force observed, a community of one hundred pioneers could be living and working on the Moon.[60] Lunar settlers would explore their new home and set up scientific instruments such as a radio astronomy observatory on the back side of the Moon, shielded from interference from Earth. Like pioneers before them, they would look for ways to make their expeditions pay. Experts were especially intrigued by the possibility of mining the Moon. Solar flares, experts suggested, had deposited on the lunar surface quantities of helium-3, which could provide a rich source of fuel for fusion reactors, should that technology ever take hold. One lunar booster argued that just sixty thousand pounds of helium-3 per year returned to Earth would satisfy the energy needs of the whole planet.[61]

To others, development of a lunar settlement seemed like a waste of time. The United States had been to the Moon. If the objective of a lunar base was to prepare for Mars, why not get on with the larger goal? In the spring of 1981 a collection of space-interest groups organized a conference at the University of Colorado in Boulder to examine whether "a manned Mars mission was a viable option for our space program."[62] Out of the meeting emerged the so-called Mars Underground, a congregation of students, space boosters, and aerospace professionals devoted to making (as the title of a series of books supporting the concept revealed) *The Case for Mars*. More books and conferences followed.[63] In 1986 NASA joined the discussion with its own Mars conference and the following year established an Office of Exploration for the purpose of coordinating agency activities and convincing Congress and the president to approve the endeavor.[64]

Much of the practical work contained in the various studies concerned the best way to get to Mars. In her 1987 report, astronaut Sally Ride proposed a series of three short sprints, with ten- to twenty-day stays on the planet's surface and an overall journey of no longer than a year. NASA engineers developed plans for more elaborate expeditions taking nearly three years. Debate led to government infighting. White House officials grumbled about conservative NASA bureaucrats and commissioned outside experts to develop more imaginative proposals. A deftly illustrated report by astronaut Thomas Stafford drew on suggestions from outside experts that nuclear propulsion could cut the one-way transit time to Mars from 224 to 160 days.[65]

Amid the clamor and debate, the long-term goal remained steady. Humans would depart Earth, settle the Moon and Mars, and eventually move to the stars. A report from the Mars Underground predicted that the first human child would be born on Mars in the year 2020, when the population of the outpost reached one hundred persons. By 2081, authors of the chronology speculated, two hundred thousand colonists would live on Mars.[66]

For all of its popular appeal, however, Mars remained a very inhospitable place. The 1976 Viking landers revealed a cold, dry desert with little atmospheric protection from sterilizing ultraviolet rays. Humans had not yet settled the Antarctic on Earth, which was absolutely balmy by comparison. How did Mars enthusiasts plan to handle the hostile environment on Mars? No problem, they replied. They would simply transform the planet into an Earth-like environment by altering conditions there. Of all of the recommendations for pioneering space, few were as imaginative as the proposals for terraforming Mars.

Terraforming was once the preserve of science fiction writers. In the April 1937 issue of *Astounding Stories*, Ross Rocklynne described a successful attempt to move fifty-two-million cubic miles of frozen water from the asteroid belt to Mars.[67] A series of stories published under the pseudonym Will Stewart in the 1940s explained the use of "paragravity generators" to attach atmospheres to previous lifeless bodies. Stewart, whose real name was Jack Williamson, gave the process the name that has remained with it since.[68] In 1950 Robert Heinlein described the terraforming of Ganymede, one of the giant moons of Jupiter, by several thousand colonists from planet Earth.[69]

In 1961 the young Carl Sagan published an article in *Science* magazine containing a plan for making Venus habitable, one of the first serious proposals for altering the environment of planets. Although Sagan's plan was flawed (he overlooked the problems posed by the density of the atmosphere), the article gave scientific respectability to the concept.[70] By 1975 NASA was ready to give its official blessing, sponsoring a study that examined the possibility of altering the Martian environment to make it more habitable.[71] The advocacy group Mars Underground was born out of the interest of people anxious to explore and transform that planet.[72]

The concept continued to draw public interest. In May 1991 *Life* magazine ran a cover story on the subject of terraforming Mars. Relying upon scientific opinion, the editors presented an ambitious 150-year scheme for transformation of the cold, dry planet. Orbiting solar reflectors would melt the polar ice caps, and Martian factories would produce greenhouse gases and ozone substitutes. As the planet warmed, nitrogen and water would seep out of the Martian soil, and the atmosphere would thicken. This in turn would cause further warming. Clouds would appear, and the color of the sky would shift from pink to blue. Oxygen for the newly forming atmosphere could be extracted by local factories from carbon dioxide, carbonate rocks, and deposits of iron oxide. Pioneers would plant tundra plants and hearty evergreens as the mean planetary temperature approached the freezing point of water. Rain would fall and agriculture would thrive, but the maturing atmosphere would need more oxygen to allow humans and animals to live outside. With enough oxygen-producing factories and vegetation, the editors predicted, the planet could be made totally suitable for human habitation by the year 2170. Streams, lakes, and oceans would cover the surface of a new moist green globe.[73] To celebrate the achievement, artist Robert McCall prepared a painting depicting pioneers deep in a Martian canyon emerging from glass-domed biospheres as cannons in the background pump greenhouse gases into a newborn atmosphere. Touches of green, tundralike vegetation appear on nearby hills as temperatures rise. The inspirational painting, titled *Terraforming Mars*, hung for many years outside the NASA administrator's office in Washington, D.C.

The cost of making Mars fit for human habitation would be high, but not beyond the reach of technologically advanced nations. *Life's* editors optimistically predicted that investment costs for their terraforming proposal would peak at $45 billion per year during the early buildup stage—an impressive sum but a fraction of what earthly nations spend annually on national defense or government-assisted health care. The technical problems would be formidable, but not insurmountable. Engineers would need to develop inexpensive rocket ships and cheap sources of energy (fusion reactors would help considerably). With much fine-tuning and the avoidance of undesirable side effects, it could be done. As Carl Sagan observed, "We need look no further than our own world to see that humans are now able to alter planetary environments in a profound way."[74]

On 20 July 1989, to commemorate the twentieth anniversary of the first landing on the Moon, President George H. W. Bush challenged the United States to commit itself to a sustained program of exploration that would lead to the settlement of space: "From the voyages of Columbus—to the Oregon Trail—to the journey to the Moon itself—history proves that we have never lost by pressing the limits of our frontiers." To begin the process, he called upon Congress to join him in supporting an outpost on the Moon and the first human expedition to Mars. In a later speech he set a goal of 2019 for the first Mars landing.[75] When NASA's cost estimate for the venture soared, members of Congress refused to fund the efforts, a response the first President Bush characterized as an act of short sightedness by people who lacked vision.[76]

President George W. Bush tried again in 2004 when he announced a program to return to the Moon and send humans to Mars. Referring to the expedition led by Meriwether Lewis and William Clark two hundred years earlier, the second President Bush noted that the United States undertook that journey "to learn the potential of vast new territory, and to chart a way for others to follow." For the same reasons, Bush announced, Americans would return to the Moon by 2020. "With the experience and knowledge gained on the moon," he continued, "we will then be ready to take the next steps of space exploration: human missions to Mars and to worlds beyond."[77] This time NASA officials struggled to fund the effort from their existing budget without exponential growth, without success.

The United States and the other nations of the world may achieve these goals. It is technically feasible and not beyond the financial reach of nations rich enough to fly in space. Yet even if it occurs, would a human presence on Mars accomplish the broader goals contained in the exploration vision?

The image of the frontier is America's creation myth. For many (but not all) Americans, it explains where they came from and why they are special among the peoples of the world. According to this story, America was essentially an unoccupied land of boundless opportunity. Hardy, independent pioneers settled the wilderness through their own ingenuity and resources and created a new civilization. Unencumbered by old traditions, they formed simple democratic communities with governments that became a model for the entire world. Their inventiveness led to the richest and most technologically advanced nation in human history. The work was hard but satisfying. Generation after generation repeated this experience until the frontier was gone. The impulse to explore and settle remained in American culture, however, waiting to be reapplied on some new frontier.[78]

For all of its cultural appeal, the image, alas, is factually wrong. It is based upon a romanticized interpretation of history as far removed from reality as the Buffalo Bill Wild West Show was from the real events it sought to portray. The American West was not an empty land waiting to be settled, as the native Americans and Hispanics who already lived there knew. Settlers depended extensively upon subsidies and capital provided by people from more settled regions for activities such as railroad transportation, dam building, and irrigation. In business enterprises such as the gold and silver rushes, failure was as common as prosperity. The economic principles that favored the formation of large companies employing low-wage workers were not suspended in the West. Territorial governments were no less corrupt nor more democratic than those in the East, and as many principles of democracy emerged from the pens of intellectuals residing in the eastern United States and Europe as from the frontier experience.

If space is like the real frontier, what might the experience tell us? As people who have labored in the enterprise already know, space travel can be hard and dangerous work. Spacecraft and space stations can be noisy, crowded, even foul places filled with temperamental equipment. Long voyages under such conditions can encourage thoughts of homicide. In that sense, human space travel may not be much different from life on an actual frontier. "Nobody wants to be a cowboy," lamented one western employment specialist. "It's hard work, it's dirty work, it's round-the-clock work." It is something most Americans want to watch from a distance.[79]

Frontiers are rarely utopian in spite of efforts of their advocates to portray them as such. Commenting on the challenges of founding a lunar colony, Thomas Paine assured his supporters that it would "sweep aside old world dogmas, prejudices, outworn traditions, and oppressive ideologies." Konstantin Tsiolkovsky predicted that in space colonies "human society and its individual members [would] become perfect." Gerard O'Neill predicted that life in his suburb-like colonies would permit "most of the human population to escape from poverty" and that the environment would "be optimized for good health."[80] By their apparent openness and lack of rules, frontiers attract utopian thinkers. History suggests, however, that new settlers bring society and all its imperfections with them. Utopian communities commonly fail because they cannot escape the human traits their founders wish to leave behind.[81]

The harsh conditions on imagined new worlds may pose additional challenges. Utopians see in new places the opportunity to construct communities free of the old. Local conditions can obstruct this dream. The extraterrestrial community into which explorers emerge after traveling through the portal in the widely viewed science fiction film *Stargate* is an oppressive despotism drawn from the ancient civilization that built the Egyptian pyramids. Luke Skywalker's foster parents are killed by Imperial Stormtroopers on the wild west and largely ungovernable planet of Tatooine. Significantly, both of these imagined spheres present dry environments like those found on the planet Mars. A particular theory of developing civilization suggests that settlements arising under acutely arid conditions give rise to despotic and bureaucratic governments because forced labor and strict rules are necessary to regulate the distribution of water.[82]

The general term for such a community is dystopia—a version of society characterized by oppression, misery, and undesirable living standards. Science fiction offers many examples, from George Orwell's classic *1984* to the failed attempt to create a utopian community in Aldous Huxley's *Brave New World* (technically an anti-utopia). The challenges of maintaining an extraterrestrial outpost on a largely airless and water-scare world might create conditions favoring highly regulated, autocratic governing bodies that are more corporate than democratic in form. In America, frontiers are thought to promote equality, independence, and freedom, but on the Moon or Mars the opposite might be true. In advance, it is hard to know.[83]

Frontiers are also a metaphor for ungrateful dependence. American colonists depended upon British troops for protection from the French and their Native American allies and were notoriously reluctant to pay their share. When the British imposed the Stamp Act as a means to recover their investment, American colonists organized the boycotts and demonstrations that led to war. One century later, eastern taxpayers and financiers invested heavily in the development of the American West. How will Earthlings react when space pioneers demand outside investment for their extraterrestrial colonies while complaining about outside rule? Will people on Earth glorify the independence of these space age pioneers or treat them like ungrateful children?[84]

In spite of the relentless attacks of classroom historians, the romantic image of the frontier endures among the public at large. Many people continue to believe in it. Space advocates call upon the popular image of the frontier to garner support for their visions, even as historians attack them. The space frontier is an appealing analogy to many people in the United States, given its pioneering history. The American creation myth provides a level of vindication for space exploration that compensates for less-glamorous byproducts. It is doubtful that Americans would pay hundreds of billions of dollars to send humans to Mars simply to gain some technology spinoff or to establish the interplanetary equivalent of an Antarctic research station. The frontier analogy, with all of its flaws, allows people to believe that space exploration will reopen one of the longest and most formative chapters in American history. Never mind that the reality of space colonization may differ considerably from the popular image of it. Space flight is a dream, and dreams do not have to be entirely real in order to motivate behavior.

The Moon and Mars and other places could be explored for the purposes of scientific understanding by robots and machines alone. To do so, however, would fail to satisfy one of the central elements of the spacefaring dream—the extension of humanity into the extraterrestrial realm. Listen again to the words of Carl Sagan, set down in the book *Pale Blue Dot*, shortly before his death. Sagan traced the spiritual erosion of modern life to two great developments. The first was the closing of terrestrial frontiers. We are all wanderers, he maintained, from Ice Age humans who crossed the Bering Straits to Polynesian argonauts in outrigger canoes and American pioneers: "This zest to explore and exploit . . . is not restricted to any one nation or ethnic group. It is an endowment that all members of the human species hold in common."[85] Humans, Sagan continued, first wandered as hunters and gatherers and continued to migrate as explorers and pioneers. Only recently, for a brief period in the lineage of the species, have humans confined themselves to established settlements. In spite of the material advantages to be found in villages and towns, humans remain restless. "For all its material advantages," Sagan professed, "the sedentary life has left us edgy, unfulfilled." Humans have not lost their urge to roam: "The open road still softly calls, like a forgotten song."[86]

Sagan suggested that the urge to move on is necessary to survival, an instinctive drive built into human behavior as a result of natural selection. Towns and villages do not last forever, and the people who crave new places protect their descendants against the catastrophes that inevitably befall those who remain behind. The human experience has been diminished by this loss of openness, Sagan suggested, not just among Americans lamenting the loss of the Western frontier but among humans everywhere.

Joining this restlessness of place, Sagan suggested, is a new desperation of spirit. For centuries humans took comfort in the knowledge that the Earth sat at the center of the universe, that the Sun and Moon and stars rotated around the Earth, and that God had created humans in His own image for a special purpose. Science devastated those beliefs. Earth is a tiny blue dot rotating around an inconspicuous yellow star on the outer reaches of one of a hundred billion galaxies. There are certainly other planets, probably housing other life-forms, and possibly other universes that operate according to different laws of nature. According to Sagan, the best available evidence does not support the need for a Grand Designer.

"Human beings cannot live with such a revelation," Sagan quoted British journalist Bryan Appleyard as saying.[87] The great demotions, as Sagan called them, have created a more mature view of nature, but they have also devastated the human spirit. Maturity is painful; it is easier to think like a child. In the past, when humans believed themselves part of a greater purpose, they could accept moral codes passed down from people presenting themselves as the worldly agents of the creator. Humans could follow the exhortations of religious and secular authorities. The apparent insignificance of the Earth in the cosmos weakened those codes. It is difficult for humans to respect strict moral codes when those doctrines are based on patently false cosmologies. The new view of the universe has undermined the leadership of religious and secular authorities and bred a sense of hopelessness.

Against this sense of desperation, Sagan believed, a new spirit of discovery could arise. The very science that created the sense of despair could create a new state of wonder: "Once we overcome our fear of being tiny, we find ourselves on the threshold of a vast and awesome Universe that utterly dwarfs—in time, in space, and in potential—the tidy anthropocentric proscenium of our ancestors." For the immediate future, Sagan thought, humans could do their investigating from the Earth. Yet a time would come when humans would move out from the planet. Their instincts and survival demanded it. "On behalf of Earthlife, I urge that, with full knowledge of our limitations, we vastly increase our knowledge of the Solar System and then begin to settle other worlds."[88]

7

Stations in Space

I think if you ask the public at large, and quite possibly most of the people within NASA, what a space station was, they would think in terms of the movie that came out fifteen or twenty years ago.

— John Hodge, 1983

Having created an imaginative vision of humans leaving Earth and settling space, devotees faced the practical difficulties of actually doing so. An essential step in practically every settlement or exploration scheme was the creation of a permanent facility in an orbit above the Earth at which humans could live and work. Such a station, as it was called, provided a mechanism for completing the activities humans imagined taking place in space. Unfortunately, those activities gathered together on a single faculty tended to conflict with one another, exacerbating the practical difficulties of actually constructing one.

In the centuries preceding the space age, nations seeking to extend lines of exploration and settlement built the terrestrial equivalent of space stations across the face of the Earth. Frontier forts, way stations, trading posts, and base camps provided convenient means for advancing human presence into unconventional territory. Colonists and pioneers, mountain climbers and polar explorers commonly employed structures such as these for the purpose of marshaling material and fashioning sanctuaries in hostile realms. By the start of the space age, the concept of the frontier fort or base camp as a jumping off point for more daunting adventures was well established in the public mind. Quite naturally, advocates of the high frontier assumed that advances into the cosmos would require similar structures above the surface of the Earth.

Elaborate plans for orbiting stations appeared. Some of the proposed facilities rotated so as to provide occupants with the comforts of artificial gravity; others moved quietly through the ether as their occupants, living in places with practically no gravity at all, studied the unique effects. Some served as navigational aids; others as observational posts. Some promoted national security; others functioned as orbiting science laboratories. Ambitious engineers designed transportation depots at which humans could board spacecraft bound for points beyond; others designed factories, hotels, and other commercial facilities. To excite public interest, some promoters portrayed space stations as very large structures, housing hundreds of specialists and supporting personnel; others proposed stations that were small.

In January 1984 President Ronald Reagan directed the National Aeronautics and Space Administration to build a permanent way station in space. "Develop a permanently manned space station," he told officials at NASA, "and do it within a decade."[1] Twenty-five years later the United States and a consortium of allies were still working to assemble what became known as the International Space Station. Constructing an Earth-orbiting space station turned out to be easier to envision than to accomplish. Once again, reality challenged imagination.

Imagining the idea of a space station was not hard. The concept was especially well set in North America, where frontier forts and way stations had speckled the face of the newly settled continent. From the garrison houses of New England to the station settlements of Kentucky and Tennessee, from the brick and earthen fortresses of the Atlantic coast to the simple stockaded trading posts of the West, Americans had built forts and stations as an initial step toward settling unfamiliar territory. Many present-day cities bear the names of the forts around which colonial settlements grew—the Spanish fortification at St. Augustine in Florida, the French forts on Mobile Bay along the Gulf of Mexico, and the English Fort Pitt at the confluence of the Monongahela and Ohio Rivers. Frontier forts offered safe havens from the dangers of the wilderness, attracted commerce and trade, and provided jumping-off points for ventures into the wilderness. The impulse to settle new lands, said one commentator, transformed America into "the most internally fortified territory in the world."[2] By the end of the nineteenth century, fort building was firmly associated with conquest and settlement in the American mind. Space pioneers did not need to explain the underlying association between frontier stations and space exploration, and rarely did, because the association was so deeply rooted in the national experience.

The necessity for establishing way stations in space was reinforced by the obsessive attempts during the first half of the twentieth century to reach the last unexplored regions on Earth. Efforts to reach the South Pole and ascend Mount Everest attracted as much attention from armchair explorers as later voyages to the Moon. In 1911 the Norwegian explorer Roald Amundsen and four companions set out for the polar plateau. They were followed along a slightly different route twelve days later by a party led by Englishman Robert Scott. Amundsen located the South Pole on 14 December; Scott and his four compatriots reached the pole thirty-five days later but perished on the return journey when their supplies ran low.

The success of Amundsen's plan, as with nearly all polar expeditions to follow, owed much to his skill in establishing an adequate base camp on the outer edge of unfamiliar territory. Amundsen's base camp was prefabricated and shipped to the Ross Ice Shelf some 750 miles short of the pole. During the year that preceded his successful journey, Amundsen used the camp as a base from which to test equipment and set up caches of food and fuel along the proposed route. The base camp also served as an observatory to study the polar environment, a workshop to rebuild equipment tested in the field, and a shelter to provide protection from the wind and cold.[3] Eighteen years later, the explorer Richard E. Byrd established Little America on the site of Amundsen's abandoned base, from which Byrd conducted the first airplane flight over the pole.

The public impression of base camps as adjuncts to exploration was further amplified by efforts to climb the highest mountains in the world. Because of the great distances and harsh conditions involved, mountaineers constructed elaborate pyramids of people and supplies designed to put a few skilled individuals on the Himalayan summits. The first successful ascent of Mount Everest took place in 1953, by Edmund P. Hillary and his Sherpa guide Tenzing Norgay. A supporting team of climbers and porters organized a large base camp at sixteen thousand feet, from which a series of nine additional camps were established at ever-increasing altitudes. Seven and a half tons of supplies were carried from Kathmandu to the base camp at Thyangboche. Those supplies supported a final cache of 650 pounds used in the final assault. It was not uncommon for Himalayan expeditions to employ five hundred to six hundred porters to move the basic loads.[4]

Space pioneers envisioned similar schemes for the exploration of the Moon and planets. Hermann Oberth, the German rocket pioneer, stated that a space station would be necessary as a "springboard" for flights to the Moon. He drew directly from Antarctic experience in framing plans for his *Weltraumbahnhof.*[5] Willy Ley, one of Oberth's disciples, argued that a "terminal in space" would be necessary to marshal the equipment needed to explore the Moon and nearby planets. Ley and other fellow advocates of human space travel supported orbital stations for a practical reason. The laws of physics, given the limited thrust of available propellants, did not permit workable rocket ships with humans on board to be propelled directly from the surface of the Earth to the Moon and back, and certainly not to the planets. By one calculation, a ship bound directly from the Earth to Mars would need to burn 105 tons of liquid propellant during the first second of flight just to heave itself a few feet off the ground. "Landing on the moon is beyond the borderline of what chemical fuels can do," observed Ley. "The direct trip to the neighboring planets is even further beyond." Some sort of transit and refueling station was obviously required. The space station, as Ley observed, would provide the much needed "cosmic stepping stone for spaceships which were too weak to reach another planet directly. . . . Trips to the moon, around the moon, and even to the other planets are no longer difficult if they are made from that station."[6]

Creators of these early images were keenly aware of the advantages to be gained by using extraterrestrial stations as the new fortresses of the space age. Forts, after all, are military facilities, and throughout history, fortifications have been used to defend settlements and protect territorial claims. In North America, the need for fortifications intensified because of the competing claims of European nations bent on controlling common territory, to say nothing of the anger expressed by the native population that already lived there. Indian wars, colonial wars, and revolutionary wars all spawned the need for fortification.

In contemplating the nature of space, many pioneers believed it would be subject to the same sort of territorial claims. Nations might lay claim to the Moon and planets, planting their flags on extraterrestrial territories just as they had done on terrestrial lands.[7] In this event, the space station would serve a valuable purpose, standing as the twenty-first-century equivalent of a fortress guarding the entrance to a harbor or waterway. Sitting at the gateway to the cosmos, a space station would fortify the claims of constructing nations to operate freely much as European nations seeking to settle the American frontier had used defensive works along colonial waterways to do the same. Like frontier forts, space stations would limit the claims of competing nations. "Whoever is first to build a station in space can prevent any other nation from doing likewise," the editors of *Collier's* magazine bluntly warned in 1952.[8] Circling the planet, the space station would cover far more territory than any fortress fixed on the Earth. Rocket pioneers devised a variety of schemes to take advantage of the new military high ground. Scientists working for the German government formulated plans for a giant solar mirror, a "Sun gun" that could incinerate enemy forces on the ground. Wernher von Braun suggested using an orbiting space station as a platform from which to drop nuclear bombs. Officials in the U.S. Department of Defense pursued the possibility of using manned orbiting laboratories (MOL) as spy stations in the sky. Oberth predicted that "whenever work on a space station is started, it will undoubtedly be for military reasons."[9]

The U.S. military never built an Earth-orbiting space station during the formative years of space flight. No space base served as the jumping-off point for the first expeditions to the Moon. Between the visions of space flight and its actual execution, a considerable gap emerged. Space travel proceeded down a course much different from that conceived by its advocates.

The concept of an orbital station as a space-age fortress received its first practical test during the 1960s when the U.S. Department of Defense received permission to move ahead with plans for a manned orbital laboratory. To MOL partisans, space was the ultimate high ground from which military campaigns of the future would be won or lost. To test the feasibility of the concept, military officers drew up plans for a small, fifty-six-hundred-pound research laboratory in which two soldiers could conduct a variety of surveillance activities and experiments. The U.S. Air Force even went so far as to recruit a corps of military astronauts.[10]

As the decade progressed, the applicability of the fortress analogy waned. Shortly after entering office in 1969, members of President Richard Nixon's administration canceled the MOL project. A number of factors doomed the idea. Technological advances allowed the development of remotely controlled satellites that could complete surveillance activities at a fraction of the cost of stations with humans on board, rendering bulky fortifications far less effective than satellites in space. During the 1960s the main combatants in the Cold War agreed to forgo territorial claims in outer space and accept unrestricted overflight of their territories by orbiting objects, decisions prompted largely by the knowledge that everyone's assets were equally vulnerable to attack. They agreed to ban nuclear weapons in space. This removed much of the motivation behind the development of large stations and maneuverable spacecraft designed to control access to the new high ground.[11]

Although these events undercut the fortification analogy, the image of a space station as a base camp for distant exploration remained strong. But as with the military plans, practical developments undermined this analogy as well. In 1962 NASA engineers decided to forgo any sort of orbital assembly or refueling point around Earth as a prelude for the trip to the Moon. Instead, they devised a radical method that substantially reduced the weight necessary to take humans to the lunar surface and back. Lunar orbit rendezvous combined with the development of engines that burned liquid hydrogen allowed the United States to reach the Moon directly from Earth with a three-stage Saturn launch vehicle weighing only thirty-two-hundred tons at launch, fully fueled. Because the method required no space station or Earth-orbit assembly techniques, it dashed the various schemes for a pyramid of equipment and supplies that had characterized exploration plans in previous decades.[12] The rocket technology necessary to achieve lunar orbit rendezvous rested easily within the grasp of NASA engineers. Space station advocates agreed to the plan reluctantly. Although they understood that it offered the only hope of reaching the Moon by the president's end-of-the-decade deadline, they also knew that the plan diminished the rationale for permanent facilities in orbit around the Earth. As NASA deputy administrator Hans Mark later observed, the price paid for Apollo was the lack of any space-based infrastructure once flights ended: "Apollo was essentially a dead-end from a technical viewpoint."[13]

As if to further undermine the supporting analogy, mountain climbers experimented with Alpine-style dashes to Himalayan summits that minimized the need for an elaborate pyramid of camps supplied by small armies of porters. As in space, lightweight equipment and new techniques transformed earthly exploration. In 1978 Reinhold Messner and Peter Habeler scaled Hidden Peak, a 26,470-foot companion to nearby K2, with only a single depot above a minimal base camp. As if to emphasize their point, Messner and a companion pulled a sled across Antarctica to prove that the new approach would work there too.[14]

    The most durable aspect of the space station image as presented to the public was its size, an aspect maintained not only by practical engineers but also by creators of fantastic stories. Both promoted the notion that any space station worthy of public support would be a massive structure—so large that people on Earth would be able to see it fly by.[15] In the minds of its advocates, creation of an Earth-orbiting space station would rank among the greatest engineering accomplishments of all time. Public interest was nurtured with images of big stations in space.[16] This created a discomforting dilemma for station advocates in search of government support for their plans. The station designs that generated the most public interest were large, multipurpose facilities that typically cost more than politicians were willing to spend, whereas the stations the government could afford to build were often too small to arouse much public interest. It was a situation guaranteed to produce anxiety. Politicians choked on impressive space station plans and sneezed at modest ones.

    In his "Crossing the Last Frontier," von Braun proposed the construction of one of the most impressive stations of all time, 250 feet wide and orbiting 1,075 miles above the Earth's surface. Shaped like a wheel, the station would make a complete turn on its axis every twenty-two seconds so as to give the sensation of gravity, one-third of the force felt by humans on Earth, to occupants in its outer rim. Chesley Bonestell painted a wide picture of the station to accompany the article. The illustration, one of the most reproduced works of twentieth-century space art, revealed a hublike wheel with graceful lines passing over the Earth near the Panama Canal far below. Astronauts in space suits worked at a nearby observatory and on the space wheel; others unloaded supplies from a winged space shuttle recently arrived from Earth. Space taxis shaped like overgrown watermelons carried astronauts between the winged spacecraft and space station and the observatory orbiting nearby.

    To simplify construction of the space wheel, von Braun suggested fabricating it out of inflatable nylon and plastic. The material could be collapsed for its trip into space, then pumped up once in orbit. Astronauts would assemble twenty sections, each an independent unit, to complete the wheel. Constructing and resupplying the orbital facility would require frequent shuttle flights. According to von Braun, "There will nearly always be one or two rocket ships unloading supplies."[17]

    Accompanying a separate article by Willy Ley in the same issue of *Collier's* magazine, a painting by artist Fred Freeman showed a cutaway of one section of the wheel. The fifty members of the space station crew worked on three decks inside the outer rim, much like sailors on a large submarine. Elevators carried workers to the station hub; regulators and pipes kept the station cool. Von Braun estimated that the space station would require five hundred kilowatts of electric power, which he proposed to generate through a condensing mirror and generator. A highly polished metal trough ran the circumference of the facility, concentrating the rays of the Sun onto a steel pipe containing liquid mercury. Heat from the Sun transformed the mercury into hot vapor, which in turn drove a turbogenerator.[18]

Stanley Kubrick improved substantially upon the image of the large wheel with the release of *2001: A Space Odyssey*. The phase of the movie devoted to modern times opens with the sight of a space shuttle preparing to dock at a very large space station. Kubrick's depiction of orbital activities set new standards for special effects. His station, with work on one of its twin hubs still under way, spun in an orbit two hundred miles above the surface of the Earth and measured an astounding nine hundred feet across, as large as a modern aircraft carrier. It rotated less forcefully than von Braun's wheel, one revolution per minute, producing a sense of gravity equal to that on the Moon. The facility housed an international crew of scientists and bureaucrats, welcomed passers-through on their way to the Moon, and offered amenities as comforting as those in a modern hotel.[19] In a further effort to equate size with significance, the creators of *Star Trek: The Motion Picture* designed an orbital facility that moved across the screen like a huge city. The facility served as a dry dock for the USS *Enterprise*, to which Admiral James Kirk was returning. A smaller orbital transfer station hovered nearby.[20]

Engineers working to design space stations for the U.S. government did not avoid the proclivity for bulk. In 1969 NASA officials asked aerospace contractors to prepare preliminary designs for a permanent orbital facility that could house between fifty and one hundred occupants.[21] The effort produced some grand designs. North American Rockwell proposed a pinwheel shaped configuration that grouped four 33-foot-wide cylinders around a common core. The core consisted of an axis 359 feet long, a requirement made necessary by the placement of two nuclear-powered generators at the far end of the central shaft. Attached by booms to the opposite end of the shaft, the cylinders stood a total of 234 feet apart from end to end. The base could be rotated so as to produce artificial gravity or dampened out for experiments requiring none at all. From the basic building blocks to the completed pinwheel, the resulting configuration was impressively large.[22]

Just as terrestrial explorers used frontier forts and base camps to open up remote areas on Earth, space pioneers envisioned orbital stations as stepping-stones for the exploration of space. Stanley Kubrick placed a space station 900 feet wide in the 1968 motion picture *2001: A Space Odyssey*. The station rotated so as to produce a sensation of gravity similar to that experienced by travelers on the Moon. This movie poster by space artist Robert McCall depicts a winged space shuttle departing the docking port in the large orbital facility. (*2001: A Space Odyssey* © Turner Entertainment Co. A Warner Bros. Entertainment Company)

In 1984 NASA finally received the political approval necessary to begin designing a real space station. The baseline design announced two years later was more generous in all its dimensions than von Braun's 250-foot-wide wheel. NASA engineers rejected rotating space stations in favor of a configuration that minimized the force of gravity. To accomplish this, designers removed the experimental labs and crew quarters from the outer edge of the station and placed them at the center of mass on a long transverse boom. Two vertical keels provided mounts for various experiment packages and the servicing of spacecraft and satellites, while the outer ends of the central boom provided attachment points for a solar-power-generating system. To illustrate the size of the impressive facility, NASA officials prepared a computer-generated illustration that superimposed the dual-keel design across a view of the U.S. Capitol. The station stretched from the House to the Senate, joining the two legislative chambers in what space enthusiasts hoped would be the political consensus necessary to complete the job.[23]

Space enthusiasts not only imagined that the station would be large but also assumed that it would last a long time. NASA engineers wanted the structure to last for at least twenty or thirty years, perhaps longer.[24] Once up, astronauts would modify the structure and eventually replace it with other, more advanced stations. Initial occupancy of the space station would mark an important milestone in history. From the time of its occupancy forward, humans would always live and work away from the Earth. The station would create a permanent human presence in space.

Station advocates carefully maintained this image. During the late 1960s, NASA officials won approval for an orbital facility called *Skylab* that was constructed out of equipment developed for the Apollo moon program. During 1973 and 1974 *Skylab* housed three crews of three astronauts each, who spent a total of 171 days in orbit.

Clearly not a permanent facility, *Skylab* plunged into the atmosphere in 1979. NASA executives were careful to call it an orbital workshop, not a space station. "The ideal station would be permanent and large," confessed one NASA reference work. *Skylab* was "a worthy precursor to a larger, more elaborate station."[25] It was not the real thing.

When Congress in 1984 attempted to steer NASA back toward an orbital edifice that would house astronauts only part of the time (a "man-tended" facility), NASA executives objected strenuously. President Reagan had committed the United States to develop a "permanently manned space station," they pointed out. "The Space Station will, of course, be a highly automated system, and it will require many advances in automation techniques and robotics," NASA administrator James Beggs argued. "It is, however, the presence of man which makes it a unique national resource."[26]

To attract public interest, a space station had to be large. To fulfill the vision of humans moving permanently into space, the station had to last a long time. To justify a large and permanent orbital outpost, advocates had to propose a facility that performed many functions. As they knew, a big space station that performed only one task would not attract the support of the various scientists, engineers, entrepreneurs, astronauts, military officers, and exploration advocates interested in space. Station advocates had to find many jobs for their crews to perform. Many functions helped to justify the government's investment in a station. It also complicated the station's design.

According to popular conceptions, a space station could serve various purposes—an observation platform, a scientific laboratory, or a manufacturing factory. Using it as an operational base, its crew could service satellites or prepare them for transfer to higher orbits. A space station could provide an orbital dry dock for the final assembly and fueling of large spacecraft bound for the Moon or Mars. The space station presented in *2001* served as a hotel, managed by the Hilton Corporation. Von Braun's wheel served as a reconnaissance platform—an orbiting "battle station" from which soldiers could track the movement of enemy forces. NASA officials suggested that the U.S. Defense Department might want to conduct missile intercept experiments from one.[27]

In few areas did the reality of space engineering clash so dramatically with the images necessary to motivate public interest. Big stations had to perform multiple missions, but those missions were often better performed on individual platforms specifically designed for specialized purposes. Station planners, nonetheless, herded them onto the main facility.

Advocates insisted that a large orbital outpost could serve as an observation platform, the primary function of von Braun's 1952 wheel. "Technicians in this space station," von Braun wrote, "will keep under constant inspection every ocean, continent, country and city. Even small towns will be clearly visible." The desire to observe all portions of the Earth drove von Braun to select a polar orbit for his space facility. With the planet turning below them, members of the crew would use specially designed telescopes attached to large optical screens and radarscopes to observe activity below with pinpoint detail.[28] Observations could be used to improve weather predictions by monitoring natural events like cloud patterns. Most of the observations von Braun discussed, however, concerned national security. Well into the 1980s, space station advocates believed the U.S. military would be an important customer for any orbital facility, given its potential to act as a spy-in-the-sky.[29]

Observations could also be made of astronomical phenomena. For years, scientists have recognized the advantages of astronomical observatories in space. Orbiting telescopes achieve unprecedented clarity from their vantage points above Earth's murky atmosphere. Photographs in visible light are substantially impaired by atmospheric fluctuations, and only a small fraction of the infrared energy emitted by other heavenly bodies reaches the ground. Scientists devised plans for space telescopes that could collect visible and ultraviolet light, radio waves, X-rays, gamma rays, and cosmic rays. Astronomers agitated for space telescopes of all sorts.[30]

Station advocates viewed scientists as a powerful group whose cooperation broadened considerably the advocates' base of strength.[31] Unfortunately for the advocates, the space-based telescopes desired by astronomers required such long exposure times and exact pointing requirements that the telescopes could not be housed on the multipurpose facility. Other station activities, from the docking of spacecraft to the simple movements of people on board, would disturb the alignment of telescopes that probed the void. Recognizing this problem, von Braun proposed placing a space observatory on a remotely controlled platform separate from the main station.

This created something of a conceptual puzzle. If the space observatory was not physically attached to the station, how could the two facilities be treated as one? Von Braun solved this problem by placing the space observatory in the same orbit at the station, hovering nearby. Astronauts from the station would service the observatory, he said. They would change its film. In the days before remote sensing, when sharp images could not be transmitted electronically, von Braun imagined that station technicians in space suits would manually load canisters of film and other special plates into the observatory. Von Braun thus linked the station and the orbiting observatory in a direct way, as two parts of the same system.[32]

NASA's Space Station Task Force, set up in 1982 to develop conceptual plans for the first permanently occupied facility, took this idea one step further. In order to win support from scientific groups, leaders of the task force included within the space station program two free-flying platforms. One would travel in the same orbit as the main facility and could be serviced by station technicians much in the same way as von Braun had envisioned. The other platform, however, would travel along an entirely different path, moving around Earth's poles. To reach it, station technicians would need to return to Florida, fly to California, and catch a space shuttle launched from the West Coast and headed in an entirely different direction. Members of the task force attached the polar platform to the multipurpose space station in a most ingenious way. They included the appropriation for the platform in the budget for the main facility. NASA's whole space station program, as a result, contained both "manned and unmanned elements." Unless NASA built a multifunctional facility, observed the leader of its task force, "it really is not going to be justifiable."[33]

In addition to its functions as an observation platform, the space station could serve as a scientific laboratory, uniquely situated but not conceptually different from other government-funded national research laboratories on Earth. Scientists in a stable or nonrotating facility could conduct research on the effects of weightlessness on materials and living things. (The technical term is microgravity—any large space station will produce a small amount of gravity from its own mass.) Microgravity is a condition impossible to reproduce on Earth for any significant period of time. Critics scoffed at the image of laboratory studies on weightlessness as "another two decades of original research on why astronauts vomit."[34] To be sure, considerable research would be done on the human response to weightlessness, but the research proposed for various space stations went considerably beyond this.

Writing well before the first humans flew into space, Hermann Oberth predicted that materials with unusual properties, such as new metal alloys, could be created in the weightless environment of space. A continuously operating laboratory in space would accelerate the research necessary to develop such materials. Experiments could be conducted on plant growth under weightless conditions. Oberth wondered, for example, if plants unconstrained by the force of gravity would transform themselves into unusual shapes and sizes. "All of our scientific and technical knowledge is based on the existence of the force of gravity," he wrote. "It is most interesting to imagine a scientific and technical world without gravitation and to attempt to picture its consequences." An orbital laboratory would also permit extended research using nearly perfect vacuums or temperatures close to absolute zero.[35]

One of the more creative uses of the space station arose from the notion of an orbital factory. During the early 1980s, a number of space enthusiasts promoted outer space as the new frontier of entrepreneurial capitalism. Preliminary experimentation had revealed a number of products that could be manufactured under the special conditions existing in space. In the near weightless environment, drugs with a standard of purity not commercially attainable on Earth could be produced using a technique called electrophoresis. Integrated circuits made from new materials such as gallium arsenide promised to further revolutionize the computer industry. A huge new industry based on the possibility of growing protein crystals in space promised advances in the battle against diseases such as cancer and in the improvement of chemical processes such as gasoline production. Materials nearly free of contaminants could be produced using the extremely high vacuum of space.[36]

Swept along by excessive enthusiasm, NASA's space station planners offered their orbital facility as the industrial park of the twenty-first century. The United States could capture this billion-dollar market by taking the lead in establishing an orbital laboratory to complete much-needed research and set up the processing facilities needed to begin production. When President Reagan met with a group of industrial executives familiar with the commercial potential of space in 1983, he asked what the government could do to speed up the commercialization of space. More than anything else, the executives replied, they wanted a space station. The promise of space commercialization figured heavily in Reagan's decision to start work on the orbital facility.[37]

Designing a multifunctional space station that could perform such activities posed a number of practical problems. Once again, the gravity problem emerged. Scientists measure the absence of gravity in micro g, where one micro is equal to one-millionth part (10^{-6}) of the force felt by a person standing on the surface of the Earth. Space-based facilities for microgravity work require levels below ten micros, approaching one micro where possible. The dual-keel design for NASA's space station was chosen because it promised to place the laboratory modules in the one-micro zone.[38] A space facility where everything is perfectly still can produce gravity reductions in this range. Machinery that vibrates will compromise the standard, as will clumsy people as they move about. As a simple matter of physics, more activities increase the number of micros likely to be felt. Under perfect conditions, a standard of one micro might be attained. Under actual circumstances, it could slip into the range of ten to one hundred micros.[39]

In part to solve this problem, spacecraft designer Max Faget proposed the construction of what he called an industrial space facility (ISF). After his retirement from NASA, Faget founded Space Industries, Inc., from which he sought support for an automated platform that could manufacture materials in space. The platform would be visited by astronauts, who would service the facility, change equipment, and set up new activities. The rest of the time, the platform would orbit quietly without people on board. Removing people from the processing facility enhanced the possibility of maintaining very low levels of gravity. It also reduced the cost of construction.[40]

In 1987 the president's Office of Management and Budget inserted $25 million in the NASA budget and instructed the space agency to investigate the possibility of supporting this commercial venture by leasing it from Faget. This caused great consternation among station advocates. At the time, the space station was fighting for its political survival. Advocates angrily viewed Faget's proposal as a diversionary tactic that would allow politicians to kill the big, multipurpose station while shifting its most promising activities to a human-tended orbital facility. NASA executives raised elaborate objections to the ISF. With support from competing industries and the National Academy of Sciences, NASA officials beat back attempts to shift commercial activities off the multifunctional space station. "It had to be killed," Faget grumbled, "and they did kill it."[41]

Complicating all these proposals was the desire to use the space station as an operational base, a vision that had captivated exploration advocates for decades as they imagined a series of interplanetary service stations spreading out toward the Moon, planets, and stars.[42] In 1980 planners at NASA's Johnson Space Center put forward one of the most imaginative versions of a station emphasizing this function, which they called a space operations center. At that time, any object traveling to geosynchronous orbit or beyond had to be launched directly from the Earth or carried partway on the space shuttle. (NASA stopped using the space shuttle to transfer payloads to higher orbits after the *Challenger* accident.) From a space operations center in low Earth orbit, the planners said, astronauts could assemble and test objects bound for geosynchronous orbit. A permanent facility would allow astronauts, working in space, to construct enormous communications platforms and high-resolution parabolic antennas too large to be launched from Earth. Using a remotely controlled orbit transfer vehicle or space tug, astronauts could push those platforms out to appropriate locations.[43]

The configuration proposed for the operations center, with its agglomeration of modules and trusses, looked like many designs of the day—with one exception. In the most widely distributed drawing, a hanger protruded from the supporting framework. Planners envisioned the large hanger as an orbital dry dock for servicing and repairing satellites, a major purpose of the operations center. The designing engineers shifted most science and industrial applications to free-flying platforms, such as automated factories or space telescopes. "Some operations actually proceed best without the disturbance of human presence," the engineers admitted. Those platforms, however, would require periodic servicing and adjustment.[44] This work could be done by having astronauts fly from the operations center to the platforms in a small space tug or by using the tug to bring satellites into the hanger for repairs. Science was not a major activity within the space operations center. Operations was.

NASA executives dismissed the proposal for a space operations center as soon as they began to seek political support for a large space station in 1982. The officials did not want scientists standing outside of any space station looking in. They pushed the station concept back toward its original premise. Any station large enough to be worth building would have to accommodate many users.[45]

In 1984, having received permission to construct a large, multifunctional space station, NASA began the process of actually designing one. This proved extraordinarily difficult to do. The image of a space station was well set in the public mind: it had to be big, it had to be multipurpose, and it had to provide a permanent outpost for future exploration. Given the budgetary realities of the 1980s, it also had to be affordable. These expectations created conflicting requirements. For nearly ten years, NASA employees and their contractors struggled to produce a workable design. They spent more time redesigning the space station than their predecessors had spent two decades earlier to reach the moon.[46]

Explaining the vision to a Senate subcommittee, NASA administrator James Beggs recited the capabilities a multifunctional facility could provide:

Properly conceived, a station could function as: A laboratory in space for the conduct of science and the development of new technologies; a permanent observatory to look down upon the Earth and out at the universe; a transportation node where payloads and vehicles are stationed, processed, and propelled to their destinations; a servicing facility where these payloads and vehicles are maintained and, if necessary, repaired; an assembly facility where, due to ample time on orbit and the presence of appropriate equipment, large structures are put together and checked out; a manufacturing facility where human intelligence and the servicing capability of the station combine to enhance commercial opportunities in space; and a storage depot where payloads and parts are kept in orbit for subsequent deployment.[47]

Engineers initially settled on what they termed the dual-keel design. The configuration featured a rectangular metal truss as large as a football field (361 feet by 146 feet) on which engineers planned to mount a number of experiments and payloads. In the plans, stellar and solar observatories on the upper boom pointed toward the heavens. Earth observatories on the lower boom pointed toward home. Along the sides of the rectangle engineers planned to mount canisters that could house scientific experiments and small factories. Two rectangular service bays provided space to store equipment and service payloads.

On and around the core space station, NASA engineers planned to deploy a variety of exotic machines. A long robotic arm crept around the truss like a spider, moving payloads and performing assembly work. Manipulators inside the service bays assisted astronauts repairing satellites. A busy little orbital maneuvering vehicle (OMV) flew around the space station like an intelligent bug, repairing faulty equipment and retrieving satellites. An orbital transfer vehicle, mated with its own OMV, carried payloads to and from geosynchronous orbit twenty-two thousand miles away. A self-powered, automated platform conducted science experiments a short distance from the main station, while another funded out of the same budget circled the poles.

Complementing the dual keels, engineers added a horizontal boom 503 feet long centered on the rectangular truss. The boom held solar collectors and photovoltaic arrays that generated the station's electric power, as well as radiators for dissipating the heat that power generation would create. Station designers planned to start with seventy-five kilowatts of electric power and expand to more than twice that level of power as the station grew.[48] Attached to the midpoint of the horizontal boom, at the center of the rectangular truss, sat the modules in which the crew would live and conduct laboratory experiments. With a useful life of thirty years, the station would evolve as its capabilities grew. It could eventually provide "a staging point for spacecraft of unprecedented size," which could take humans back to the Moon or on to Mars. Engineers built "scars and hooks" onto the framework of the station on which future ambitions could be placed.[49] They involved the European Space Agency, Canada, and Japan, which would contribute their own facilities.

The space station proposed by NASA and approved by President Ronald Reagan in 1984 abandoned the wheel-shaped design in favor of a facility that produced practically no gravity at all. NASA engineers set living quarters and laboratories at the center of a long traverse boom. Two vertical keels provided room for observatories, experiments, satellite repair, and preparation of spacecraft bound for missions deeper into space. Though not wheel-shaped, the resulting configuration was large—more than 500 feet long and 360 feet tall. (NASA)

On paper, the space station looked extraordinary. In practice, it substantially exceeded the original cost estimate that planners had advanced when the project was approved. In a premature and somewhat accidental response to White House inquiries, station planners had released an $8 billion estimate of the facility's cost. The estimate set well below the formal cost studies under way within the space agency and the true cost of a complex, multifunctional facility. NASA executives spent years attempting to reconcile their desire for a multifunctional space station with the modest cost ceiling imposed over it.[50]

As ambitious as the design appeared, it still fell short of expectations from the past. The station, for example, would not house a large crew. The living quarters contained room for no more than eight astronauts, cramped in a fourteen-foot-wide habitat module.[51] John Hodge, the leader of NASA's Space Station Task Force and later director of the development program, admitted at a congressional hearing that the public at large and "quite possibly most of the people within NASA" thought of a space station "as a very large rotating wheel with 100 people on it and artificial gravity." Writing for the popular *Science 83* magazine, Mitchell Waldrop observed that the actual station "will look more like something a child would build with an Erector set." It would not resemble a wheel, it would not rotate, and it would not soar very high. "2001 it's not," the magazine complained.[52]

As design work continued, the station shrunk. It grew smaller and smaller, moving back from the expectations created by a century of wondering about outposts in space. Within a year of announcing their intention to construct the large, multipurpose, dual-keel space station, NASA officials scaled that design down. In April 1987, confronting the fact that the proposed facility would exceed budget guidelines, NASA abandoned the vertical keels. That eliminated the capability of the station to operate as a service station in space. As the servicing capability disappeared, so did the co-orbiting platform. NASA officials hoped to resurrect these elements during a secondary construction phase, but deferrals became permanent as cost and design problems swelled.[53]

By eliminating the vertical truss structure, NASA officials sharply reduced the functions the station could perform. Instead of a large, multipurpose facility, NASA was left with an orbital research lab whose residents would spend most of their time conducting experiments and making observations of the Earth and heavens. In submitting her report on the future of the U.S. space program to NASA administrator James Fletcher, Sally Ride correctly identified the revised configuration as a "laboratory in space." Without much discussion, many of the functions that had motivated space pioneers to dream about orbital stations disappeared. Ride could only look to the future: "Other capabilities, such as an assembly station or a fueling depot, will not be included in the initial phase, but could be accommodated later if a need for those functions is clearly identified."[54]

Any hope that even those capabilities could be patched onto the initial facility evaporated when the station contracted again. In 1991 NASA officials issued plans for what they called the "restructured space station," also known as *Space Station Freedom*. The remaining horizontal truss was snipped back to 353 feet, and the habitation and laboratory modules were reduced in size. The station was reconfigured, in the words of the advisory committee recommending the changes, "with only two missions in mind." The four-person crew would carry out life-science experiments and conduct microgravity research. Gone were the hopes for a large, multipurpose outpost in space.[55]

With a change in U.S. presidents, the space station contracted once more. Upon entering office, the Clinton administration requested another space station redesign. Not unexpectedly, this produced a new configuration with less volume, less electric power, and a smaller crew.[56] The new *Space Station Alpha*, as it was dubbed, disappeared quickly as the United States and Russia agreed to join their two space stations efforts into one. Like a dieter rebounding after a long fast, the new configuration gained volume. The crew quarters where the Russian modules appeared grew a few feet larger and the central truss grew longer in order to accommodate a more extensive electric-power-generating system.[57] Russian participation halted nearly a decade of configuration decline, but it was not sufficient to restore the large, multipurpose facility from which planning began.

The design to emerge from these efforts met few of the criteria established by people who had tried to imagine what a space station would be. It was not round; it did not rotate. It was not a fortress in space; it could not serve as a jumping-off point for voyages beyond. It would not perform many functions, being primarily a research laboratory. It did not have a large crew.

Why did the space station prove so difficult to design? NASA spent nearly ten years and more than $10 billion preparing station designs, an amount roughly equal to the 1984 cost estimate for the entire facility.[58] The continuing agonies of the space station program did not arise solely from engineering complications and cost overruns. They also appeared as a consequence of deep-seated conceptual contradictions in the idea.

Station advocates sought to build an orbital facility that was both cheap and grand. As NASA officials learned, it was hard to design a station that simultaneously met public expectations and budgetary constraints. The pioneers of space flight envisioned a station that would rival the other great engineering feats of the twentieth century, in both scope and cost. Scientists and engineers called together to advise *Collier's* estimated that their 250-foot-wide orbiting wheel would cost $4 billion, a huge sum in the currency of the day.[59] Four billion dollars then, when adjusted for inflation, equals $21 billion in 1984 dollars. Yet when pressed for a cost appraisal during the White House review process, NASA officials backed away from an engineering project of that scale. They issued a preliminary estimate of $8 billion in 1984 dollars for the development of the orbital facility, a number based largely on the political reaction to various cost estimates. "I reached the scream level at about $9 billion," said John Hodge, leader of NASA's Space Station Task Force.[60]

The $8 billion cost estimate was based on a number of assumptions that were technically feasible but difficult to fulfill. To construct a space station that fit within the budgetary estimate, NASA executives in Washington, D.C., had to restrain the desire of officials at the agency's field centers to drive up the price with Earth-based infrastructure, overhead, and cost reserves. These extras contributed little to the capability of the orbiting facility but let the NASA field centers thrive. As the centerpiece of the human space flight program for the 1990s, the space station provided a number of center directors with a lucrative opportunity for strengthening their institutions. Managers in the field pushed financial requirements higher; headquarters executives fought back. Field officials eventually won.[61]

One important feature that allowed NASA to issue a small estimate for a large project backfired. NASA officials believed that a space station could be constructed in such a way as to grow incrementally. Using a modular design, NASA could deploy a small space station in the beginning and add more parts as funding allowed. "Space stations are the kind of development that you can buy by the yard," James Beggs pointed out in his effort to justify the low cost estimate.[62]

A space station that can grow "by the yard" can also shrink in the same way, as station planners learned. Every cost review produced an incrementally shrinking design. Modules, air locks, satellite servicing bays, solar collectors, experiment mounts, space tugs, orbiting platforms, and scientific equipment fell off as cost reviews sped by. A 1985 letter written by station overseer Philip Culbertson that initiated the first redesign came to be known as "scrub mother," the parent of all reductions to come.[63]

NASA executives premised the development of the space station on the availability of cheap and easy transportation to space. In Chesley Bonestell's famous painting of the rotating space station prepared for *Collier's*, a winged space shuttle hovers nearby. In *2001: A Space Odyssey*, people arrive at *Space Station One* on a winged passenger shuttle.[64] President Ronald Reagan took the first step toward approving the space station at the 4 July 1982 landing of the space shuttle *Columbia*. James Beggs planned to fund much of the cost of the space station through funds freed as the space shuttle began to pay its own way. The orbital facility, he frequently said, was the "next logical step" after completion of the shuttle development program. As late as 1985, NASA officials were planning to launch as many as twenty-four shuttle flights per year. Construction of the space station depended upon the availability of cheap, easy access to space.[65]

All of that changed with the explosion of the space shuttle *Challenger* on 28 January 1986. Concerns about the cost and reliability of the space shuttle forced NASA engineers to make a number of important changes in station design. Engineers cut back on the number of shuttle flights needed to construct and maintain the facility, both to reduce costs and to enhance safety. They added the cost of space transportation to the space station budget. Members of the original station task force did not include transportation in the original $8 billion estimate because the cost of space transportation was forecast to be so low.[66]

Worries about the reliability of the shuttle raised concerns about the safety of the space station crew. With shuttles always ready to fly, the crew had access to a rescue vehicle should some catastrophe occur. In a serious emergency, NASA engineers anticipated that the crew could retreat to what they called a "safe haven" within the station until a shuttle arrived. Problems with the shuttle encouraged engineers to prepare alternative plans for a "lifeboat" stationed on the facility that could drop the crew back to Earth following a major malfunction. The lifeboat had to be ready to fly with little preparation. Space station planners considered a number of options—a winged X-38 that could hold up to seven crew members, two Russian Soyuz spacecraft with three seats in each, and a special version of the capsule-shaped *Orion* spacecraft with room for six. Safety concerns, joined with the limited capacity of possible return vehicles, restricted the size of the crew. The need for lifeboats drove up costs once more.[67]

People imagining the space station had little difficulty envisioning the technologies needed to operate it. Engineers knew how to prepare livable environments in pressurized spacecraft; they knew how to generate electric power in space. Experts knew how to stabilize large structures in orbit and write computer programs that could regulate the automatic functions of such craft. Even the 1952 editors of *Collier's* assumed that the technology for the space station was well understood. "Our engineers can spell out right now," they wrote, "the technical specifications for the rocket ship and space station in cut-and-dried figures."[68]

Technological specifications might have been understood, but NASA engineers did not want to build a space station with old technologies. They did not want to reconstruct *Skylab*. They wanted to build a complex, multipurpose facility with new technologies, taking a quantum leap into the twenty-first century and thereby producing knowledge that could be applied to bolder projects such as interplanetary spacecraft and bases on Mars.[69] Imagining new technologies for the space station additionally complicated its design.

Those who had to operate the space station, including members of the astronaut corps, grumbled that engineers were creating a technically infeasible and potentially unsafe facility. The assembly phase was too complicated, they said. The complex remote manipulator system would break down. Assembly and upkeep of the station would require too much extravehicular activity. Planned maintenance would require frequent spacewalks, and that did not include unanticipated repairs. Without frequent maintenance, the station would not last long.[70]

Planners worried that a large, multifunctional facility would not work well. As Sally Ride wrote in her 1987 report on the future of space exploration, "a laboratory in space, featuring long-term access to the microgravity environment, might not be compatible with an operational assembly and checkout facility." In an initial facility, with a small crew and a limited number of activities, competing uses could be compressed within a single structure. The crew could take turns performing different functions. But as the size of the station grew, "branching" would surely occur, meaning that the government eventually might need to build separate space stations for each function. Station advocates faced enough opposition to just one orbital facility and chose not to broadcast the fact that future evolution might require more.[71]

From its beginnings, the space station was a concept at war with itself, built on terrestrial analogies with limited applicability to space. Even on Earth, the importance of base camps and forts diminished as technology progressed. Space station advocates sought to combine on a single, relatively low-cost facility a large number of potentially conflicting functions. Advances in technology drove down the cost of performing those functions on separate, often-automated facilities. The attractiveness of the multipurpose, centralized station faded as experience accumulated. NASA executives, once the strongest partisans on behalf of the construction of an orbiting space station, wondered if they should withdraw from the program and move on. Various officials proposed that NASA abandon the space station just a half-dozen years after completing it.[72]

The large space station that the world community cooperated to build performed fewer functions than the space station that President Reagan approved. Conflicts between ambition and cost prompted cutbacks in station design. Even so, assembly of the orbital facility marked what exploration advocates hoped would be an important milestone in the advance of human space flight—the first permanent human presence in space. (NASA).

Even the appeal of the space station as a depot for voyages beyond declined. In 1990, when members of the so-called Synthesis Group released their report on human exploration of the solar system, they scarcely mentioned the station. Members of an Advisory Committee on the Future of the U.S. Space Program took a similar position. "We do not find compelling the case that a space station is needed as a transportation node for planetary exploration," they wrote. "First, many promising flight profiles do not appear to require such a node and, second, if they did, the need in our judgment is sufficiently far in the future that we would hardly know today what to ask of such a terminal." NASA's plans to return to the Moon envisioned Earth-orbit rendezvous, but no assembly and checkout at an Earth-orbiting space station.[73]

Yet even as the concept of the space station shrunk, its significance grew. Some two dozen orbital workshops with human crews circled the Earth before orbital assembly of the big, International Space Station began in 1998. They included the U.S. *Skylab*, six versions of the Soviet *Salyut*, one Russian *Mir* space station, and a succession of missions involving the European-built *Spacelab* modules flown on the U.S. space shuttle. As impressive as they were, none of these facilities achieved the intended goal imagined by people promoting the space station concept—a commitment to a permanent human presence in space. The big space station did. On 2 November 2000, with the arrival of the Expedition 1 crew, the commitment to a permanent human presence in space was fulfilled.[74]

Members of the NASA task force that helped win approval for the actual space station wanted the facility to mark such a milestone in human history. They looked forward to a point in time at which humanity would no longer be confined to the surface of the Earth. If humans could learn to live in a space station, they would learn how to live in spacecraft on long voyages as well as surface shelters on other orbs.

The technology required to develop a lunar habitat, fly living beings to other planets, or build an outpost on the surface of Mars is the same technology needed to construct and operate an orbiting space station. It is the technology of power generation, artificial atmospheres, propulsion, communication, heat dissipation, onboard data management and flight control, fire and hazard suppression, life control mechanisms that recycle waste, and the systems integration techniques that allow all of these components to work in a coordinated fashion. The most devoted advocates of space stations envisioned station-like facilities throughout the solar system. They would be needed not just around the Earth but around the Moon and Mars and on other places. Danny Herman, the chief engineer for the NASA task force that helped win approval for what became known as the International Space Station, spoke of these technological innovations in the following way. "A manned Mars mission," he said "is a space station that is going to leave earth's orbit." Building a space station that could not encourage those innovations, he concluded, "is not exciting to me."[75]

The space station built by sixteen nations, like other great human structures, will eventually deteriorate and fall. Its legacy, however, may persist in the structures that come after it. If the lessons learned from the movement to construct a large space station lead to future structures in Earth orbit and beyond, space flight advocates will have achieved an important element in their dream. They will have created the means by which their own species can live permanently beyond the confines of its home planet. No one yet knows whether this will occur, but if it does, future generations will celebrate the achievement. They may even laugh over how much it cost.

8

Spacecraft

It has proven to be much more complicated than I thought.
—Hermann Oberth, 1985

Advocates of the dominant approach to space flight fostered an enticing expectation. Under the terms of their vision, large numbers of people would board extraordinary spacecraft and travel well beyond the planet's confines. Space flight, once the domain of test pilots and other brave individuals, would receive ordinary citizens not much better prepared for travel than automobile drivers or airline passengers back on Earth. Everyone would fly into space. The experience would transform humanity.

This vision drew its inspiration from the history of aviation in the first half of the twentieth century. The advent of aviation spawned an outpouring of prophecy and public enthusiasm similar to that accorded space flight. At a time when only the brave and foolhardy flew, prophets of aviation predicted that air travel would become so commonplace that ordinary people in large aircraft unaffected by weather would travel from city to city across the face of the Earth—a notion unimaginable to a populace stuck on muddy roads and horse-drawn carriages. Aviation fascinated the general public, encouraging even bolder claims that became the basis for what air-minded people called "the winged gospel."[1] The fact that so many of the prophecies came true encouraged advocates of space travel to believe that a similar future awaited them beyond the Earth.

For both aviation and space travel, the organs of popular culture promoted the popular view. Hollywood, which had churned out scores of aviation films, released movies and television programs depicting the effortless quality of space travel. Invariably, filmmakers and screen writers portrayed space flight as less difficult than it actually was.[2] Spacecraft of the popular mind covered vast distances at remarkable speeds and were affordable, reliable, and not much more complicated than a modern jet aircraft. Passengers suffered few ill effects, underwent little preparation, and encountered modest risk. Science joined fancy in fostering the chorus of anticipation. Otherwise pragmatic rocket engineers issued extraordinary claims about the future of powered spacecraft, especially the NASA space shuttle. Promoters predicted that the shuttle would cut the cost of space flight by a factor of ten. It would allow ordinary citizens to fly into space. It would be so easy to launch that a fleet of five could make fifty round trips every year.[3]

In 1970 the president's Science Advisory Committee assessed the status of human space flight. Impending developments, they predicted, would determine future capabilities. The prevailing costs of human space flight imposed by existing expendable rocket technology "do not justify the conduct by astronauts of space science or space applications," committee members wrote. If rocket scientists succeeded in developing low-cost, reusable spacecraft, then that situation would change. Such a development, the committee members predicted, could "usher in a new space era" in which human activities in space would expand significantly, coming to resemble aviation access "to hazardous and remote environments on earth."[4] However, if rocket scientists failed in this endeavor, then a different scenario would emerge. Robotic spacecraft and remotely controlled satellites might replace human beings as the dominant travelers in the solar system. Space travel under the latter conditions would not resemble the model supplied by aviation, which rose from the sands of Kitty Hawk into a worldwide transportation network in hardly sixty years. It would resemble something else. The outcome, members of the committee cautioned, would depend to a large extent upon NASA's ability to build and fly its proposed space shuttle.

Since the beginning, advocates of the spacefaring creed have embraced the notion of a future in which practically anyone would be able to travel in space. They were not content to picture a few astronauts planting flags on outlying orbs or a handpicked crew surviving in a small research facility on some distant place. They wanted large numbers of people to live and work in space. In order for the full vision to come true, ordinary people had to move into the cosmos, traveling in spaceships accessible to the common folk. That, in turn, required cheap and undemanding transportation.

The anticipation of easy transport clashed with the imagery used to promote the early space program. The first flights into space were conducted under conditions so stressful that the government would allow only test pilots to fly. Although the courage of early astronauts boosted public interest in space travel, it distracted from the ultimate aim of human flight. For its most devoted advocates, space travel had to move beyond the personal confrontation of high danger—not to diminish the bravery of astronaut pioneers but so that ordinary people could imagine themselves flying in space.[5]

In clinging to this vision, advocates found comfort in the aviation experience, a powerful model for imagining that ordinary people could fly beyond the atmosphere. Many of those who promoted the U.S. human space flight effort were earlier involved in aviation, in both industry and government research labs, achieving the milestones of atmospheric flight.[6] They knew that air transport, a familiar experience to millions of Americans by the second half of the twentieth century, had overcome the same sort of skepticism that affected the early space program. Like promoters of space flight after them, the pioneers of aviation had struggled to convince a skeptical public that mass transport above the surface of Earth was attainable. Newspapers and magazines had greeted the earliest attempts at powered flight with extreme disbelief. Turn-of-the-century stories about airplanes were lumped into the same category as reports of perpetual-motion machines and messages from Mars. Humans would fly, reported one magazine, just as soon as they could get the laws of gravitation repealed. The historic flight of Orville and Wilbur Wright near Kitty Hawk, North Carolina, on 17 December 1903 went virtually unreported in the mainstream press. Said Orville, "I think this was mainly due to the fact that human flight was generally looked upon as an impossibility, and that scarcely anyone believed in it until he actually saw it with his own eyes."[7]

To spread the gospel of flight, aviation pioneers took their evidence to the public at large. Wilbur awed Europeans with his flights over the French countryside in 1908, while brother Orville attracted attention in the War Department trials at Fort Myer, Virginia. The following year, one million people watched Wilbur pilot one of their airplanes up the Hudson River from Governors Island to Grant's Tomb and back.[8] In 1910 the brothers formed the Wright Exhibition Company, whose newly trained pilots toured the country performing at county fairs and specially organized shows. Glenn Curtiss, a rival of the Wrights, formed a similar company that same year.[9] After the World War I, barnstormers took up the cause. Flying surplus military aircraft, barnstormers often worked alone, dropping into farmer's fields in search of passengers and flying stunts wherever a crowd could be found. "By decade's end," wrote historian Joseph Corn, "flying missionaries had exposed nearly every hamlet and crossroads in the land to the airplane."[10]

The message they carried was relatively simple: airplanes would become as commonplace as trains or steamships and would transform the lives of ordinary people. A statement by Orville Wright was unequivocal: "I firmly believe in the future of the aeroplane for commerce, to carry mail, [and] to carry passengers." Alexander Graham Bell, another air enthusiast, predicted that the next generation would "see the day when men will pick up a thousand pounds of brick and fly off in the air with it."[11]

These predictions, so skeptically viewed, eventually came true. The rapid development of aviation became one of the most impressive technological achievements of the twentieth century.[12] An airplane manufacturing industry rose up in response to the call of the Joint Army and Navy Technical Board in 1917 for the production of twenty thousand military aircraft. Airline entrepreneurs prospered after federal officials in 1925 established contract mail routes, providing a secure financial base for the business of carrying freight and passengers through the air. In 1936, only thirty-three years after the Wright brothers' 1903 flight, American Airlines put the first DC-3s into regular service. This workhorse of air transport allowed investors to make a profit simply by hauling passengers from town to town. Three years later, Pan American Airways began transatlantic passenger flights with the Boeing Clipper. In 1954, with scarcely fifty years of aviation history behind them, workers rolled out the first prototype of the Boeing 707, inaugurating the era of jet transport. Fourteen years later Boeing introduced the 747 jumbo jet, each capable of carrying 490 passengers across the world.[13]

How many could have thought, standing at the dawn of the air age, that scarcely seventy-five years later powered aircraft owned by American firms would carry in a single year more passengers than the entire population of the United States? To commemorate the seventy-fifth anniversary of the Wright brothers' first flight, the Air Transport Association reported that in 1978 American carriers conveyed 275 million passengers through the air. Flying had become as commonplace as the pioneers of aviation had predicted. Indeed, anyone could fly.[14]

Not all of the prophecies prevailed. High expectations encouraged wilder claims, some of which could not be achieved. Among the more extreme was the belief that personal aircraft would become as common as the family automobile. Not only would everyone fly, but nearly everyone would own their own instrument for doing it. Executives would conduct business from the air and commuters would take to the sky, leaving soot and traffic behind. "An Airplane in Every Garage" became the rallying cry for efforts to fully democraticize the experience of flight.[15]

This belief gave rise to experiments both amusing and sad. In 1926 Henry Ford introduced the "flying flivver," a small single-seat aircraft that attracted considerable interest until pilot Harry Brooks killed himself during a demonstration flight over Miami Beach. During the 1930s engineers at the government's aeronautical laboratory at Langley Field, Virginia, developed a foolproof two-seat aircraft that would not stall or spin. Unfortunately, the engineers gave up versatility in exchange for safety, a feature that earned the enmity of experienced pilots. Built for land lovers accustomed to driving automobiles across the two-dimensional space of the road, the aircraft could not perform the most rudimentary maneuvers necessary to move precisely through the three dimensions of the air. During the 1930s the U.S. Department of Commerce offered incentives to anyone who could develop an affordable personal aircraft that the average person could fly. The competition encouraged the submission of a number of hybrid designs, half aircraft, half automobile. One prototype arrived like a car at the front door of the Commerce Department on Pennsylvania Avenue in Washington, D.C. The pilot then motored to the National Mall and after a few adjustments lifted off for nearby Bolling Field. The prototype performed as poorly in rush hour traffic as it did in the air.[16]

The contribution of these inventions to the aircraft industry was less significant than the expectations they generated. Promoters of flivvers and other simple aircraft sought to break down the barriers of cost and complexity separating aircraft technology from ordinary citizens. Visionaries such as Henry Ford had successfully integrated the citizen and the automobile, creating an unprecedented degree of personal freedom. Anyone with a minimal amount of training could drive a car. The prophets of aviation sought to expand this freedom to include the air, and if it could be done in the atmosphere, imagine what would be possible in space. The drive to make aircraft and eventually space travel accessible would alter the lives of common citizens in ways un-imaginable to their earthbound ancestors. The effort to let every middle-class American family own its own aircraft failed, but the expectations it generated lived on.

For human space flight advocates, realization of their dream required the development of spacecraft that could carry ordinary people. Unfortunately, early design choices stood in the way of this goal. In rushing to put the first Americans into space, the U.S. government shunned efforts under way during the 1950s to apply aircraft analogues to outer space. During that decade, the government began work on a number of hybrid aircraft capable of flight in both air and space. Flight engineers experimented with test aircraft that could rocket to the edge of space and studied airfoils called lifting bodies that could return from orbit like conventional airplanes. In 1958 the first X-15 experi-mental aircraft arrived at the Flight Research Center at Edwards Air Force Base in Southern California. Preparations ensued. With the assistance of a B-52 drop plane, the X-15 rocketed to an altitude of fifty-nine miles, nearly half the distance attained by Alan Shepard's 1961 suborbital flight. Unfortunately for advocates of aircraft technology, the X-15 performed this feat more than one year after Shepard flew. By then, John Glenn had already orbited the Earth. Anxious to push Americans farther into space, government officials rejected the aircraft model for space travel, instead adopting the capsule approach.[17]

If a group of people set out to design a spacecraft contrived to discourage public confidence in the ability of ordinary citizens to fly, that group could not have made a better choice than the space capsules designed for the early NASA flight program. The first astronauts traveled inside conical objects that re-sembled warheads used to deliver nuclear bombs, from which their shape was actually derived. An image further removed from conventional winged aircraft would be hard to imagine. Like nose cones, the space capsules sat atop modified intercontinental ballistic missiles (ICBMs). Such missiles were never designed to take people anywhere and exploded with statistical regularity. Concerned with meeting their own deadlines, air force officers resisted changes in missile design that might have satisfied NASA safety standards.[18] In response, NASA engineers designed escape rockets that could separate a space capsule from a malfunctioning ICBM. Once in orbit, astronauts faced the difficult task of plowing back into the atmosphere at speeds in excess of seventeen thousand miles per hour. To prevent a human barbecue, NASA engineers installed an ablation shield that enveloped the capsule in a raging fireball as hot as the surface of the sun.[19] Back in the atmosphere, the astronauts could not fly home. Parachutes slowed their descent until the capsules dropped into the ocean, adding the possibility of drowning to the dangers of death by fire. This was not a craft designed to inspire confidence among ordinary ticket holders.

Tom Wolfe called the experience the human cannonball approach to space flight. Astronauts were launched into space by the brute force of military missiles; they splashed into the ocean in a capsule they could hardly control. Lacking even the simple ability to guide their craft to a landing field, astronauts had to depend upon armadas of ships and aircraft to find them. In the space business, as Wolfe observed, recruitment was confined to those "whose profession consists of hanging their hides, quite willingly, out over the yawning red maw."[20]

In an attempt to put the Gemini astronauts on solid ground, NASA officials and their contractors tried to suspend the returning spacecraft from an inflatable paraglider. This would at least allow the bell-shaped capsule to skid back to Earth in what vaguely resembled an emergency landing. The paraglider looked and behaved like a paper airplane. Tests proved so unsuccessful that NASA officials abandoned the system in 1964 and told the Gemini astronauts to prepare for a watery return.[21]

The methodology of early space flight was not contrived to make space travel a pleasant experience. Winged spacecraft capable of controlled landings and new propulsion technologies to ease the discomforts of launch were not available on the time line dictated by the Cold War. Once again, the history of aviation suggested that such obstacles could be overcome. Aviation had surmounted similar dangers as it matured. Danger was a close companion on the early exhibition circuit. Of the four pilots who signed a two-year contract with the Wright Exhibition Company, only one lived to fulfill it. Orville Wright injured himself seriously at the Fort Myer army trials after his propeller cracked, forcing a crash landing. His passenger, Thomas Selfridge, broke his skull in the impact and died. Dogfights and stunts did little to inspire public confidence in flying. "These accidents are dreadful," said aviation pioneer Glenn Curtiss.

"Every time a flyer dies in a crash, we not only lose a precious life, but public confidence as well. How can we ever sell the idea that flight can be a safe means of travel?"[22]

Air flight improved as technology advanced. By the late 1920s airlines were carrying passengers in transport aircraft like the Ford Trimotor. The fully enclosed cabins were a significant improvement over open-cockpit, two-seat biplanes. Even so, flying conditions remained primitive by modern standards. Like other unpressurized aircraft, the Ford Trimotor flew close to the ground, where turbulence inevitably occurred. Airlines kept a variety of implements on board to minister the needs of airsick passengers. In an emergency, customers could ultimately hang their heads out movable windows. "Patrons immediately behind an airsick passenger learned to keep their windows closed," observed historian Roger Bilstein. "It was not unusual to hose out the entire interior of a plane after completing a turbulent flight."[23]

Bit by bit, flying conditions improved. The Boeing 307 entered service in 1940, introducing passengers to the blessings of pressurized aircraft that could overfly troublesome storms. Tricycle-type landing gear improved takeoff efficiency and allowed passengers to find their seats without walking uphill. Piston-engine technology increased flight ranges eightfold while cutting costs in half. Gas turbine (jet) engines further increased speed and range.[24] Pioneers of space flight wanted to apply similar technologies to spacecraft. They wanted to put wings and landing gear on vehicles bound for the new frontier. The pioneers wanted ordinary people to fly and be physically altered by the experience. As with aviation, this would mark an end to the "barnstorming" era of space flight and the advent of technical maturity. It was a formidable challenge.

As if the desires of practical space flight engineers were not enough, the promise of easy space flight was ground into the public consciousness by people producing works of fiction. Throughout the twentieth century, effortless space travel was no further away than the bookshelf or the television tube. By and large, producers of imaginative literature assumed that obstacles to space transportation would fall away in the face of modern technology. They certainly would be no more difficult to overcome than the challenges of aviation. Writers of fiction and popular science anticipated most of the important technological breakthroughs in space flight before those developments occurred. Famous spaceships traveled through the minds of ordinary people well before they ever soared in space.

To producers of imaginative literature, space transportation was frequently an afterthought to other, more important purposes. When H. G. Wells dispatched Martians to engage Earthlings in *War of the Worlds*, he gave little attention to the practical details of transporting them here. Flashes of hydrogen gas on the surface of Mars announce their departure; the first Martians arrive about ten days later.[25] Wells was more concerned with describing the clash between different cultures than explaining the mechanics of interplanetary transport. The ultimate dismissal of transportation obstacles occurred when Edgar Rice Burroughs allowed John Carter to be transported to Mars as the result of a trance. Teleportation neatly avoided the technological obstacles to space travel and allowed Burroughs to get on with his story.

The first widely read writer of imaginative literature to deal seriously with the practical challenges of space travel was Jules Verne. Verne's cannonball approach to space flight anticipated NASA's capsule system by nearly one hundred years. To dispatch his crew toward the Moon, he employed a large cannon capable of firing a nine-foot-wide aluminum shell. Verne understood that the force of the cannon blast would disturb the occupants of his bullet-shaped spacecraft and therefore installed a water bed in the spacecraft to absorb some of the shock. In fact, the crew's situation was much worse than Verne estimated. The force required to launch such a projectile toward the Moon would have subjected the occupants to g forces far in excess of those humans could endure. As Ron Miller, a chronicler of spacecraft history observed, the g forces produced by the cannon blast would have reduced the spacecraft occupants to thin smears on the cabin floor.[26]

Having committed himself to a realistic treatment of space flight, Verne found himself in a literary quandary. Lacking similar artillery on the lunar surface, he could not imagine any realistic way to return his space travelers from the Moon. He therefore concocted circumstances that forced his lunar explorers to approach the Moon without actually landing on it. Finally, Verne faced the difficult problem of a safe landing on Earth. He adopted a method whose credulity was bolstered only by the fact that NASA eventually adopted it: Verne's capsule splashed down in the Pacific Ocean to be recovered by the American navy.

Most writers who followed Verne treated space transportation as an afterthought to more pressing literary objectives. Where practical transportation barriers stood in their way, most writers simply dismissed them. This gave rise to inventions both practical and implausible. Writers of imaginative literature anticipated real developments such as the NASA space shuttle, but they also invented mysterious forces produced by devices such as antigravity machines and jumps into hyperspace.

The sum effect of these literary inventions helped convince an uncertain public that space transportation was a resolvable challenge that did not impose significant obstacles on the more exotic enterprise of interplanetary exploration. To persons growing up during the first half of the twentieth century, this was not an altogether unreasonable idea. Forty years after the Wright brothers' 1903 flight, pressurized four-engine aircraft shuttled passengers routinely between Europe and America. Was it so hard to imagine that similar changes might transpire by 2001, forty years after the first human flight into space? And what about the centuries beyond?

In the 1929 film *Frau im Mond* (*By Rocket to the Moon*), producer Fritz Lang presented plans for a multistage rocket capable of flying humans to the Moon and back. The Lang rocket ship resembled in many details the multistage Saturn V rocket that delivered American astronauts to the surface of the Moon forty years later. It is hard to tell, given the history of the movie, whether fiction influenced science or science influenced the film.[27] In preparing the script, Lang drew on the technical advice of German rocket pioneer Hermann Oberth. Oberth in turn influenced Wernher von Braun, who directed the work that produced America's Saturn V. Von Braun then used his charismatic expertise to promote the development of multistage rockets in popular outlets such as *Collier's* magazine, the movie *Conquest of Space* (1955), and the three-part Disney series released as the space age began.

The most impressive anticipation of near-term technology appeared in Arthur Clarke and Stanley Kubrick's *2001: A Space Odyssey*. The space shuttle in that classic science fiction film flew across the movie screen just four years before President Nixon gave NASA the directive to start building a real one. Like many of the structures to fly from the minds of imaginative writers, Clarke's winged spaceship was impressively large. It measured two hundred feet from wing tip to wing tip. The space shuttle NASA eventually built had a wingspan of seventy-eight feet, still imposing compared to the thirteen-foot-wide Apollo command module. To boost the imaginary shuttle into orbit, Clarke employed a two-stage reusable system similar to the one NASA engineers hoped to build. After carrying the wing-shaped vehicle to the edge of space, the lower stage glided back into the atmosphere and landed at the Kennedy Space Center. The lower stage would be as easy to maintain as a modern jet liner, Clarke predicted: "In a few hours, serviced and refueled, it would be ready again to lift another companion."[28] NASA engineers set similar goals for their proposed space transportation system.

Features of the Clarke space plane made the trip into space nearly as comfortable as a passage on a commercial jetliner. Passengers suffered little sense of discomfort as the craft's steady source of propulsion generated only two gravities of force at its maximum point of acceleration, which was a vast improvement over Jules Verne's cannonball ride. On a second spacecraft, coasting toward the Moon, Clarke described an onboard toilet that spun like a centrifuge to provide relief for passengers during periods of microgravity flight. The spinning lavatory generated the equivalent of one-fourth g, enough to "ensure that everything moved in the right direction."[29]

Clarke predicted that a reusable space shuttle would reduce the cost of space flight substantially, a goal that proved easier to attain in works of fiction than in real life. Each flight of the Earth-to-orbit shuttle in *2001* cost a little more than $1 million, Clarke wrote. That worked out to $50,000 for each of the twenty seats, not exactly economy fare. Compared to the old Saturn V rocket, however, it was a huge savings. In the uninflated dollars of its day, each launch-ready Saturn V rocket cost $185 million and never carried more than three people per flight.[30]

Although expert advice helped determine the shape of imaginary spacecraft, so did cultural trends. During the era of the underwater boat, many imaginary spacecraft resembled flying submarines. When automobiles developed fins, so did spaceships. Reflecting the fashion preferences of his day, Flash Gordon flew a spaceship designed in the art deco style. Buck Rogers piloted a similarly streamlined spacecraft, with sweeping windows and symmetrical fins.[31] That cultural trends created engineering problems hard to overcome is easily illustrated by the history of spacecraft design after World War II.

Contemplating rocket technology in the decade after the Second World War, filmmakers and story writers employed the familiar shape of the German rockets that had crashed down upon European cities during that great conflict. The terror of military rocketry had seared the image of the V-2 into the public consciousness, intensified even more by the thought that nuclear warheads would be launched on vessels of similar design. When George Pal produced the classic *Destination Moon* in 1950, he dispatched his crew to the lunar surface in a rocket that looked like the German V-2. The most famous rocket shape of its time, the V-2 helped establish an image of aerodynamic grace among a

From Buck Rogers's comic strip to 1950 motion picture *Destination Moon*, space enthusiasts portrayed space travel as a relatively simple, effortless affair. This Chesley Bonestell painting of a spindle-shaped rocket ship with sweeping fins graced the cover of Bonestell and Willy Ley's 1949 book *Conquest of Space*. The rocket's profile copied the German long-range A4b, a launch vehicle derived from the German V-2. (Reproduced courtesy of Bonestell LLC)

generation of would-be rocketeers. The spacecraft was sleek and tall, shaped like a spindle with large fins at its tail, and rose from the Earth in a single stage so as not to compromise its graceful style.

During the Second World War, Germans had begun work on an advanced version of the V-2 called the A4b. Designed to extend the range of the V-2, the A4b looked like a V-2 with the added advantage of two medium-sized wings. A single-stage rocket shaped like an A4b lifted Earthlings from their doomed planet in the motion picture *When Worlds Collide*. In that film, scientists construct a space ark to carry forty humans and a menagerie of livestock to a planet where they can begin life anew. The technological challenges of such a planet hop are severe. The rocket must develop sufficient thrust to escape the gravity of Earth, decelerate through the atmosphere of the new planet Zyra, and deposit its heavy load on alien soil. Engineers decide to boost the space-craft by propelling it along a mile-long slide using a rocket-driven undercar-riage. Although the project directors worry about fuel, the rocket ship flies successfully between the two orbs.

Both spacecraft designs, although aesthetically pleasing, were technically infeasible. In its only successful test, the German A4b reached an altitude of eighty kilometers. A fully fueled, single-stage V-2 could attain an altitude of about one hundred miles, if launched vertically, thousands of miles short of any extraterrestrial body, to say nothing of the propellant needed for landings or a return voyage.[32] Screenwriters for *Destination Moon* overcame these limi-tations in an ingenious way. They developed an atomic power plant for the single-stage rocket, drawing upon the most promising energy source of their day. NASA actually tested a nuclear rocket engine nineteen years later, begin-ning in 1969. The NERVA (Nuclear Engine for Rocket Vehicle Application) project used a very hot nuclear core through which liquid hydrogen was pumped. The rapidly expanding gas provided the necessary thrust to propel the accompanying spacecraft. NASA canceled the project in 1972, citing bud-getary cutbacks, but continued to study nuclear power as a source of interplan-etary and translunar propulsion.[33]

Nuclear power seemed to hold the most immediate promise for advancing propulsion technology. In 1969 NASA proposed the use of a nuclear-powered spacecraft to shuttle humans between a proposed Earth-orbiting space station and a proposed transit station circling the Moon. When White House officials sought new ideas on the best way to speed humans to the planet Mars, they called on a group of outside experts, who in 1990 again laid out the virtues of nuclear power.[34] Popular culture helped reinforce the notion that more effi-cient propulsion technologies lay just over the horizon if only the government would take the risk and invest in them.

The vision of effortless space travel was expeditiously advanced with the 1977 release of George Lucas's *Star Wars* (later retitled *Star Wars Episode IV: A New Hope*), one of the most popular science fiction films of all time. Of the numerous spacecraft depicted in this motion picture, none did more to rein-force the vision of accessible flight than the *Millennium Falcon*. Hoping to es-cape his frontier outpost on Tatooine and travel to the Alderaan system, Luke Skywalker is introduced to Han Solo in a rear booth of the Mos Eisley Cantina. With the assistance of the Wookie Chewbacca, Solo works as an independent freighter pilot and owner of this wondrous spaceship. The character of Han

Solo, played adroitly by Harrison Ford, combines the bravado of a mercenary pilot with the technical skill of a garage mechanic, the latter recalling the "hot rodders" of the 1950s who worked to improve the performance of their own machines, stock automobiles whose mechanical alteration could vastly improve the capabilities of engine and drive train. Such alterations permitted unlicensed drag racing and legendary flights from law enforcement authorities.

As portrayed in *Star Wars*, personal spacecraft technology is so simple that Solo can maintain and improve the *Millennium Falcon* himself. "She may not look like much," Solo admits, "but she's got it where it counts, kid. I've made a lot of special modifications myself." Given its patchwork appearance, Luke Skywalker doubts that the custom-modified spacecraft can even fly. "What a piece of junk," he says. Solo is indignant. "It's the ship that made the Kessel run in less than twelve par seconds!" he responds. "I've outrun Imperial starships and Corellian cruisers."[35]

The success of the *Star Wars* tales was due in large part to Lucas's ability to take familiar images from American popular culture, such as stock-car bravado, and place them in a galactic setting. Audiences instinctively respond to such imagery. The film helped solidify the notion that the space frontier would provide a tableau for the completion of old fantasies. In this sense, the *Millennium Falcon* represents the ultimate fantasy of aerospace pioneers—a personally accessible spacecraft that is relatively easy to operate and maintain.

In one major respect, the *Millennium Falcon*'s capabilities exceeded the hopes of contemporary rocket engineers. Solo's ship speeds through space by jumping to hyperdrive, a faster-than-light-speed form of travel. As Solo points out, "Traveling through hyperspace isn't like dusting crops."[36] To cross the great distances imposed by galactic travel, writers of imaginative literature have adopted devices that circumvent the laws of physics as understood by modern scientists. According to Einstein's physics, velocity and mass are interrelated. The faster one travels, the heavier a spaceship becomes. This effect is not noticeable at modest speeds but begins to intrude upon galactic travel as one approaches the speed of light. At such velocities, spacecraft mass grows rapidly. As it becomes heavier, the spaceship requires more energy to push it. An infinite amount of energy is eventually required to accelerate a spacecraft to the speed of light—about three hundred thousand kilometers per second. Simply put, there is not enough energy in the universe to propel a spacecraft that fast. Nature appears to have established a cosmological speed limit for spacecraft traveling between the stars.[37]

No fictional spacecraft did more to heighten expectations about easy space travel than the Starship *Enterprise*. First introduced in 1966 as part of the *Star Trek* television series, the various configurations of the USS *Enterprise* use antimatter engines to race through the galaxy at "warp" speeds. ("Star Trek" ©1966 by Paramount Pictures)

In works of imagination, however, this constraining limit disappears. Visionaries have developed a variety of techniques for approaching and even exceeding the speed of light. To many, the speed of light is no more insurmountable than prior obstructions to human flight like the sound barrier. E. E. Smith presented his classic *Skylark of Space* only one year after Charles Lindbergh's 1927 crossing of the Atlantic. Chuck Yeager would not break the sound barrier for another nineteen years, but the hero of the *Skylark* tale crosses 237 light-years in a mere twenty-four hours. *Skylark* began its serialization in a 1928 issue of *Amazing Stories* that also featured the first Buck Rogers tale.[38] Both served to launch the modern "space opera" with its vast interplanetary settings, formidable weaponry, and awesome spacecraft. Such spacecraft crossed portions of the galaxy as easily as airplanes spanned the oceans.

Images of interstellar travel in fictional literature helped generate scientific interest in advanced propulsion systems. A figure no less imposing than NASA's associate administrator of Manned Space Flight argued that humans would eventually develop the means to traverse the galaxy. "The future of Mankind," George Mueller argued in 1984 after his retirement from NASA, "lies in populating first, the Solar System, from there developing the technology to visit the stars and to begin populating the Universe as we now know it." Mueller called it the human destiny and maintained that it could be done. He insisted that history would vindicate his view. In the 1930s most people treated serious proposals for flights to the Moon as science fiction. Thirty years later humans had accomplished that feat. Now experts were contemplating the technical requirements for interstellar flight. "Once again," Mueller observed, "the idea will be widely regarded as science fiction. But I am convinced that the more we learn, the more the question becomes not *whether* we can go to the stars, but *when*."[39]

Mueller believed that interstellar travel would be expedited by propulsion technologies that could be developed soon. Such technologies crossed the boundary between science fact and science fiction. Some of the proposals were primitive, such as the notion that an interplanetary spacecraft could be propelled by exploding small atom bombs behind it. Following the landings on the Moon, members of the British Interplanetary Society proposed the use of nuclear fusion. The promoters of their Project Daedalus, named after the mythical Greek craftsman who built wings for himself and his son Icarus to escape imprisonment in the Labyrinth, envisioned a large reaction chamber into which small pellets of deuterium and helium-3 would be thrust. Beams of electrons would cause the pellets to fuse, producing the power that drove the spacecraft. Although only a small percentage of the pellet mass would fuse, the power produced would be sufficient to drive a large spacecraft to just over 12 percent of the speed of light. With such a propulsion system, a robotic spacecraft could conduct a flyby of Barnard's Star in forty-five years flight time.[40]

One of the most promising methods of propulsion technology involves antimatter drive. While this sounds as fictional as an antigravity device, the concept is physically real. Antimatter possesses physical properties the opposite of normal matter. The protons in the nucleus of an ordinary atom carry a positive electrical charge, whereas antiprotons carry a negative one. When introduced into the same chamber, protons and antiprotons annihilate each other. All the antimatter (and accompanying matter) is converted to energy, producing a spectacular burst of power.[41]

Substantial but not insurmountable obstacles exist to the development of actual antimatter engines. Small quantities of antimatter are routinely created in large research laboratories. Once produced, antimatter is notoriously difficult to store because it reacts violently with ordinary substances. Storage is not impossible, however. Research scientists have suspended antimatter in magnetic fields where it does not touch ordinary matter. As a rocket fuel, antimatter could take the form of small balls of frozen antihydrogen channeled through a magnetic nozzle. According to one calculation, a two-thousand pound payload could be propelled using an antimatter drive into an orbit in the Alpha Centauri system (our nearest stellar neighbor) in just twenty-five years.[42]

If this sounds like science fiction, it is. Although hundreds of scientific papers have been written on the technology of interstellar travel, most people learn about exotic propulsion systems through works of imagination. The most famous spaceship to fly in modern science fiction uses an antimatter drive. Publications directed at the infinite curiosity of *Star Trek* fans describe the operation of various *Enterprise* spacecraft in delicious detail. The technical manuals for these Galaxy-class spacecraft may be prophetic in some particulars. According to the writers, antimatter is generated with relative efficiency at major Starfleet fueling facilities using a combination of solar and fusion power. It is produced as antihydrogen in slush form. Magnetic fields continue to be the preferred method of storage. These details are indistinguishable from known scientific principles. The combustion chambers on the *Enterprise* spacecraft, however, are not. According to its creators, the combustion nozzles use a fictitious material known as "dilithium crystal," which antimatter passes safely through, eliminating the need for magnetic injectors.[43]

From these brushes with reality, *Star Trek* writers venture into a familiar fantasy. As in earlier galactic sagas, writers plot paths through the speed-of-light barrier that scientists cannot find. In the *Star Trek* series, the pathway takes the form of "warp" speed. Theoretical work underlying superluminal flight was completed by the middle of the twenty-first century, according to the writers. The first prototype capable of exceeding the speed of light for a brief instant was produced in 2061. The key to avoiding the limits of physics, the writers speculate, lay in the process of nesting layers of acceleration against each other with exponential effects. The first practical engines to use this method helped establish the Alpha Centauri colonies. The spacecraft crossed the span of 4.3 light-years in just four years. The fifth *Enterprise*, commanded

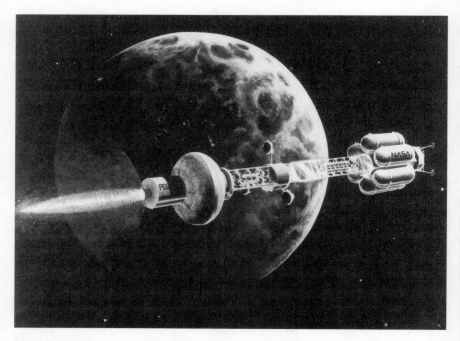

Rocket scientists contemplated fantastic forms of propulsion. Some studied electric propulsion vehicles that used magnetoplasmadynamic thrusters such as the one depicted here as a means for moving cargo to Mars. Others experimented with engines that pumped liquid hydrogen through very hot nuclear reactor cores, lending hope to enthusiasts who dreamed of speeding into the solar system and beyond. (NASA)

by Jean-Luc Picard and commissioned in 2363, can maintain a cruising speed of warp 6, or 392 times the speed of light. For brief periods of time it can accelerate to velocities four times that fast. As the Milky Way is 100,000 light-years across, a complete transit at cruising speed would require some 250 years. Limits on the velocity of the *Enterprise* arose more from story lines than technology, the writers explained. Making the ship go too fast "would make the galaxy too small a place for the *Star Trek* format."[44]

Who knows how much of this will come true? At the outer reaches of scientific investigation, inquiry merges with imagination. Faster-than-light spacecraft and antigravity machines violate the known laws of physics. Antimatter engines, ion drive accelerators, and hyperspace do not. The casual reader might be forgiven for confusing the real with the fictional.

Before their liftoff in an imaginary rocket to Mars, visitors to Disneyworld were informed that the speed at which the spacecraft travels was considered science fiction only a few decades earlier.[45] Imaginative voyages help create challenging expectations. As John Mauldin observes in his serious review of interstellar travel, "When a theme as common as travel to the stars is taken for granted in a large but fictional literature, the public tends to assume that the concept will become certainty, even that particular methods will be developed fairly soon."[46]

With the era of human space flight hardly a decade old, NASA officials took the first step toward the development of spacecraft in which ordinary people could fly. NASA officials proposed and won approval to build a fleet of four space shuttles that could travel to and from low Earth orbits "in an airline-type mode." The shuttles were envisioned as part of a larger Space Transportation System that would expedite travel deep into space. For use beyond the range of the space shuttle, NASA hoped to deploy a chemically fueled space tug and a nuclear-powered transit vehicle.[47] Only the shuttle was approved.

Expectations for the new spacecraft were high. "Toward the end of the Seventies you will no longer have to go through grueling years of astronaut training if you want to go into orbit," predicted Wernher von Braun. "A reusable space shuttle will take you up there in the comfort of an airliner." The shuttle was designed to make space flight accessible to the common folk. "I'm convinced that by 1990 people will be going on the shuttle routinely—as on an airplane," said NASA planner Robert Freitag.[48] Although the shuttle would take off like a rocket, it would land like an airliner. This was an important milestone in making space travel seem familiar. "When I was a kid reading *Buck Rogers*, the spacecraft all looked like bullets or saucers, with sweeping fins and fancy tail skids," said astronaut Michael Collins. "We are beginning to see Buck's dream emerge in the squat but elegant space shuttle."[49]

Significantly, NASA planners meant to give the shuttle wings. They sought to take NASA out of the cannonball business and put it back into the realm of conventional flight. For those who began their careers in the NACA, this was particularly important. As the story goes, Johnson Space Center director Bob Gilruth approached spacecraft designer Max Faget and told him to "get off this blunt-body, parachute stuff. It's time we thought of landing on wheels."[50] Wings would allow the spacecraft to land in the manner familiar to all pilots. Faget wanted to put real wings on the shuttle, the type that conventional airliners use. At lower altitudes, conventional wings would give the spacecraft substantial lift. Such a spacecraft would land practically like a glider, at the relatively low speed of 150 miles per hour. A shuttle with conventional wings could land at conventional airports, like New York's Kennedy or Washington's Dulles. As a safeguard against trouble, it could glide around and land slowly.

Airplane-type landings would help make the space shuttle accessible to ordinary people. American astronauts falling back from space in blunt-bodied capsules faced deceleration forces in excess of seven g's, to say nothing of the indignity of landing in a large body of water. Passengers on the shuttle would experience forces less than three g's and would depart the spacecraft on a concrete runway. John Young, who flew the first shuttle into space, announced, "If you're alive and breathing, you can fly on the shuttle."[51]

Most important, the shuttle would be reusable. The need to drop rocket ships into the ocean after their first and only use frustrated efforts to make space flight more accessible. "There's no way that you can make a railroad cost-effective," observed one NASA executive, "if you throw away the locomotive every time."[52] The space shuttle, as designed, could be used again and again, cutting the cost of space transportation significantly. Cost expectations for reusable transportation systems approached hyperbole. NASA administrator Thomas Paine predicted that "by 1984 a round trip, economy-class rocket-plane flight to a comfortably appointed orbiting space station can be brought down to a cost of several thousand dollars." The cost of a trip to the Moon would fall "to the $10,000 range."[53] Officially, NASA committed itself to reduce the cost of Earth-to-orbit transportation "by a factor of ten." At that time, people needing to deliver payloads to low Earth orbit on conventional rockets paid about $1,000 per pound. The nonreusable Titan IIIC, for example, delivered its twenty-three-thousand-pound payload to low Earth orbit at a 1972 cost of $24 million.[54] To do an equivalent amount of work at one-tenth the cost, NASA engineers sought to develop a shuttle with a fifty-five-thousand-pound payload that could be launched for no more than $5.5 million per flight in 1972 dollars. Officially, NASA committed itself to an estimate of $10 million per launch—still impressive compared to the expense of launching larger rockets such as the Saturn IB.[55]

Given its low cost, NASA expected that the space shuttle would become the preferred carrier for delivering scientific, commercial, and military payloads to space. Various publications called it a "cargo plane" and a "space truck," and an aerospace industry publication predicted that the shuttle "will eliminate the need for expendable launch vehicles."[56] With prompting from the NASA leadership, officials at the U.S. Department of Defense agreed to use the shuttle as their primary vehicle for placing spy satellites and other payloads in orbit. NASA officials tried to convince European nations to forgo development of the Ariane rocket, a competing technology, on the grounds that the shuttle would make nonreusable rockets obsolete.[57]

As demand for the shuttle grew, NASA officials predicted, the cost per flight would fall. Other government agencies, including the Department of Defense, commercial firms, and other nations would pay NASA to put payloads in space. The money NASA received from these customers would help defray the cost of developing and operating the shuttle. As the cost per flight fell, more customers would appear. The shuttle would pay for itself. To reach the point of economic efficiency, NASA had to conduct at least twenty-five launches per year. One study set the figure as high as fifty-two flights annually, or one launch every week.[58] On paper, the plan looked convincing.

Critics of the shuttle program failed to construct a persuasive rebuttal, and President Nixon approved development of the system in 1972.[59] The shuttle vision seemed reasonable, given common beliefs at the time. Popular science and works of imagination had encouraged the public to expect cheap, reusable spacecraft. The airline industry had made enormous advances by 1972, and optimists wanted to believe in a similar future for space. NASA's recent record of performance in sending humans to the Moon, a task deemed infeasible only a few decades earlier, encouraged confidence in the ability of the agency to make the program work. NASA officials played on those popular expectations by setting extraordinary goals for the new machine.

Expectations melted away in the face of reality. By the mid-1980s, a few NASA watchers had concluded that the shuttle would not meet its original goals. The explosion of the space shuttle *Challenger* in January 1986 communicated that message to the nation at large.[60] Flight procedures were anything but routine. The shuttle orbiter, although reusable, was much too complicated to permit a launch every week or two. The largest number of launches NASA achieved in any one year was nine in 1985, far short of the expected level of twenty-five to fifty-two. Flight costs remained exceptionally high. The $10 million per launch estimate escalated to $57 million and then to $225 as flight experience accumulated. In 1992 NASA set the average recurring cost of each flight at $412 million, roughly ten times the original inflation-adjusted goal of $10 million per launch. Simply put, the spacecraft conceived for the purpose of reducing the expense of space flight by a factor of ten cost as much to fly as the vehicles it had replaced.[61]

In one respect, the shuttle did meet expectations. It flew like a spacecraft and landed like an airplane, an impressive achievement. Its airplane-like configuration allowed ordinary people to fly in space. Unfortunately, the first civilian to accept this challenge, Christa McAuliffe, boarded the space shuttle *Challenger* for its final voyage. NASA officials abandoned the civilian-in-space program shortly thereafter, along with their commitment to fly military and commercial payloads on the space shuttle.

From the start, the effort to build an easy-to-fly spacecraft encountered obstacles. NASA rocketeers originally sought to build a fully reusable shuttle, mating the shuttle orbiter to a winged booster as large as a Boeing 747 jumbo jet. With a crew in each vehicle, the winged booster would lift the orbiter to an altitude sufficient to allow the latter to proceed into space. As the orbiter raced away, the crew in the booster would return to a landing strip near the launch site, with workers waiting to refurbish the booster for its next flight. The second crew in the orbiter would eventually return in the same manner, landing its craft like an ordinary airplane.[62]

The fully reusable design was very expensive. One internal NASA memo set initial development costs at $10 billion to $13 billion. The estimate did not fit under the spending ceiling that White House aides imposed on NASA during the early 1970s. In order to get the shuttle program approved, NASA executives felt obliged to propose a shuttle design with startup costs estimated at only $5.5 billion, which required rocket engineers to substitute two liquid-fueled boosters for the reusable first stage. The boosters would carry the orbiter to an altitude of about forty miles and fall away, after which parachutes would break their descent to Earth. NASA officials planned to reuse the boosters after retrieving them from the ocean. Flight engineers were understandably nervous about reusing rocket engines that had been dunked in salt water and were discouraged that they had lost their airplane-like first stage.[63]

Worse news followed. Under increasing pressure to cut shuttle costs further, NASA executives abandoned the use of boosters powered by liquid propellants and substituted solid fuel. Whatever their shortcomings, liquid-fueled boosters had one major advantage: they could be shut down if problems occurred. Solid-fueled boosters, on the other hand, would burn continuously for the first two minutes of flight. Once the solid rocket boosters were lit, experts observed, the shuttle was going to go somewhere for two minutes.[64]

Even the original design for the shuttle wings disappeared. Engineers at the Johnson Space Center preferred conventional wings, which would provide more lift during the crucial landing phase. Defense Department officials, however, favored delta-shaped wings. Landing a spacecraft with delta wings is tricky. The spacecraft comes in much faster, dives in order to maintain speed, then flares up seconds before landing. A delta-winged orbiter needs longer runways and cannot land at conventional airports; its higher landing speeds put extra pressure on failure prone brakes. Critics complained that a delta-winged shuttle would possess the gliding characteristics of a pair of pliers. Nonetheless, the Defense Department insisted on delta wings, and NASA needed their support in order to get the project approved.[65]

During the 1960s, NASA flight engineers understood that going to the Moon would be difficult to do. With sufficient funds, they managed the risks and completed the mission on time. The space shuttle created a different challenge. Funds were short and space advocates announced that a reusable spacecraft would be easy to fly. Making such an announcement was relatively easy to do, given the image of spaceflight in the popular mind. Executing it proved much harder. In reality, cheap and easy were on a collision course with each other. Development of safe, reliable flight technologies for orbital flight requires substantial subsidies, experimentation, and the ability to replace mistakes with new designs. Those elements cost money, which NASA officials did not receive.

Rather than announce that they needed larger development outlays, space advocates perpetuated the myth of cheap and easy space flight by announcing that the era had arrived. Dissent was largely ignored. Optimism prevailed. The public, as a result, was unprepared for a catastrophe. "When the shuttle prangs," one realist warned, "it will be the media event of the decade."[66]

In producing a low-cost spacecraft, NASA officials took a risk. The size of the risk was not difficult to calculate. In the wake of the *Challenger* accident, NASA officials announced that the space shuttle contained approximately 750 objects that, if any failed, could endanger the life of the crew. Experts set the overall probability of a catastrophic failure at between 1 and 2 percent, a prediction confirmed by actual events.[67] After the loss of the *Columbia*, NASA leaders reconfirmed the risk. The chances of a shuttle accident during any one flight ranged from one in eighty to perhaps one in three hundred. Flying the shuttle is risky, not something an average person would want to do every day. It is clearly not as safe as flying a commercial airliner. "Anyone who has lived with large rocket engines," observed astronaut Michael Collins, "understands that . . . a thin and fragile barrier separates combustion from explosion."[68]

Practical experience showed how hard the development of advanced spacecraft could be. NASA executives wanted to develop a reusable Earth-to-orbit shuttle that could be flown frequently for one-tenth the cost of conventional launch vehicles. Although the space shuttle was a technological marvel, being the first rocket ship to reenter the atmosphere and land like an airplane, it proved neither cheap nor easy to fly. (NASA)

Throughout the early 1980s, NASA officials and their allies promoted the shuttle flight experience as something increasingly routine. Expressions of confidence increased as one rose up the government hierarchy. After hearing an estimate that set the probability of a catastrophic failure at only one in one hundred thousand, physicist Richard Feynman asked incredulously, "What is the cause of management's fantastic faith in the machinery?" Feynman was a member of the commission that investigated the *Challenger* accident. As he pointed out, probabilities on that order would allow NASA to launch one shuttle each day for three hundred years expecting to lose only one.[69]

Confidence in its reliability returned as the space shuttle resumed work in the post-*Challenger* years. Encouraged by a succession of successful flights, NASA officials made new attempts to economize. They reduced the shuttle work force by more than one-third. They resumed flying the shuttle seven and eight times per year. A special committee headed by flight control veteran Christopher C. Kraft characterized the space shuttle as "a mature and reliable vehicle . . . about as safe as today's technology will provide."[70] On the basis of assurances such as these, NASA executives injected private-sector cost saving techniques into the launch procedures and transferred much of the operational responsibility for shuttle flights to a private aerospace contractor. These actions stabilized the cost of flying the shuttle, and NASA officials considered plans to fly the vehicle well into the twenty-first century.[71]

Loss of the space shuttle *Columbia* in early 2003 restored a sense of reality. Members of the Columbia Accident Investigation Board (CAIB) criticized NASA's confidence in the vehicle and the organizational culture it had spawned. NASA administrator Michael Griffin characterized NASA's thirty-two-year-old commitment to the space shuttle as "not the right path."[72] Agency leaders set a firm date for retiring the remaining shuttles and, when flights resumed, reduced the number of shuttle missions to half of what they had been before. They allocated a total level of funding that, when divided by the reduced number of flights, approached $800 million per launch. According to the CAIB investigators,

The increased complexity of a Shuttle designed to be all things to all people created inherently greater risks than if more realistic technical goals had been set at the start. Designing a reusable spacecraft that is also cost-effective is a daunting engineering challenge; doing so on a tightly constrained budget is even more difficult. . . . In the end, the greatest compromise NASA made was not so much with any particular element of the technical design, but rather with the premise of the vehicle itself. NASA promised it could develop a Shuttle that would be launched almost on demand and would fly many missions each year. Throughout the history of the program, a gap has persisted between the rhetoric NASA has used to market the Space Shuttle and operational reality.[73]

In spite of these setbacks, the rhetoric of cheap and easy space flight did not disappear. The inability of NASA employees and their contractors to meet the original cost and schedule goals set for the U.S. space shuttle did not deter people charged with maintaining the spacefaring dream from seeking new and more advanced designs. Buoyed by government funding, workers at the Mc-Donnell Douglas aerospace company tested the DC-X (also known as the Delta Clipper) during the 1990s. A cone-shaped rocket ship, the DC-X resurrected the science fiction fantasy of vehicles such as the one in *Destination Moon* that could rise gracefully in a single stage and land with their noses pointed toward the stars. The single-stage-to-orbit design eliminated the need for cumbersome booster rockets, lower stages, or external fuel tanks and marked an advance in technology promising easier access to space.

A few years later, NASA joined with the Lockheed Martin aerospace company in an effort to produce the X-33, a reduced-scale test version of a larger vehicle to be called VentureStar that would propel astronauts and cargo to orbital velocities. Like the Delta Clipper, VentureStar was envisioned by its designers as a single-stage-to-orbit vehicle that would depart and return like a lone airliner. Project engineers built a powerful linear aerospike engine and attempted to fabricate lightweight internal fuel tanks in an effort to produce a workable configuration. Designers added two small wings in order to permit aircraft-type landings at the end of each voyage.[74]

In one of their more creative efforts, NASA planners contemplated the advantages of a MagLifter single-stage-to-orbit vehicle. The concept tackled one of the most persistent barriers to cheap and easy spaceflight—the practice of placing all of the propellants needed to accelerate a rocket ship in fuel tanks or booster rockets attached to the main craft. MagLifter drew on the principles presented in the classic science fiction film *When Worlds Collide*, in which a spacecraft accelerates along an extensive metal slide. In the film, acceleration is provided by rocketry. NASA engineers suggested magnets. Once ready for flight, the MagLifter would employ repelling magnets to levitate above and accelerate along a long metal track. The spaceship would attain supersonic speeds before release at a point where the track curved upward and two large engines finished the work of dispatching the spacecraft and its crew into orbital space.[75]

In an even more advanced version of the concept, engineers outside of NASA proposed a launch track called StarTram. Using superconducting magnetic levitation, a bullet-shaped spacecraft would accelerate through a long tunnel, pumped free of much of its air to reduce drag. Through the use of advanced materials, designers suggested that the tunnel and its accompanying track could be constructed from the ground to an altitude of seventy thousand feet. From that height, the spacecraft would emerge with sufficient velocity to reach outer space. "When StarTram is built," its advocates professed, "passengers will take a trip into space with the same confidence that they now have when they fly anywhere on Earth."[76]

All four of these designs, like many more, worked better on paper than on the assembly plant floor. Money for the Delta Clipper prototype disappeared after a 1996 flight test fire; NASA executives canceled government support for the X-33 in 2001 following difficulties with the lightweight, internal composite fuel tank. Artistic conceptions of MagLifter and StarTram appeared on various Web sites but not in the budget for a real prototype.

Determined to overcome setbacks such as these, aviation pioneer Burt Rutan with financial help from Microsoft founder Paul Allen developed *SpaceShipOne*. The two accepted no government funding and spent only $25 million, a modest sum by comparison to the $1.3 billion exhausted on the unsuccessful X-33, which like *SpaceShipOne* was designed for suborbital flight. *SpaceShipOne* utilized a unique movable wing that the pilot raised to stabilize the craft for reentry as it reached the top of its arc and turned to come home. On 4 October 2004 the Rutan group won the Ansari X Prize, awarded to the first privately funded group to fly a reusable rocket ship into space twice in fourteen days. To celebrate the achievement, pilot Mike Melvill climbed onto the fuselage of the vehicle and held up a sign that read "SpaceShipOne, Government Zero."[77]

None of these concepts motivated NASA's response to the anticipated retirement of the space shuttle in 2010. Instead of moving toward a more advanced configuration, NASA engineers fell back on an orthodox design. The tall, slim Ares I rocket and cone-shaped *Orion* crew compartment on top of it looked like the missile-and-capsule configurations presented during the first decade of flight. Although the underlying technology improved significantly on the missile-based rocketry of the 1960s, the overall configuration looked nothing

Difficulties with the space shuttle did not divert human space flight enthusiasts from their ultimate vision. Rocketeers experimented with a number of
advanced prototypes, such as the X-33, pictured here. Technical obstacles
blocked completion of the X-33 program and hindered other efforts to make
space travel as accessible as atmospheric flight. (NASA)

like the futuristic designs of which visionaries dreamed. Nor did the configuration promise cheaper flight. According to a special presidential commission,
the Ares rocket would cost more than the launch systems it was designed to
replace.[78] It would not make space flight more accessible to the common folk,
and it would not land on wheels, relying instead upon the undignified experience of a parachute-assisted reentry.

Many factors contributed to the presidential decision that followed, including the usual collection of cost overruns and technical obstacles. Yet one could
hardly avoid the impression that the government personnel developing the
Ares I/*Orion* configuration spent a great deal of money to produce a launch
system that looked nothing like the space plane images just proffered. In one
unfortunate but revealing comment, a NASA administrator characterized the
new transport vehicle as "Apollo on steroids."[79] In 2010 President Barack Obama
responded. He proposed that the government cancel the Ares I project, trans-

fer to the private sector the task of building inexpensive earth-to-orbit launch-
ers, and let NASA concentrate on the development of large, exotic rocket ships
that could reach the planets and asteroids. The arrangement, he promised,
would "accelerate the pace of innovations" in space transportation.[80]

NASA provided seed money; entrepreneurs responded with conventional
and creative designs. The Boeing Company proposed a seven-person crew capsule similar to the *Orion* configuration on which NASA engineers had been
working. Blue Origin offered its *New Shepard* spacecraft, a bullet-shaped vehicle
capable of suborbital flight that drew inspiration from the defunct Delta Clipper program. The SpaceX company recommended a blunt-nosed ballistic capsule, named *Dragon*, that might carry seven astronauts to and from the International Space Station. The Sierra Nevada Corporation advanced one of the
more imaginative configurations. Engineers at that company recommended a
small shuttle-shaped spacecraft, called *Dream Chaser*, one-fourth the length of
the NASA shuttle orbiter with room for seven passengers, that could be launched
on an Atlas V rocket but land as an airplane does.[81]

The vision of cheap and easy space flight simply will not die. Aviation enthusiasts, aerospace entrepreneurs, and people who want to believe in the
cosmic dispersion of humanity promote it. The cost and reliability standards
achieved by the airline industry suggest it. Works of imagination continuously
expose the public to it. If rocket scientists someday attain the dream, the slow
and grinding progress through working rocketry may seem like a minor diver-
sion in the realization of a much larger goal. Thousands of people cheered the
SpaceShipOne team on the day its developers flew into space and claimed the
X Prize.[82] Promoters of the vision believe that thousands more will want to fly
into space once the experience becomes cheap, easy, and relatively risk free. No
one yet knows whether this will happen. If it does, a prophecy will come true.

9

Robots

Never send a human to do a machine's job.

—Agent Smith, 1999

The effort to build spaceships and explore the universe left unresolved one important question. Exactly what mix of humans and machines would do this work? At the beginning of the space age, the answer seemed clear. Humans would be in charge. Technology and culture required it.

When Arthur C. Clarke presented his famous article explaining how humans could establish a global communications network in space, he made an important but often overlooked assumption. A human crew would be needed to operate the receiving and transmitting stations, he suggested. "The station would be provided with living quarters, laboratories and everything needed for the comfort of the crew." When Wernher von Braun suggested that humans construct a large orbiting telescope, he correctly surmised that the facility would be automated. The movements of a human operator, he observed, would distort the delicate task of pointing the instrument toward its objects of study. Yet humans would need to be stationed nearby, von Braun wrote, and an accompanying illustration showed what the humans would do. The illustration portrayed the precursor of what would become the Hubble Space Telescope orbiting alongside a large space station. Von Braun explained why the station needed to orbit nearby. Astronauts from the space station, he said, would need to change the film in the space telescope.[1]

When the United States began to deploy orbiting reconnaissance platforms, it included plans for stationing soldiers on them. The U.S. Air Force recruited an astronaut corps of fourteen individuals to do this work. Air Force officers planned to equip the platforms, called manned orbital laboratories, with large optical instruments that the astronaut-soldiers could use to spy on enemies below. A few humans with powerful binoculars, it was thought, could do a better job of orbital reconnaissance than machines alone. The Soviet Union launched two such platforms with cosmonauts on board, called *Almaz*.[2] The cosmonauts were provided with cannons to repel prospective attacks.

Other proposals for humans in space played on public misconceptions. Radio waves bounce off the ionosphere, a phenomenon well known to radio operators. Hollywood producers of the classic science fiction film *Rocketship X-M* used this common understanding to suggest that radio waves from Earth would not penetrate the cosmos. Hence, a rocket ship bound for Mars, like a seventeenth-century sailing ship, would depend entirely upon its human crew because it could not communicate with the Earth. The movie used this misconception to promote the necessity for human exploration. Experts raised similar concerns with respect to radio waves to and from the Moon.[3] In fact, the phenomenon affects only low-frequency radio waves.

All of these early conceptions were misconstrued. Humans are not needed to operate communication satellites or change the film in space telescopes. Reconnaissance satellites work perfectly well without humans on board, a realization that prompted government leaders in both the United States and Soviet Union to abandon plans for astronaut-driven orbiting military platforms. Radio signals in appropriate wavelengths travel easily between Earth and space, allowing operators at ground stations to control spacecraft throughout the solar system.[4]

The ability of machines to work in space without human operators physically present has been expedited by a number of interrelated technologies. The development of the charge-coupled device (CCD) allows machines in space to capture and transmit images at resolution levels exceeding that of the human eye and in wavelengths both visible and invisible to human beings.

The development of Cassegrain antennas and cryogenically cooled amplifiers, technologies supporting NASA's Deep Space Network, allows humans to receive faint signals from distant spacecraft. Lightweight computers and digital solid-state recorders permit machines to store vast amounts of information for its eventual transmission back to Earth. Solar panels and radioisotope thermo-electric generators (RTGs) produce electric power. Integrated circuitry ensures that the power needs of the machines remain small. Together these technologies support the science of remote sensing, the ability to collect large amounts of information using devices such as satellites that are in immediate physical contact with neither the object of their studies nor the people operating them.

Developments such as these have led a number of people to question the overall need for humans in space. Opposition to human space travel is centered in the scientific community, although scientists are not unified in this respect. Many biologists favor human travel; physicists and astronomers are its most persistent critics. "The histories of the space programs of the United States, the Soviet Union and all other countries provide overwhelming evidence that space science is best served by unmanned, automated, commandable spacecraft—the obvious and only important exception being the study of human physiology and psychology under free-fall or low-g conditions." The statement was made by James Van Allen, a physicist, designer of the scientific instruments on America's first artificial satellite, and one of the strongest proponents of machine flight. Alex Roland, a space historian, agrees: "For virtually any specific mission that can be identified in space, an unmanned spacecraft can be built to conduct it more cheaply and reliably."[5]

A journalist who specializes in space policy summarized the opposition in the following way: "NASA's fixation on man in space has actually become the curse of the space program." The statement represents the view commonly held among supporters of robotic flight that money spent on human flight is wasted because it steals money from space science. Humans were "a costly nuisance in space, admittedly unexcelled in fixing orbiting toilets, but orbital plumbing wouldn't be there to need fixing if (the humans) weren't on board." The commitment to human flight, Van Allen complained, "has been elevated in some quarters to the quasi-religious belief that space is a natural habitat of human beings." Was there anything that a human could do in a spacecraft that a machine could not? "Yes," Van Allen mused, "but why would anyone wish to do it at such a high altitude?"[6]

In the final years before his death, Van Allen sought to calm the shrillness of the controversy with a measured statement. He answered his own poignant question, "Is human space flight obsolete?" with a series of further queries: "My position is that it is high time for a calm debate on more fundamental questions. Does human space flight continue to serve a compelling cultural purpose and/or our national interest? Or does human space flight simply have a life of its own, without a realistic objective that is remotely commensurate with its costs? Or, indeed, is human space flight now obsolete? . . . Does anyone have a good rationale for sending humans into space?"[7] The visions of previous decades, he concluded, "look more like delusions in today's reality." Exuberant hopes for the commercial and scientific benefits of an international space station "have been all but abandoned," he professed. The facility that became the International Space Station was approved in 1984, among other reasons, for the purpose of developing new metal alloys or pharmaceutical products that could be manufactured only in the microgravity conditions of space. Twenty-five years later, the United States and its international partners were still preoccupied with the challenges of station assembly and maintenance. Five years after approval of the space station, advocates of human flight convinced President George H. W. Bush to propose an expedition to Mars. Burdened with the staggering cost, the first proposal "disappeared with scarcely a trace." Only a tiny fraction of the Earth's inhabitants, Van Allen observed, directly experience space flight. For the rest, human space travel was a vicarious adventure, "akin to that of watching a science fiction movie."[8]

Van Allen suggested that humans stay home, abandon their vision of a human diaspora in space, and let machines do the hard work of extraterrestrial exploration. Although machines appear frequently in the popular presentation of space travel, it is generally not in terms that support the arguments of Van Allen and other supporters of machine flight. In fact, the most popular image of machines in space—as robot companions to human beings—undercuts the argument that humans should stay home and allow machines to do the work. That image supports the official position of the U.S. government: when exploration occurs, it will be done with humans and machines working together.

In one of his earliest robot stories, Isaac Asimov introduced Robot QT-1, known by his human creators as "Cutie." This was Asimov's second robot story—and the first set in space. It appeared in April 1941, when Asimov was perhaps twenty-one years old. (He was never certain of his exact age.) Cutie operates a solar-power station, directing energy beams to a home planet. Regulations require that the robot be supervised by two humans, although the latter are far less competent in the delicate work of directing energy beams than Cutie. Given his superior capabilities, the robot begins to doubt that humans could have assembled him and concludes that he must have been created by a superior being.

The story presents an important lesson in the traditional human-robot relationship. The robot remains subservient to its human overseers even though it feels superior to them. "What's the difference what he believes," asks one of the humans in the QT-1 story. The only fact that matters is that "he can run the station perfectly." The story led Asimov to explain why a machine with superior capabilities—even one that doubted its creators—would remain subservient. The following year Asimov elucidated his three laws of robotics.[9] These famous statements create a system in which robots are programmed to eternally serve humankind.

> One, a robot may not injure a human being, or, through inaction, allow a human being to come to harm. . . .
>
> Two, . . . a robot must obey the orders given it by human beings except where such orders would conflict with the First Law. . . .
>
> Three, a robot must protect its own existence as long as such protection does not conflict with the First or Second Laws.[10]

Asimov's robots are programmed to instinctively follow these laws. They cannot help doing so, any more than a human can mentally overpower his or her genes. Any failure to follow the laws, Asimov believed, represented an error in programming, easily remedied by refining the rules that guided their behavior. "I saw [robots] as machines—advanced machines—but machines. They might be dangerous but surely safety factors would be built in. The safety factors might be faulty or inadequate or might fail under unexpected types of stresses, but such failures could always yield experience that could be used to improve the models."[11]

Asimov's cooperative image dominated a large number of robot stories that followed. Robbie the Robot appeared in the 1956 movie *Forbidden Planet*. Created to assist the survivors of an expedition to the Altair star system, the seven-foot-tall robot responds to voice commands and generally acts in benevolent ways. Its creator, Dr. Edward Morbius, is a far greater threat to the astronauts sent to rescue him than his robot servant.[12] The producers of *Lost in Space*, a television series that ran from 1965 to 1968, featured a large nameless robot that warns an astronaut family of various dangers and plays with the children. In a modern retelling of the classic *Swiss Family Robinson* tale, the robot assists the Robinson family when its space expedition is blown off course after a base doctor, the nefarious Dr. Zachary Smith, temporarily reprograms the robot to sabotage the mission.[13]

Three adorable service droids named Huey, Dewey, and Louie serve as assistants to the botanist in charge of an orbiting greenhouse in the science fiction classic *Silent Running*. The botanist rescues the greenhouse, set up to preserve forest specimens from a desiccated Earth, while the droids help to keep the botanist and his ultimate vision alive.[14] The image of robots as cooperative companions inspired George Lucas to create C-3PO and R2D2 for the original *Star Wars* movie. The high-strung C-3PO is a protocol droid programmed to provide translations and advice on etiquette to diplomats and other high-ranking officials. R2D2 is an astromech droid, equipped by its manufacturer with a variety of tool-shaped appendages for use in the maintenance and repair of spaceships. The pair escapes from a captured spaceship to the planet Tatooine, where Jawa traders sell them to a farmer, whose nephew, Luke Skywalker, is destined to become the central character in the tale.[15] Lucas and his fellow writers invented no less than four dozen types of droids, who appear in the films and perform a variety of activities including surgery and battlefield work.

Robots such as these are as friendly and devoted as the family dog. When Asimov wrote his first robot tale, the vision of robots as cooperative companions fit easily into the prevailing model of industrial-age thinking. Frederick Taylor's *Principles of Scientific Management* was less than thirty years old, while Elton Mayo's classic study on *The Human Problems of an Industrial Civilization* had appeared only six years earlier. Norbert Weiner's groundbreaking work on control and communication in the animal and machine would follow in less than ten years.[16] Taylor and his disciplines viewed workers and machines as essentially interchangeable, their performance a matter of correct engineering and discovery through scientific experimentation of the "one best way" of performing every job. Mayo and other leaders of the human relations movement recognized that fatigue and anomie, the dislocation and isolation sensed by factory workers, interfered with their ability to follow prescribed procedures. Weiner developed the concept of cybernetics—the use of feedback loops to control human and machine behavior. The concept helped lay the foundation for much of the automation that followed.

All three thinkers struggled with problems of control, the same challenge that Asimov addressed with his three laws. How does one get any entity performing complex tasks to follow directions? For Karel Capek, the Czech playwright who invented the word robot, the solution seemed obvious. Invent a human substitute with the obedience of a machine and the intelligence of a person. Capek presented the concept in his 1921 play *R.U.R. (Rossum's Universal Robots).*[17] The play depicts human substitutes forced to serve as inexpensive factory workers. Capek called them *robota*, a term derived from the medieval practice of requiring peasants to work for a few days in the fields of nobles in whose districts the peasants lived. Capek relied upon biological engineering to create his robots; Asimov dropped this in favor of mechanics. To Asimov, robots were machines with brains. He had no concept of the "as yet uninvented computer" when he introduced the concept of the "positronic brain" that permitted his robots to think. Asimov admitted that the concept "was just gobbledygook but it represented some unknown power source that was useful, versatile, speedy, and compact."[18] Nonetheless, it foretold the central mechanism for installing intelligence in an artificial being.

In general, the term robot refers to a machine under human control that can perform tasks similar to those of which human beings are capable. To fully fit the definition, a robot needs some sort of central control mechanism allowing it to react to its surroundings (a computer attached to various sensors will do) and the ability to visualize objects, manipulate through touch, and move around. Treads or wheels work as well as legs for the latter. A robot that possesses human features (eyes, head, arms, and legs) is called an android, and a mechanism that combines organic and synthetic parts is called a cyborg. An avatar is a graphic representation of a human being, usually in electronic form. The term is taken from Hindu scripture where it means the manifestation of a deity come to visit humans on the Earth.

Only a fraction of the remotely controlled machines dispatched into space during the first fifty years of space travel met the general definition of robot. The NASA rovers *Sojourner, Spirit,* and *Opportunity* that began their work on Mars between 1997 and 2004 qualify. So did the lunar rovers *Lunokhod 1* and *Lunokhod 2* that the Soviet Union landed on the moon between 1970 and 1973. Robotic space flight technology is still developing, albeit rapidly. Entities performing "unmanned" space activities generally possess some but not all of the characteristics of robots. The Surveyor lander that preceded Americans to the Moon carried imaging devices (eyes) and motorized arms, but once on the lunar surface it could not move. Many satellites carry instruments for capturing images and possess jets that allow them to change position but lack arms for grasping. Such devices are often referred to as "proto-robots," meaning that they possess some but not all of the characteristics of the pure type.[19] The term robotic may be used in a general fashion to characterize any mission without astronauts on board, from rovers to space telescopes. Though broad, it avoids the obvious gender bias of terming such missions "unmanned."

NASA's official position regarding the relationship between humans and robots rests upon the mechanical view contained in the Asimov-type literature on the subject—helpful companions working in cooperation with humans under constant human supervision and control. An elaborate statement of this philosophy appeared in the guiding principles established for NASA's 2004 Vision for Space Exploration.

> NASA will send human and robotic explorers as partners, leveraging the capabilities of each where most useful. Robotic explorers will visit new worlds first, to obtain scientific data, assess risks to our astronauts, demonstrate breakthrough technologies, identify space resources, and send tantalizing imagery back to Earth. Human explorers will follow to conduct in-depth research, direct and upgrade advanced robotic explorers, prepare space resources, and demonstrate new exploration capabilities.[20]

Throughout NASA's history, people promoting human space travel have treated robots as partners. The list of "mission target dates" proposed as part of the agency's first internal long-range plan in 1959 reflected this philosophy. According to this plan, robotic spacecraft would circumnavigate and land on the Moon before humans did the same. An ad hoc panel reporting to President Dwight Eisenhower in 1960 insisted that "a great part of the unmanned program for the scientific exploration of space is a necessary prerequisite to manned flight."[21] Artist Pat Rawlings captured the spirit of this philosophy with a much-reproduced painting of an astronaut in a space suit on the surface of Mars kneeling to dust off and retrieve the frozen, windblown *Sojourner* rover that preceded humans to the planet in 1997.

NASA officials implemented this philosophy with the Surveyor program of lunar expeditions, which took place between 1966 and 1968. Approved shortly before the presidential decision to land Americans on the Moon, the Surveyor project placed five robotic spacecraft safely on the lunar surface. (Two more landers arrived but crashed.) Each Surveyor craft was a three-legged robotic

While scientists questioned the need for humans in space, NASA officials insisted that humans and robots would explore space together. In this Pat Rawlings painting, one of the first astronauts to reach Mars retrieves the *Sojourner* rover that explored the planet in 1997. (Artwork by Pat Rawlings; courtesy of NASA)

lander containing a variety of instruments designed to test the suitability of the lunar surface for human occupancy. Some scientists feared that a thick mantle of lunar dust covered the Moon's surface and that any approaching spacecraft or astronaut would disappear into it. The Surveyor missions proved that the lunar surface, while dusty, was sufficiently solid to accommodate human occupancy. The first humans from Earth arrived nineteen months after the last Surveyor spacecraft touched down. The second human crew landed less than two hundred meters from *Surveyor 3*. Astronauts Charles Conrad and Alan Bean visited the spacecraft and retrieved parts to be brought back to Earth for analysis.

Oran Nicks, an aeronautical engineer who directed NASA's first robotic missions to the Moon, Mars, and Venus, articulated the agency's early views on this partnership when he attacked the distinction between robotic and human missions. The distinction, Nicks insisted, was a false one. "The truth is that there were no such things as unmanned missions; it was merely a question of

Human space flight advocates viewed robots as precursors to exploration conducted by people, a relationship that left humans clearly in control. Before astronauts reached the Moon, robotic spacecraft tested local conditions such as the density of the landing surface. Apollo 12 astronauts landed within walking distance of the *Surveyor 3* spacecraft that had preceded them to the Moon. (NASA)

where man stood to conduct them." Sometimes humans sent machines into space while controlling them remotely; sometimes part of the control was transferred to humans on board, who were "equipped with suitable life support systems."[22] In both cases, humans at flight stations on Earth directed various phases of spacecraft operations from the ground. As in Asimov's three laws, personal control remained a centerpiece of human-machine relationships in space.

The extent of human control over both piloted and robotic spacecraft neatly fit into the more optimistic presentations of technology offered in the mid-twentieth century. Campaigns to promote public acceptance of new technologies consistently emphasized human mastery. Most new technologies, such as the jet engine, were difficult to understand. Others, such as early computer software, proved exasperating to operate. Some were powerfully frightening, such as the technology of atom bombs and the generating plants that used nuclear power to produce electricity. Many commentators worried that technology was out of control. To sell Americans on the notion of progress through technology, advocates extolled the concept of human mastery.

Marketing experts advertised human mastery in a number of ways. One of the most effective was the use of the helmsman, a figure who used technology to master a challenging environment and did so in association with a product the advertisers wished to sell. The figure of the helmsman was used to promote a number of postwar products, none more successfully than Marlboro cigarettes. Advertisers presented the Marlboro Man as a master of technology—a pilot, race car driver, captain of boats, and, most effectively, a cowboy, an autonomous individual prospering in a harsh environment. By smoking Marlboro cigarettes, the ads suggested, consumers could become masters of their environments, too.[23]

Advertisers and manufacturers also utilized technological jargon to convey a sense of mastery. Technojargon filled advertising campaigns in the 1950s, most notably in the automobile industry, where advances in transportation technology accelerated past consumer understanding. Dealers discussed vehicle features such as "torsion-Aire" suspensions and "HC-HE engines" in an effort to convince baffled customers that they actually understood how those features worked.[24]

In the process of selling the newly created space program to American voters, advocates of space flight invariably made use of these techniques. The astronaut provided the figure of the helmsman, the lone eagle of Lindbergh fame recast as the space cadet of tomorrow. Presenting the original astronauts as moral supermen, which the press corps eagerly did, ensured that the public would grow attached to the mission objectives that the astronauts endorsed.

As with advertising campaigns, astronauts and flight controllers employed technojargon to make rockets and spacecraft appear more controllable than they really were. Ready launch systems were "A-OK," not just okay, suggesting a standard of reliability that exceeded normal expectations. Astronauts in orbit "initiated a retro-sequence" with their "retrograde package," a delicate way of announcing that they were prepared to ignite rocket engines whose failure to start would condemn the occupants to a cold and asphyxiating death. The language of space flight engineering suggested total control. NASA used "systems analysis" to assure "operational control." The dangers of "liftoff" were managed by "Launch Control," and the lives of the astronauts were protected during their journeys by "Mission Control." In space, astronauts kept their rocket ships from tumbling by the use of "10 reaction control system engines" and provided "attitude control about 3 axes."[25] In an insightful essay on the subject, Michael L. Smith characterized the efforts of flight advocates to sell Americans on the virtues of space travel as "a triumph of commodity scientism." "The space race was consummately other-directed, revealing a curious mixture of unsurpassed power and deep insecurity among American leaders. In constant doubt of their global technological superiority, and unsure how to apply it, they rushed to outdistance their geopolitical rivals in every measurable contest for prestige."[26] Early advocates presented space travel to the American public as one might advance any product based on a new science or technology. Become a better person, the campaign suggested, by mastering this task.

Under the circumstances, the presentation of a space program in which humans did not play the central role was unthinkable. This realization led to a series of statements extolling the importance of humans in space. Nicks expressed a commonly held sentiment in human space flight circles when he suggested that "if the mission was manned, people cared deeply, and if only instruments flew, interest was lessened and somewhat remote." When asked to characterize the central purpose of sending humans to the Moon, astronaut Neil Armstrong replied that the flight provided its own justification. "The objective of this flight is precisely to take man to the moon, make a landing there, and return. . . . The primary objective is the ability to demonstrate that man, in fact, can do this kind of job."[27] Comparing flights into space with the efforts to climb tall mountains, an advisory committee headed by aerospace executive Norman Augustine later observed that "there is a difference between Hillary reaching the top of Everest and merely using a rocket to loft an instrument package to the summit."[28]

Statements such as these exasperated many space scientists. Echoing broader attacks on public gullibility, Van Allen assailed the claim that human space adventures were necessary to maintain public interest and political support. Such an assertion, he said, "is an insult to an informed and intelligent citizenry."[29]

NASA officials continued to insist that the objectives of space flight were best maintained by a combination of humans and machines. Engineers would push automation as far as it would go, but at some point human intelligence and versatility necessarily intervened. "The more complex the mission, and the farther from the Earth it must be carried out, the greater will be the need for that human versatility," NASA's director of Space Sciences argued when the space program began. NASA administrator James Fletcher explained eight years later that automated systems were appropriate where a detailed definition of the space mission could be specified in advance. "But when the objectives and opportunities cannot be fully defined in advance, as in the case of exploration, or when the required operations are exceedingly complex . . . the presence in space of man with his unique intelligence and versatile physical capabilities can be an essential advantage."[30]

Nonsense, replied space scientists in an increasingly contentious debate. The notion that humans were needed to compensate for the shortcomings of machines, in the view of many scientists, had "very limited validity." To the contrary, they asserted, humans were a hindrance on most space flights. They used up precious space. They displaced instruments that could be installed to gather information with equipment needed to keep the occupants alive. What would humans do on a rocket ship bound for the outer planets, or a spacecraft investigating the surface of Venus? Humans could not stand on the gas giants even if they got there, nor could they survive the nine-hundred-degree temperatures on Venus "underneath its veil of sulfur rain."[31]

Advances in technology slowly transformed the human versus robotic space flight debate. When asked in 1971 whether machines could substitute their judgment for that of human beings, the director of NASA's Institute for Space Studies, Robert Jastrow, argued that the machines could not. Jastrow noted that a machine with the cognitive powers of a single human brain, as of the year in which he made his statement, would cost $10 billion, weigh one hundred thousand tons, occupy eight million cubic feet, and consume one billion watts of electric power. "It seems a safe bet," he concluded, "that in space exploration man will be superior to the machine for all difficult investigations, from now to the end of the 20th century."[32]

The final words in Jastrow's observation were prophetic. The twenty-first century arrived, and with it came great advances in computer technology. In 1971 spacecraft computers were not very sophisticated. The Apollo block computer, which guided astronauts to the surface of the Moon, was one of the most advanced calculating machines of its time. Apollo astronauts utilized two—one in the command module and another in the lunar lander. The computer was developed at the Massachusetts Institute of Technology Instrumentation Laboratory and assembled by Raytheon. It occupied one cubic foot of space, weighed seventy pounds, and had a 38K word memory. Its computing power and relatively small size allowed engineers to transfer much of the flight control activity that previously had been confined to massive computers on the ground into the spacecraft.[33]

In an extraterrestrial version of Moore's Law—Intel cofounder Gordon E. Moore's 1965 prediction that the number of transistors that can be crammed on an inexpensive, integrated circuit will double approximately every two years—computer capability steadily improved. About twenty-five years later, scientists assembling the Mars exploration rovers placed each spacecraft's computer in a small, insulated box slightly above the wheels. The computers, which reached the planet in 2004, each possessed 128 MB of random access memory.[34]

Extrapolating such trends, various thinkers have imagined points in the future when computers become as smart as human beings. At some future point in time, humans will probably push technology to the level that an inexpensive personal computer attains the raw computing power of a human brain. The next step is more speculative. As the computing power of machines continues to increase, humans might learn how to code algorithms that simulate brain activity. For people attempting to visualize the future, the latter point is known as "the last invention" or, more popularly, as "the singularity." Writing in 1965, Irving Good speculated on the consequences of developing such a machine: "Let an ultraintelligent machine be defined as a machine that can far surpass all the intellectual activities of any man however clever. Since the design of machines is one of these intellectual activities, an ultra-intelligent machine could design even better machines; there would then unquestionably be an 'intelligence explosion,' and the intelligence of man would be left far behind. Thus the first ultraintelligent machine is the *last* invention that man need ever make."[35]

At this stage, the ability of humans to foresee the innovations that intelligent machines might make breaks down. Past experience and the limited cognitive abilities of less intelligent beings would be insufficient to anticipate what ultraintelligent entities might do. Drawing a term from physics, Vernor Vinge described this point of transition as the singularity.[36] The term refers to the point on the event horizon of a black hole at which the quantities associated with a gravitational field become infinite. In lay terms, nothing beyond the singularity happens in the manner that creatures on a more conventional world experience.

The significance of the singularity has been popularized by Ray Kurzweil, who views it as a time of exceedingly rapid and unpredictable technological change. "You will know the Singularity is coming," he joshes, "when you have a million e-mails in your in-box." According to Kurzweil, the singularity will occur during the twenty-first century. At that point, inexpensive computers will vastly exceed the computing power possessed by thousands of human brains. Simultaneously, humans might learn how to enhance their reasoning powers by connecting themselves more directly to computers. The eventual result, Kurzweil reasons, will be the emergence of new life-forms: machines that claim to be conscious and cybernetically augmented human beings. The implications for the future of space exploration are profound. Although no human can predict exactly what form the machines might take (the singularity prevents that), one possibility is the production of ultraintelligent entities no longer limited by the conventional arguments favoring human beings for space travel. The entities would be better at it—stronger, smarter, and more versatile.[37]

Science fiction writers have for some time contemplated the consequences of producing sentient machines—manufactured entities that learn to think for themselves. Sentience refers to the state of consciousness. It describes an entity with a mind of its own. Isaac Asimov introduced one of the most famous sentient machines in the 1976 novella, "The Bicentennial Man." In the story, a household robot named Andrew Martin begins to produce pieces of art. His manufacturers explain that some error in the plotting of his positronic pathways has produced an artistic sensitivity. Robotics is not an exact science, they point out, offering to replace Andrew with a less deficient model. Andrew's owner politely declines. The robot is valuable in all other respects. "He performs his assigned duties perfectly," the owner attests.[38]

Over time, the artistic sensitivity manifests itself as a desire to wear human clothes, make human gestures, and to undergo a series of surgeries that replace metal parts with synthetic tendons and skin. Further surgeries implant human body parts. Andrew wishes to become a complete human being. A member of the World Legislature points out an important obstacle to Andrew's quest. The human brain eventually dies. The robot's brain is mechanical. It has lasted more than two hundred years "and can last for centuries more." The eventual deterioration of the organic human brain, says the legislator, creates a barrier between it and the robot's central processing unit that is "a mile high and a mile thick."[39] Relentless in his desire to become fully human, Andrew undergoes a final procedure that allows his positronic pathways to deteriorate and die. Asimov called the story "my most thoughtful exposition on the development of robots."[40]

The story follows a very familiar narrative line. It is the story of the slave seeking its freedom. An entity, previously viewed as someone else's property, acquires capabilities that prompt it to request the full rights and privileges accorded to human beings. In Andrew's case, courts rule that "there is no right to deny freedom to any object with a mind advanced enough to grasp the concept and desire the state."[41]

The story addresses a conundrum presented by one of the most popular characters in the fictional presentation of space exploration. On *Star Trek: The Next Generation*, Lieutenant Commander Data is the chief operations officer on the USS *Enterprise*. Data is an android, a robot in human form, constructed to resemble the doctor who created him but possessing an artificial brain with extensive computational power. The brain makes him ultrarational, like the previous half-human, half-Vulcan character Spock in the original series. Data's form makes him curious about human behavior and draws him to think of himself in human ways. In the episode "The Measure of a Man," Data resists an order from a resident cyberneticist who wants to perform experiments that could lead to the robot's demise. To the cyberneticist, Data is nothing but property and has no more right to resist the order than a toaster. Data appeals to the judge advocate general and wins a legal judgment declaring him to be a sentient being that possesses the rights and privileges of any citizen in the

United Federation of Planets. "He may be a machine," the judge rules, "but he is owned by no one and has the right to make his own decisions regarding his life."[42]

Like many other robots in the science fiction realm, the characters Andrew Martin and Lieutenant Commander Data show intelligence. They do not possess organic brains, but they can think on their own. Is this strictly fiction, or could it actually occur? No less a figure than the physicist Stephen Hawking suggests it could come true. "Some people say that computers can never show true intelligence whatever that may be. But it seems to me that if very complicated chemical molecules can operate in humans to make them intelligent then equally complicated electronic circuits can also make computers act in an intelligent way. And if they are intelligent they can presumably design computers that have even greater complexity and intelligence."[43]

Were this to occur, it could alter the course of space exploration in astonishing ways. The conventional vision of space exploration holds that human beings, with robot helpers, explore the solar system. The vision is based on the assumption of human superiority, a presumption that leaves human beings, if not more capable, at least in control. Reverse that presumption and a multitude of strange opportunities occur. They have been imagined, but their influence on the conventional practice of space exploration has been small.

Consider for a moment the challenge of interstellar flight. Astronomers have already identified a variety of planets orbiting nearby stars; it seems inevitable that humans will eventually discover a planet with liquid oceans and an atmosphere. People will want to know more about the object. The desire to explore such a planet will be strong. To transit the distances involved, even with the most advanced propulsion technologies, will probably exceed the life-span of the normal human being. Visionaries have imagined various solutions to the challenges posed by the desire to transport human beings to such objects, including wormholes through space and multigenerational space ships on which descendants of the original emigrants complete the voyage. Intelligent robots provide another solution, one not much discussed when people set out to imagine such things.

Andrew Martin, the lead character in "The Bicentennial Man," lived two hundred years. He could have easily lived two hundred more. Long-lived, intelligent robots provide a means for interstellar exploration that avoid many of the challenges imposed by solutions like the multigenerational spaceship, with its requirements for a sufficiently diverse genetic pool and the enforced consent of persons born during the voyage. The producers of the television docudrama *Alien Planet* dealt with this challenge in an intriguing way. To investigate a life-supporting planet six and one-half light-years from Earth, humans send robots. Three robotic explorers, shaped like small space shuttles, traverse the distance in an interstellar spacecraft the size of a nuclear attack submarine. The robots will not return; they are on a one-way mission to explore the planet and its life-forms and report their findings to Earth. Upon their arrival at the planet, the explorers face a special problem. The first explorer disintegrates as it enters the planet's atmosphere and attempts to land.

Under the procedures utilized for earthly flight, the dispatching government would assemble an investigating committee. Members of the committee would review the malfunction and meet until they determine a cause. The undamaged spacecraft would remain in place until a solution could be implemented, perhaps a period of two to three years. Adding in the period for radio transmissions—six and one-half years each way—the remaining spacecraft would need to wait fifteen to sixteen years for new instructions. If one imagines the robots to be sufficiently intelligent, this is clearly unnecessary. The explorers can conduct their own investigation and learn from it. They can, as the show's producers say, "think on their own." In this work of fiction, the remaining two explorers proceed and successfully land.[44]

Having relinquished control of the mission to the robot explorers, the humans who dispatch them are in a poor position to govern their activities. The explorers could decide to go somewhere else. Their sentient qualities and superintelligence would permit it. "It is curious that this point is so seldom made outside of science fiction," Irving Good noted. "It is sometimes worthwhile to take science fiction seriously."[45] Humans could attempt to maintain control. Using their intellectual powers, they might devise ways to control more intelligent entities. They might "hard wire" obedience into the machines at the point of their creation. The positronic pathways in Isaac Asimov's fictional robots are plotted so carefully that they cannot be disturbed. In Asimov's imagined world, robots always obey the three laws, no matter how intelligent or sentient the robots become. Perhaps humans will simply hope that the machines will be docile enough to tell humans how to maintain control.

The social foundation for the vision of robots permitting human control follows a well-established tradition, drawing on the historic treatment of servants, including their participation on expeditions of discovery. The use of servants, or domestic workers, was widespread in the decades preceding the presentation of Karel Capek's *R.U.R. (Rossum's Universal Robots)*. The practice of keeping servants extended from the upper classes, whose houses provided living quarters for butlers, housekeepers, valets, maids, cooks, nursemaids, and chauffeurs, to the upper reaches of the middle class, where a single servant might commute to and from the home. In Europe and America, the system reached its zenith around 1901, a period marked in the United Kingdom by the death of Queen Victoria and the short-lived but fondly remembered Edwardian period, which lasted until Edward VII died in 1910.

The servant system was held in place by a rigid class system, limited social mobility, and a clearly understood distinction between the status of masters who lived "upstairs" and servants who worked "downstairs." It was expedited by a ready supply of labor produced by denying education and alternative vocational opportunities to individuals, especially women, from rural areas and the lower social classes. Families compensated servants for their work, but not much. Unlike the slavery system, servants were free to quit their places of employment, although few alternative vocational opportunities existed.[46] Like the slavery system, the widespread use of servants was eventually undermined by governmental legislation that provided schooling, employee benefits, and political rights to persons confronting such situations.

Servants to upper-class families traveled with them on tours and accompanied explorers on a variety of expeditions during the late nineteenth and early twentieth centuries. Henry Morton Stanley's seven-hundred-mile expedition to locate Scottish missionary David Livingston left Zanzibar in 1871 with some 200 porters. A porter is a servant employed to assist travelers or carry baggage. Theodore Roosevelt's ill-advised expedition into the headwaters of the Amazon River employed 15 Brazilian porters. Roosevelt almost died from an infected leg wound and had to be guided out by itinerant rubber workers. Edmund Hillary's famous Himalayan expedition in 1953 was supported by 362 porters and 20 Sherpa guides. To his credit, Hillary shared the honor of attaining the summit of Mount Everest with his Nepalese Sherpa, Tenzing Norgay, insisting that the two had reached their objective simultaneously.[47]

One of the most interesting episodes in the servant-explorer relationship occurred in 1909 during the race between Robert Peary and Frederick Cook to the North Pole. Each man insisted that he had reached the North Pole before the other, a controversy that continued throughout the century that followed. Peary was accompanied by Matthew A. Henson, originally employed by Peary to serve as a valet. An African American store clerk from Washington, D.C.,

Henson became an accomplished sled driver and traveler through two decades of polar work. Henson was a strong walker. On their expedition to reach the pole, he reached their destination forty-five minutes ahead of Peary. "I think I'm the first man to sit on top of the world," he confessed to Peary. "Oh, he got hopping mad," Henson recalled. "He didn't say anything, but I could tell."[48]

The controversy that ensued focused on Peary's and Cook's contesting claims, ignoring almost entirely the possibility that Peary might have been preceded to the North Pole by his African American valet.[49] Given the social conventions of the time, it was unthinkable that a servant would be given credit for the accomplishments of his or her master, even if the servant did the work. Under the strict social codes that governed master-servant relationships during this period, servants were expected to defer—even though they were sentient, thinking beings.

In many respects, robots were presented as a means for extending the more fondly recalled aspects of the traditional servant system into a future increasingly influenced by science and technology. In the same manner that servants followed slaves, robots followed servants. As each group achieved the ability to leave its particular situation, the wealthier classes found acceptable substitutes. It is no accident that Isaac Asimov's first robot story dealt with a robot named Robbie that worked as a nursemaid for an upper-middle-class couple and their daughter Gloria. Machines such as the airplane, radio, and snowmobile—beginning with Richard E. Byrd's expedition to Antarctica in 1928-30—were replacing the work done by human porters and other servants on earlier expeditions of discovery. The robot provided a means to continue the master-servant relationship in a machine age.

The proposition that intelligent machines will be as content to accept the oversight of human beings is in many ways as naive as the assumption that servants in the Edwardian age contently served their masters. We know how that story turned out, just as we know the outcome of the slavery debate. Servants and slaves given the opportunity for alternative situations will accept them over servitude. The concept of the contented servant, like that of the happy slave or "Uncle Tom" deployed to work in the master's home, is a fiction maintained by the people in charge and not generally shared by the other side. The concept is socially outmoded, making it hard to imagine that humans will long embrace a body of visionary literature portraying robots as contented slaves.

People who reject this social tradition more readily envision an alternative future. Many science fiction writers have embraced this alternative, imagining a multitude of rebellious, often murdering, robots. In his classic tale *Do Androids Dream of Electric Sheep?*, science fiction writer Philip K. Dick describes an understandable reaction among sentient robots faced with the prospect that humans might shut them down. The story reached a wide audience after director Ridley Scott transferred it to film in the 1982 movie *Blade Runner*.[50]

Dick draws directly on the concept of slavery to tell his robot tale. To encourage humans to settle Mars, the government grants each emigrant a personal servant. The servant is an android, made entirely of organic parts and nearly indistinguishable from a human being. Commentators on the government-owned television channel publicize the incentive by announcing that the practice of using custom-made androids "as body servants or field hands . . . duplicates the halcyon days of the pre–Civil War Southern States."[51]

Life on Mars is hard, given the primitive nature of settlement conditions there. It is not much better on Earth, where a nuclear war has poisoned much of the biosphere and forced humans who want to own animals to purchase artificial, electric ones. For the androids, life on a toxic Earth as a free human being seems preferably to life on Mars as a slave. Humans realize this. In order to discourage androids on Mars from overpowering their owners and escaping to Earth, humans ban them from the home planet. The book and film deal with the efforts of bounty hunters—notably the character Rick Deckard (played in the movie by Harrison Ford)—to locate androids loose on Earth and terminate them.

The story presents concepts typically associated with the darker side of robotics. Humans invent technologies, despoil the Earth, and fight with machines for control of what is left. One of the gloomiest portrayals of this narrative appears in the widely viewed science fiction film *The Matrix*. The title refers to the simulated reality in which humans think they exist. Humans believe that they are living in a vibrant metropolitan community in the year 1999. In fact, they are living two hundred years in the future, in 2199, in a world desecrated by conflict between the remaining group of free human beings and their machines. Free humans create thick black clouds to cut off the machines' access to solar power. The machines place most of the human population in pods, in a reddish goo, for the purpose of harvesting their biological electricity and bodily heat as an alternative energy source. Using neural implants, the machines force the captured humans to imagine they are living in a different, happier time.[52]

The Agents against which the free humans do battle look like human beings. They could pass for androids if one wanted to imagine them that way. Their physical appearance is an illusion, however, created within the virtual reality world of the machine. *The Matrix* takes the dark side of technology to a new level, one in which superintelligent entities no longer need to occupy physical bodies but can appear to do so if they wish. The Agents are manifestations of an artificial intelligence machine. To the humans who encounter them, they appear to have superhuman powers. Yet their existence is as artificial as the characters in a computer game, their powers being limited by boundaries in the software programs that establish them.

As *The Matrix* suggests, a superintelligent entity need not take a robotic form in order to pose a danger to human beings. One of the most famous rogue entities in science fiction, the HAL-9000 computer in the movie *2001: A Space Odyssey*, is nothing but a big, smart CPU. It controls robotic devices, such as the ship's space pods, and operates various subsystems on the spaceship *Discovery*, but it has no robotic form save a well-distributed red eye.

HAL stands for heuristically programmed algorithmic computer. He (the machine has a male voice) is an artificial intelligence computer programmed to operate the spacecraft and lie to the crew about the true nature of its top-secret mission. The astronauts are chasing evidence of alien life; scientists are afraid that astronauts aware of this objective will reveal it over open radio channels to the enemies of the Western world. The need to lie causes the HAL-9000 computer to make an error. Rather than admit his fallibility and risk being shut down, the computer commences an effort to murder the crew.

The sole surviving astronaut in this conflict, Dave Bowman, dismantles the HAL-9000 computer in an intriguing scene. Bowman enters the computer's Logic Memory Center, a large area containing numerous book-sized modules. As Bowman turns keys, the modules emerge and the computer's memory disappears. "My mind is going," the computer announces plaintively. "I can feel it. I'm afraid."[53]

Is this only fiction? A number of technology experts think not. "When pocket calculators can out-think humans, what will a big computer be like?" asked Hans Moravec in 1988. "We will simply be outclassed."[54] Moravec, a computer scientist and university professor, predicted that artificial intelligence computers would evolve into a new type of species far more intelligent than human beings. Bill Joy, an American computer scientist who cofounded the Silicon Valley firm Sun Microsystems, wrote one of the most widely read warnings about the dangers of this possibility in the April 2000 issue of *Wired* magazine. In the article, Joy observed, "Biological species almost never survive encounters with superior competitors." Whether in the natural world, where nature pits one species against another, or in the social world, where cultures clash, the competition for space and energy favors the more capable entity. Is there any reason to think that this would not occur between humans and their machines? "Our most powerful 21st-century technologies—robotics, genetic engineering, and nanotech—are threatening to make humans an endangered species," Joy announced. Robots might allow humans to exist in the same way that humans keep domestic animals, but the machines would clearly be in control.[55]

Ray Kurzweil, as the back cover of one of his books emphasizes, believes that these developments will produce a "breathtakingly better world." In *The Age of Spiritual Machines*, Kurzweil predicts that humans and machines will begin to merge within the next one hundred years. Humans will acquire cybernetic implants, including ones that allow their minds to interact directly with artificial intelligence machines. Machines will learn to think and sense their surroundings much like human beings. The conventional distinction between humans and machines will disappear. At that point, Kurzweil asks, why would anyone want to remain in a vulnerable, squishy organic body with limited powers of intelligence that ages and dies?[56]

HAL physically occupies a specific space in the *Discovery* brain room. Kurzweil suggests that the artificial intelligence machines of the future will not have a permanent physical presence as might be represented by a single computational processing unit. Like the Internet, for example, they might be distributed. They might even approach some incorporate form.[57] Like gods, incorporeal beings do not rely upon physical material for their existence.

In presenting the screenplay for *2001*, Arthur Clarke suggests that any sufficiently advanced entity will eventually evolve into an incorporeal life-form. Physical entities are represented by the HAL-9000 computer and the astronaut Dave Bowman, who occupy specific spaces in machine and biological form. Incorporate forms are represented by the aliens they are pursuing, who have long since left their physical presence and never appear as specific entities in the film. "Into pure energy," Clarke explains, they "transformed themselves." As the *2001* series of four novels progress, the HAL-9000 computer and astronaut Dave merge and download themselves into a computational medium. Clarke, who formulated the first steps for the commencement of human space flight, imagined its last consequences as well. Transformations in the form of intelligent beings remained a constant theme in Clarke's writings, from *2001: A Space Odyssey* to other novels such as *Rendezvous with Rama*.[58]

Kurzweil foresees a similar history of human-machine mergers, with the resulting entities eventually moving out into the universe in search of new material in which to store their thoughts. What travels through space is neither human nor machine, but computational power. In a feat of astonishing imagination, Kurzweil envisions intelligence radiating into space, breaking down whole planets, moons, and asteroids and utilizing their constituent elements to think. Kurzweil suggests that the intelligence will be preceded by swarms of nanobots that "take root" in other planetary systems and prepare those systems for the arrival of information dispatched at the speed of light. "Sending biological humans on such a mission would not make sense," Kurzweil observes. At any rate, he concludes, "human civilization by that time will be nonbiological for all practical purposes."[59]

Clarke and Kurzweil approached these possibilities with unbroken optimism. To both, the emergence of new worlds and new life-forms assures the attainment of utopia through technology. Humanity will face dangers from these technologies, even possible extinction, but the alternative is worse. The alternative to technological progress and space travel, Clarke believed, was eventual extinction of the species, because *Homo sapiens* huddled on the Earth would ultimately encounter some human-made or natural disaster. Hans Moravec agreed. "If, by some unlikely pact, the whole human race decided to eschew progress, the long-term result would be almost certain extinction. The universe is one random event after another. . . . The bigger, more diverse, and competent a culture is, the better it can detect and deal with external dangers."[60]

During the earliest phases of space exploration, the human-machine issue took the form of a dichotomous debate. Would humans dominate the exploration of space, as von Braun believed, or would machines do most of the work, as Van Allen proposed? Officials in the U.S. space program suggested a compromise, a cooperative venture of men, women, and machines. As visions and experience accumulated, the dichotomy began to break down. In the minds of people thinking about machine technology and space travel, new entities emerged—part human, part mechanical, or even something beyond. Such entities might be far better suited for space travel than human beings or the earliest machines sent from Earth to explore the cosmos. The new visions are highly speculative, of course. They contemplate a future far different from the traditional vision of human beings in a never-unchanging biological form happily exploring the cosmos with a little help from their mechanical friends. The new vision provokes humans to contemplate alternatives to conventional models of exploration.

The first practical effort to apply this new type of thinking to space travel occurred immediately after the launching of the first Earth-orbiting satellites. In 1960 two research scientists, Manfred Clynes and Nathan Klein, discussed the issue of human transformation through space travel at a government-sponsored symposium organized by the U.S. Air Force School of Aviation Medicine. Clynes and Klein suggested that the emphasis upon building artificial environments for transporting humans through the cosmos was misplaced.

Suppose that someone wanted to prepare a fish to live on land, the two scientists proposed by way of analogy. The fish could be surrounded by a bubble of water, as in an aquarium. Such a device would be fragile and would not allow the fish to interact directly with the land. Over millions of years, nature devised an alternative path. It changed species of fish so that they acquired legs and the ability to breathe air naturally. If the fish had been sufficiently resourceful, with access to adequate funding and research laboratories, they might have developed techniques that would have sped up the transformation. This, the research scientists surmised, is the situation that humans contemplating space travel face. "Artificial atmospheres encapsulated in some sort of enclosure constitute only temporizing, and dangerous temporizing at that, since we place ourselves in the same position as a fish taking a small quantity of water along with him to live on land. The bubble too easily bursts. . . . If man attempts partial adaptation to space conditions, instead of insisting on carrying his whole environmental along with him, a number of new possibilities appear."[61]

Clynes and Klein devised a word to describe such an adaptation—*cyborg*, a portmanteau of cybernetics and organism—and proposed a number of potential methods to achieve them. Although the conference at which they presented their paper dealt with the psychological aspects of space travel, their insights proved far too radical for government officials charged with the sensible expenditure of public funds. The concept exploded within the realm of popular science and science fiction, however, emerging through presentations such as the Borg on *Star Trek* and the television series *The Six Million Dollar Man*.[62]

During the preliminary stages of the real space program, the skirmishes between humans and machines took place at a far different level from the one contemplated by Clynes and Klein. Rather than investigate how humans and machines might merge, the people working on the early space program returned to a more conventional issue. To what extent would the astronauts who traveled inside spacecraft control the machines within which they flew? While the public witnessed debates that pitted advocates of human flight against scientists favoring robotic missions, the more revealing battles within the space program set humans against themselves. The engineers designing spacecraft had to deal with the astronauts assigned to fly in them. In the United States, at least, the interests of the astronauts prevailed.

The original Mercury space flight capsule, the first spacecraft to carry an American into space, had no centerline window. The absence of a window was symbolic of a larger, more technical controversy taking place within NASA and among its contractors over the degree of astronaut control. Engineers designed the earliest space capsules to be flown automatically. Sometimes the capsules flew empty and sometimes they flew with monkeys and chimpanzees in the passenger seat. NASA conducted more than ten test flights of its Mercury capsules without humans on board. Engineers wanted to fully certify that the autopilot and its related systems worked adequately before qualifying the spacecraft for human occupancy on an orbital flight. "This vehicle does not behave like an airplane," explained a group assigned to analyze the layout of the instrument panel in the capsule. Even an experienced pilot would have difficulty controlling one. As a consequence, the engineers explained, "primary control is automatic." According to one of the human engineering experts hired to address the problem of pilot control, "Serious discussions have advocated that man should be anesthetized or tranquillized or rendered passive in some other manner in order that he would not interfere with the operation of the vehicle."[63]

Under pressure from NASA executives and the Mercury Seven astronauts to give the occupants more control, the engineers proffered a compromise. Primary control of the spacecraft would be automatic, as planned. But astronauts would monitor the flight and make adjustments as necessary. In the evolving capsule, engineers installed a trapezoidal window, a three-axis hand controller, and a redundant fly-by-wire system that allowed an astronaut to interact with the automatic controls. A set of sticky fuel values prompted astronaut John

Glenn to override the automatic system and pilot his *Friendship 7* spacecraft around the Earth. Said Glenn after his return: "Now we can get rid of some of that automatic equipment and let man take over."[64]

While supportive of human flight, engineers on the ground continued to express doubts about the capacity of astronauts to diagnose spacecraft malfunctions and execute a proper response. A multitude of potential malfunctions could occur and busy astronauts tended to make judgment errors regarding events for which they had not practiced in advance. During his orbital flight, John Glenn observed specks of yellow-green light that hovered like fireflies around his capsule. The mysterious lights fascinated Glenn and excited public speculation about the possibility of an alien encounter. The NASA psychiatrist who later interviewed Glenn about the lights gave a deadpan response. "What did they say, John?" Glenn failed to correctly identify the flakes as steam expelled from the life-support system condensing and freezing during its passage through the Earth's shadow, an incident that raised questions about the cognitive limits of humans in space. Encased in a space suit far from home and preoccupied with the business of staying alive, an astronaut may be in a poor position to make complex decisions. Most of the work astronauts perform is highly rehearsed and leaves little room for improvisation.[65]

When Soviet engineers developed their version of the winged space shuttle, called *Buran*, they constructed it so that it could be flown without cosmonauts at the controls. Cosmonauts objected to this decision, but a special government commission ruled against them. The first test flight was conducted on 15 November 1988. The spacecraft completed two orbits of the Earth and landed in a completely automated mode with no humans on board.[66]

The Soviet approach contrasted sharply with NASA's approach to its reusable space shuttle. Astronauts participated extensively in the design of the space shuttle cockpit. NASA officials promoted the image of the space shuttle as an airplane-type vehicle that humans would fly. Before qualifying the Mercury, Gemini, and Apollo spacecraft for human occupancy, engineers had test flown the capsules in an automated mode. Yet in the manner of its design, certification of the shuttle for human flight required humans at the controls. On 12 April 1981, in a significant departure from prior practices, NASA engineers allowed astronauts John Young and Robert Crippen to pilot the space shuttle *Columbia* into orbit on its very first voyage. Qualifying the spacecraft for human flight by putting humans at the controls was a risky but ultimately successful move.

NASA planned to use the space shuttle for a wide range of missions in which astronauts would check out, deploy, repair, and return otherwise automated equipment. This represented a new level of human-machine interaction and one that NASA officials hoped would dominate future missions. The new policy was dramatically illustrated on 24 April 1990 when five astronauts in the space shuttle *Discovery* delivered the Hubble Space Telescope to an orbit roughly 380 statute miles above the surface of the Earth. NASA officials designed the telescope so that it could be launched and retrieved in the shuttle's payload bay. On five subsequent servicing missions between 1993 and 2009, astronauts repaired and maintained the large telescope. They replaced cameras, spectrometers, gyroscopes, tape recorders, computers, solar arrays, thermal protection blankets, and optical equipment designed to overcome a flawed primary mirror.

Restrictions on the use of the space shuttle following the *Challenger* accident prompted NASA officials to forgo extensive use of the vehicle for the delivery and repair of subsequent observatories. The next two telescopes in the Great Observatories program—the Compton Gamma Ray Observatory and Chandra X-ray Observatory—were delivered by astronauts in the NASA space shuttle, but no efforts to maintain them in space were made, even though their orbits would have allowed it. Scientists disconnected the fourth Great Observatory—the Spitzer (Infrared) Space Telescope—from the human flight program entirely. The Spitzer Space Telescope was launched on an expendable Delta II rocket into a heliocentric orbit in which the instrument trails the Earth around the sun, well beyond the range of shuttle astronauts.

NASA officials wrestled with the role that astronauts would play in the fifth and final serving mission of the Hubble Space Telescope. Automation advocates suggested that a robot launched into space could do as good a job repairing the telescope as an astronaut in a space suit. Restrictions following the loss of the space shuttle *Columbia* prompted NASA officials to consider the proposal seriously.

The mission allowed a rare head-to-head comparison of humans and machine capabilities for a single mission. "Our confidence is growing that robots can do the job," NASA administrator Sean O'Keefe announced. The leading candidate to represent the machines was a two-armed robot named Dextre being developed for the International Space Station. Dextre would ride into space on an expendable launch vehicle. In one scenario, the robot would be accompanied by a smaller robotic arm that would guide Dextre to the telescope, where the larger robot would install fresh instruments, gyroscopes, and batteries.[67]

For year, scientists had argued that robots could work in space at a fraction of the cost of human beings and at less risk. NASA officials submitted the issue to a special board of the National Academy of Sciences. The board, headed by Louis Lanzerotti, carefully examined the options. Its findings stunned advocates of robotic flight. "The immaturity of the technology involved and the inability to respond to unforeseen failures . . . make it highly unlikely that NASA will be able to extend the scientific lifetime of the telescope through robotic servicing." Although cost estimates varied widely, the probable expense of a robotic servicing mission was equal to if not more than the full cost of a single shuttle servicing mission. A shuttle mission would involve obvious risks to the crew, but a clumsy robot could produce more risks to the telescope. "A shuttle servicing mission is the best option for extending the life of the Hubble Space Telescope and ultimately deorbiting it safely," the group concluded.[68] Astronauts on the space shuttle *Atlantis* completed the repair mission in 2009. Once again, the human approach and NASA's reliance upon it prevailed.

A small bit of related symbolism illuminated NASA's continuing philosophy regarding machines. When astronauts deployed Dextre on the International Space Station during three spacewalks in 2008, they praised the capabilities of their mechanical friend. Dextre (short for special purpose dexterous manipulator) is a twelve-foot tall, thirty-four-hundred-pound robot with eleven-foot-long arms and delicate grippers that allow it to perform tasks outside the space station previously assigned to astronauts. NASA officials presented the device as a giant "robot repairman" with arms, legs, a torso, and a head. Characterization of the latter appendage was misleading. The dish-shaped object on the top of the robot's torso is not a head, as illustrations suggest. It does not contain a positronic brain. The headlike object is in fact a coupling device known as a grapple fixture used to attach the robot to a longer space station arm while Dextre does its work. Dextre does not have a head or a mind of its own. It is operated by astronauts working inside the space station. Had Dextre been used to repair the Hubble Space Telescope, the robot would have been guided in its motions by an operator on the ground, most likely at the Johnson Space Center outside Houston, Texas.[69]

The decision to deprive Dextre of a brain was a small technical decision, but it said a great deal about NASA's basic approach to the interface between hu-

Robotic science advanced more rapidly than human flight technology during the first half century of space travel. Although NASA officials dispatched astronauts to conduct the fifth and final servicing mission of the Hubble Space Telescope, agency leaders seriously considered sending a robot such as Dextre, developed for use on the International Space Station, to do the job assigned to human beings. (NASA)

mans and machines. NASA officials are committed to use of robots in the traditional mode, as servants that prepare the way for human beings and remain strictly under human control. Dextre is not allowed to roam free on the superstructure of the International Space Station, any more than the various rovers dispatched by the space agency to Mars are allowed to choose their own destinations. They may possess the capability to maneuver on their own, but they cannot decide what they will choose to explore.

The space age began with a collective presumption that humans would carry out practically all activities in space, from the maintenance of satellites to conduct of planetary exploration. Advances in robotics and related technologies undercut that assumption, prompting various scientists to question the need for humans in space at all. NASA officials adopted the position that humans and machines would explore space together, a viewpoint well represented by the large body of popular literature presenting robots as helpful mechanical servants. The debate between advocates of human and robotic flight became more complex as people contemplated the potential development of ultra-intelligent machines. Pessimists warned of the consequences of blindly pursuing the vision of machines as human substitutes. Optimists predicted a utopian future. NASA continued to follow the classical model, putting humans clearly in charge of the machinery with which they flew. What began as a two-sided debate has been transformed into a multidimensional vision of considerable complexity.

Some people imagine humans and their machines happily toiling together as they jointly set out to investigate the cosmos, a vision attractive to those who favor human preeminence and a conventional view of machines as human servants. With a robot in every garage, humans would work with their mechanical companions to explore places that neither had ever been before.

Some people imagine a dominant role for machines in which robots will perform a wide range of tasks, from fighting wars and driving combat vehicles to exploring hostile realms. If a task is too dangerous or distant for a human to complete it, increasingly intelligent machines will do it.

A third vision has now emerged. If humans are really serious about expanding earthly intelligence into the galaxy—and if machine technology continues to advance as rapidly as it has in the past—then the conventional distinction between humans and robots may disappear. Advances in biotechnology and artificial intelligence may allow humans to merge with their machines. New entities, part human and part machine, would appear, hybrids far better suited for cosmic existence than either humans or machines in their present form. Humanity would extend itself into the cosmos, but humans in their present biological form would not.

No one can predict which scenario will dominate the future realm. Imagination presents possibilities, not certainties. If humans continue to invent new technologies and pursue space travel for thousands of years, the exact course of future events will be hard to predict. Yet one can comment on the social traditions that support these imagined points of view. The first—humans and mechanical servants together—is inherently unstable. It depends upon a vision of servitude that is mythical at best and offensive at worst. It requires its advocates to embrace social relationships fast receding in influence among polite individuals. Servitude and slavery do not provide a good social foundation for a long-term vision of space flight.

The second tradition—robots alone—may be more technologically plausible, given recent developments. Yet it lacks a certain inspirational quality. If humans undertook space exploration strictly for reasons of scientific inquiry, then a program organized around machines would be acceptable. On the basis of past experience, the same could be said for national security and commercial activities in space. Machines can do that sort of work just fine. Yet space exploration involves more than science, security, and commerce. Humans undertake it in large measure as a means of extending humanity into the cosmos, as part of a search for a certain measure of immortality that will allow life from Earth to persist even when the Earth does not. Any vision of space exploration that does ignore this element, powerfully represented in the social traditions of science fiction, will be ultimately unsatisfying.

This leads to the third alternative, certainly the most difficult of the three to accept as a factual reality. Might a *Homo cosmos* emerge that is as capable of space travel as human beings have been at terrestrial migration? Of all of the visions motivating space exploration, this is the most imaginative. It may never happen. Yet, if it does, the entities that accomplish it will be able to point back to a supporting strong social tradition of cyborgs and incorporeal beings. The entities will be able to say: humans imagined it before it occurred.

10

Space Commerce

The Congress declares that the general welfare of the United States requires that the National Aeronautics and Space Administration (as established by title II of this Act) seek and encourage, to the maximum extent possible, the fullest commercial use of space.

—*The National Aeronautics and Space Act*

The first generation to experience space travel portrayed the extraterrestrial region as *outer* space, as if to emphasize the otherworldly nature of the realm.[1] Space was somewhere else, out there, away from the Earth. In reality, the most transformational effects of early space flight happened here, on Earth, to people who experienced the impact of cosmic travel without actually visiting the place.

Critics of the U.S. space program accused the government of "throwing money into space" without regard for the problems needing solutions on Earth. Through at least the first half century of space travel, however, not a single dollar was spent in space. All of the money—and most of the effects—stayed on Earth.

On Earth, space is big business, from the industries that provide goods and services under contract to government agencies to the firms that sell their own space-related products to consumers. In the largest sense, space exploration has helped to affirm the mid-twentieth-century vision of a consumer society, an economy devoted to an ever-expanding cornucopia of products generated by the intersection of capitalism and technology. The dream of this consumer society had its roots in the Great Depression and the desire of capitalists to fashion a technology-based alternative to socialism. Space travel became synonymous with consumer benefits through technology. To the nations that pursued it, advocates of space travel promised the advantages that motivated earlier efforts at terrestrial adventure: trade, new products, and great wealth. Space advocates promised commercial benefits, delivered them, and promised more.

In the future, advocates of space travel pledged, money would be spent in space. Humans and their machines would move into space, establish settlements, and engage in trade. Local economies would appear on the Moon and Mars. Commerce in space, the most fervent advocates maintained, would produce economic opportunities as profound as those previously generated by the personal computer and the Internet.[2]

Getting there will cost a great deal of money. Space exploration, especially when governments do it, is expensive. At the peak of the eight-year race to the Moon, NASA spent more money on space exploration than the federal government spent on education, health care, or international affairs. A government-run expedition to Mars, should any nation choose to undertake the venture, would cost as much as a modern war. Over three decades, NASA will have spent as much to fly its fleet of space shuttles as it spent to put human footprints on the lunar landscape. Civil space spending declined after the United States won the race to the Moon, from 4 to less than 1 percent of total federal expenditures. Individual projects like the International Space Station, nonetheless, remained big, expensive, and vulnerable to criticism from people who wanted to spend the funds on other priorities.[3]

Advocates of space exploration take a broader view. Rather than being a drain on the national treasury, they believe that space exploration has and will continue to be an engine for economic progress. It will create new products, encourage new industries, free dependence on limited natural resources, and ultimately raise the standard of living of people everywhere. Lyndon Johnson once remarked that U.S. space activities returned the money spent on them "tenfold," a reference to a specific effect on national security but one that was equally applied to its commercial value.[4]

The promise of economic benefits is an essential part of the spacefaring vision. The people who imagined space travel concurrently anticipated that humans would take their economies with them. It would begin with spinoffs, lead to space industries, and ultimately produce extraterrestrial trade.

Demonstrating spinoffs from space exploration provided an important justification for the newborn U.S. space program. Spinoffs are products and technologies developed for use in government programs that find their way into the commercial marketplace. During World War II, military research for combat-related activities led to the development of an assortment of new products that found commercial application. This included the nuclear-powered generating plant, the jet aircraft, radar, and penicillin. Fighting world wars as a means of enlarging consumer choice was a hard way to stimulate economic progress. Space exploration, by contrast, offered a method that might produce similar goods while doing much less harm.

NASA officials annually produced a book-sized publication that listed forty to fifty products and technologies expedited by space exploration that had been adopted by business firms. After more than a quarter century of tracking these spinoffs, NASA officials had identified in excess of fifteen-hundred space-related technologies that had "benefited U.S. industry, improved the quality of life and created jobs for Americans." The race to the Moon accounted for a substantial number. According to NASA, Project Apollo contributed to the development of computer-aided tomography (CAT) scan machines, kidney dialysis, cordless power tools, athletic shoe designs, freeze dried foods, and the cool suits worn by NASCAR race drivers. "Every time someone operates a computer, makes a long-distance call, watches television or uses an automatic teller machine, the benefits of space technology are being felt," one NASA administrator observed.[5]

Additionally, space spending by NASA and the Department of Defense provides direct economic benefits to the private firms that produce space hardware and associated services. Close to 90 percent of NASA's budget flows to aerospace contractors. Keith Glennan, NASA's first administrator, established the policy that NASA would rely upon private contractors instead of government-owned plants to produce rockets and spacecraft. The policy guided the U.S. civil space program in the decades that followed and provided a broad base of private support from companies that directly benefit from spending on space.[6]

Space entrepreneurs and their advocates imagined commercial opportunities that moved well beyond government contracts and technology spinoffs. They foresaw "a whole new world of commerce built on doing business in space." Writing at the dawn of the twenty-first century, CNN television journalist Lou Dobbs predicted that "in the frontier of space we will create entirely new forms of technology, new forms of manufacturing, new forms of recreation, and even new materials." It was, he enthusiastically proclaimed, "the next business frontier."[7]

The commercial space movement received a substantial boost in the summer of 1983 when business executives met with President Ronald Reagan to discuss the government's space policy—or, as the executives told Reagan, the lack of one. The business executives asked the president to develop a set of policies that would encourage private firms to invest in space. The most important initiative the government could undertake, they said, was the construction of a large, permanently occupied space station that could be used as a platform for developing commercial products. In approving the government-built facility, Reagan announced that "a space station will permit quantum leaps in our research in science, communications, in metals, and in lifesaving medicines which could be manufactured only in space." NASA's Space Station Task Force identified a dozen major commercial activities that could be performed on a properly designed station, from the creation of pharmaceutical compounds to the production of fiber-optic cables.[8]

In depictions of the future, writers of science fiction frequently present characters who do more than fly, explore, or fight inner-galactic wars. The characters perform economic activities. The receptionist who welcomes Dr. Heywood Floyd to the large, fictional space station in *2001: A Space Odyssey* is employed by the Hilton Corporation. The astronauts that fly Dr. Floyd to the space station work for the now-defunct Pan American Airways airline company. The robot QT-1 in Isaac Asimov's "Reason" operates a solar-power station directing energy beams to planetary receiving stations. Luke Skywalker works as a moisture farmer on the desert planet Tatooine before he becomes a Jedi knight. Crew members on the spacecraft *Nostromo* operate a cargo ship that transports mineral ore from Thedus to Earth in the movie *Alien*.[9]

This is fiction, but the business opportunities associated with space exploration are thought to be real. Advocates of space travel present a vision of enormous commercial opportunities waiting in the decades ahead. Participants in a "lunar underground," an informal network of individuals committed to the establishment of a colony on the Moon, suggested various commercial uses. The low-gravity, perfect-vacuum environment on the lunar surface, some said, would provide conditions conducive for the manufacture of silicon-based semiconductors and metal-ceramic materials known as cermets. Lunar advocates invited people with physical disabilities to contemplate the advantages of moving about on a low-gravity world. They imagined a variety of sporting activities on the Moon. In one iconic painting, artist Pat Rawlings visualized an enlarged athletic field under an enormous lunar dome that would house the Lunar Olympics. The pole vault bar reached nearly to the top of the transparent dome. Anticipating the first winter Olympics on the Moon, a NASA publication announced "it's only a matter of time."[10]

Historically, promoters of distant settlements have raised funds by promising to deliver precious resources such as gold and silver to their sponsors. Advocates of lunar settlements did the same. The lunar surface is dusted with helium-3, deposited over billions of years by the solar wind. The substance could be used to power fusion reactors on Earth. Fusion plants, a futuristic technology, use a controlled process similar to that which occurs inside the Sun to produce vast quantities of relatively clean electric power from small amounts of material. Helium-3, a promising fuel, is inaccessible on the Earth, having disappeared into the planet's mantel. Space entrepreneurs suggested that mining companies could produce enough lunar helium-3 to power the Earth for many thousands of years.[11]

Robert Zubrin, one of the most persistent advocates of human settlements on Mars, included business opportunities in his calculations. A fully functioning Martian economy would not be self-sufficient, Zubrin estimated, until millions of people lived there. Until then, Martian settlers would need to compensate the Earthlings who were subsidizing the planetary outpost. What goods and services might the Martians dispatch? Zubrin suggested deuterium, thought to be five times more plentiful on Mars than on Earth. Like helium-3, deuterium could be used as a fuel for fusion reactors. Zubrin calculated that a two-hundred-thousand-person Mars colony could produce enough deuterium to satisfy all of the Earth's current electric power needs. Through resource extraction, Zubrin insisted, "the economics of Mars colonization can be made to work."[12]

The 1986 National Commission on Space, headed by ex-NASA administrator Thomas Paine, extended this logic to the asteroids. "The Solar System is rich in raw materials," Paine's group announced. "In the long run, the abundance of material in the main asteroid belt is enough to support a civilization many thousands of times larger than Earth's population." Extensive mining of asteroids would not happen soon, the group agreed, "but it is good to know that they are there waiting for our descendants in future centuries." Paine's group also extolled the commercial benefits of sending self-replicating factories into space, a concept credited to the mathematician John von Neumann. The "seeds" for such factories would contain the capacity to automatically manufacture themselves from local materials after landing at a distant site. The commission additionally noted the potential benefits of solar-power satellites, a basic concept as old as the age of space travel. "It is far too early to predict that solar power satellites can undersell nuclear power," the group advised, "but the possibility is significant enough that we endorse a strong continuing program of research."[13]

The thought of all these commercial space activities spurred business executives to imagine a huge market for transportation services. With private entrepreneurs conducting so many activities in space, the potential market for rocket launches would expand. A growing launch market would attract private capital, which in turn would encourage private entrepreneurs to develop their own rockets and spacecraft. Multiple lines of development would help humans achieve an important element of the spacefaring dream—cheap, reliable, frequent access to space. As cost fell, the number of launches would continue to grow. Anticipating that growth, entrepreneurs and governmental bodies sunk money into a succession of rocket dreams, including the Delta Clipper, Conestoga, VentureStar, and the amusing Rotary Rocket Roton.[14]

With access to space assured, ordinary people could sign up. Tourism is a trillion-dollar industry, and space advocates visualized a cut of that pie. Burt Rutan and Richard Branson ordered the construction of a privately financed *SpaceShipTwo* that could take paying customers on a suborbital ride into space. Entrepreneur Robert Bigelow prepared plans for a series of inflatable modules that, among other uses, could serve as space hotels housing tourists. Space Adventures, a Virginia-based firm, promised that it "would fly more people to space than have made the journey since the dawn of the Space Age. Our clients will fly on suborbital flights, on voyages to Earth orbit and on historic expeditions that circumnavigate the moon."[15]

The optimistic vision of space commercialization contains more than an economic theory. It represents a way of life that came to dominate American society in the latter half of the twentieth century. Just two decades before the era of space travel began, to a generation of Americans struggling through the Great Depression, the economic future indeed looked grim. Thirteen million Americans were out of work. Net income for American manufacturers had fallen by more than 66 percent. The gross national product rested at a miserly $75 billion, and federal tax revenues had dipped below $2 billion annually. This was hardly a scenario capable of inspiring confidence in large technological ventures such as a voyage to the Moon.

Many people believed in socialist reform as the most hopeful solution to economic depression. Socialists promised to spur recovery by wresting economic control from private entrepreneurs and placing it in the hands of government planners. In the United States, Franklin Roosevelt offered an economic New Deal that featured social security, price stabilization, and public-works employment—not socialism per se but sufficient to frighten capitalists of the day.

To its critics, socialism and its various manifestations proffered a dreary future. Writers such as George Orwell and Aldous Huxley helped popularize the view that government-created utopias would retard individual initiative and prevent economic growth. Orwell's *1984* depicts a totalitarian government ruling an ever-increasing population struggling to divide an economic pie of diminishing dimensions. In Orwell's novel, socialist reform offers little more than equality of misery.[16]

To combat government interference in its various forms, business leaders fashioned an alternative to socialistic doctrines. They promoted a consumer society blessed with an abundance of personal goods in which economic growth would be propelled by corporate initiatives satisfying a constant desire for new products. Technology and redesign would create the goods. The private sector would manufacture them, generating the payrolls that paid the bills. During the 1930s business leaders set out to sell this alternative to a Depression-weary public. Acceptance of the vision created the culture of consumption that dominated the U.S. economy in the decades that followed, and elements of that culture nurtured support for the U.S. space program.

Advocates of this consumer society had to sell the American public on the idea that people should buy goods they probably could not afford and often did not need. Business leaders employed advertising principles and product-display techniques to promote this idea. Previously, new products such as electric lights, telephones, and industrial machines had been displayed at a succession of expositions and fairs. Building on this tradition, corporate and civic leaders promoted the vision of the consumer society at the 1939–40 New York World's Fair: "The fair's streamlined buildings rose from the ground like a utopian dream come true. . . . Its dazzling array of technological marvels attracted

To encourage Americans struggling through the Great Depression, designers presented products with streamlined forms, symbolizing the hope that technology would speed consumers toward a more luxurious future. Promoters displayed this vision at the 1939–40 New York World's Fair, producing a "world of tomorrow" that included an exhibit in which visitors entered a "City of Light" through a forty-two-foot-tall wall of water. Twenty years later, advocates of space travel embraced a similar justification for the infant space program, emphasizing technology spinoffs and the economic benefits of exploration. (Library of Congress)

millions of Americans, eager to replace their memories of poverty and bread-lines with visions of prosperity and material comfort."[17] Fair organizers displayed the latest consumer goods in theatrical fashion: automobiles, refrigerators, dishwashers, cameras with color film, and the most marvelous innovation of all, the television set. Exhibits entertained fairgoers with the vision of a consumer society in which the average family could live like royalty.[18]

To entice Americans to consume, corporations reshaped products in creative ways. Market specialists had learned during the 1920s that product appearance could be manipulated in such a manner as to affect sales. Corporations hired industrial designers, who employed artistic techniques and new materials to recreate consumer products. In an effort to entice Americans to shop their way out of the Depression, designers altered the shape and color of toasters, vacuum cleaners, radios, and other consumer goods. Invariably the items became sleek, with flowing curves and shiny surfaces, designs derived from the appearance of airplanes, boats, and trains. "The streamlined form," author Donald Bush notes, "came to symbolize progress and the promise of a better future."[19] It implied the ability to speed ahead, away from the adversities of the Depression, toward a more luxurious life-style based on modern technology. Industrialists further learned that they could render perfectly useful products obsolete simply by changing their outward form, without the introduction of any new technology. Planned obsolescence, often based on nothing more than stylistic changes, became part of the engine that ran the new economic machine.

As a social phenomenon, the U.S. space program adopted the messages that drove the consumer society in mid-twentieth-century America. This was a key factor in its early popularity. The space program promised more than the exploration of space; it offered better living through technology. Speaking of the organization he led, NASA administrator Daniel Goldin asserted that the agency was the "the one organization in American society whose whole purpose is to make sure our future will be better than our past."[20]

The space program appeared at a time when optimists wanted to believe that technology could create a better world. In formulating plans for his first theme park, Walt Disney sought to reinforce the prevailing notion that the future would be better than the past and that progress generally worked to further the common good. Disneyland opened to the public in 1955 with as much public interest as any other social occurrence of that decade. The Main Street exhibit through which visitors passed after entering the park represented a sanitized version of America's past, the way that Disney wanted to remember his boyhood in Marceline, Missouri, shortly after the twentieth century began. The Tomorrowland exhibit symbolized the future of America as the twentieth century neared its close. The intervening period marked the passage of time between two visits of Halley's Comet, a major astronomical event. Disney had seen the comet's close passage as a young boy in Marceline in 1910. What would America look like when Halley's Comet returned in 1986? "Tomorrowland was conceived with the unbridled optimism of a thriving post-war industrial society," Disney's associates wrote. "In Walt's vision, the future of 1986 would be a time of automation, a time of leisure and a time of limitless opportunity. Anything you dreamed could be created." Humans would fly to the Moon in giant rocket ships, travel freeways in sleek new sports cars, and navigate waterways in plastic boats. A Monsanto "House of the Future" displayed the latest interior design techniques. In 1967 Disney added the "Carousel of Progress," in which the General Electric Company traced the effect of ever-improving electric appliances on an imaginary American household.[21]

Disney had run out of money while constructing his magic kingdom. To finish Tomorrowland, he called on corporate sponsors to prepare exhibits depicting the future. The message that Disney and his corporate sponsors transmitted was unambiguous: progress through technology would better the lives of average Americans. "Progress is our most important product," one corporate sponsor proclaimed. No references to the darker side of technology appeared, such as the nuclear-arms race or industrial pollution. The tiny gasoline-powered Autopia sports cars underwritten by the Richfield Oil Company chugged along Disney's miniature freeway at eleven miles per hour, adding to the California smog. Little publicity was given to the problems that plagued the guidance system for the plastic boats, a fault that caused park managers to remove the ride after the 1955 Christmas season. Walt Disney did not want to scare park visitors with foreboding visions of technology. For thrills, visitors could try the 146-foot Matterhorn Mountain bobsled ride.

To entice consumers to purchase new products, manufacturers had to portray friendly technology. No snapping toasters or exploding microwaves appeared to frighten prospective customers. Such metaphors were left to critics of the consumer society. Through advertising and product display, American industrial leaders revealed gadgetry that was designed to beguile rather than frighten consumers. Early advocates of space flight reinforced this promise of benevolent technology. Cutaways of space stations and rocket ships revealed a dazzling array of technical gadgets, many of which became consumer goods. An October 1952 article in *Collier's* magazine displaying a prospective flight to the Moon revealed astronauts at work inside a four-story lunar spaceship. In the kitchen area, astronauts prepared meals by removing a precooked delicacy from a freezer and inserting it in a "short-wave food heater," a forerunner of the modern microwave oven.[22]

When the space age began, corporate leaders were obliged to deliver the goods. They began by presenting spinoffs, products based on space technology. Critics ridiculed the spinoff argument, holding up as examples the $23 billion frying pan and Tang, the drink that went to the Moon. Teflon, a substance designed to smooth the path of objects reentering the atmosphere, was developed as part of the military space effort and later used to coat kitchen utensils, a major improvement over the iron frying pan. Tang was a powdery substance that when mixed with water produced a thin orange drink. While useful for space voyages, the product was in fact developed before the start of the civil space program. Friends of the space program sought to refute the critics by presenting hundreds of products arising from space-age technology.

Had the space program merely produced tangible products, it would not have been as popular as it became. The spinoff argument certainly enhanced the rationale for government investment, but it was not the only message propelling interest within a consumer-based society. Like other products at that time, the American space program had to be sold. It competed for public attention with a variety of initiatives, both public and private. In part the early space program attracted attention because it embodied so well one of the subliminal messages motivating the consumer society.[23]

Presentation of the early space program followed a pattern already well established by American merchants when the space race began. After the post–World War II spending binge had died down, and American manufacturers had found themselves with huge inventories and lagging demand, they fashioned a creative strategy to encourage further consumption. "Our whole economy is based upon planned obsolescence," one industrial designer observed. "We make good products, we induce people to buy them, and then next year we deliberately introduce something that will make those products old fashioned, out of date, obsolete."[24] Television and radio allowed manufacturers to communicate product changes with unprecedented speed. By the late 1950s, when the space race began, Americans had become accustomed to the frequent introduction of products that only appeared to be new.

Inadvertently, the early space program adopted this pattern, regularly replacing rockets and spacecraft with new models every few years. In 1961 Alan Shepard rode a Redstone rocket on his suborbital flight from Cape Canaveral; the following year John Glenn traveled into orbit on a modified Atlas launch vehicle. The Titan III rocket made its human space flight debut in the spring of 1965 with the first piloted Gemini flight, followed in 1968 by the Saturn IB and later that year by the launch of the first Saturn V with humans on board. The single-seat Mercury capsule was replaced by the two-person Gemini spacecraft, which in turn gave way to the more spacious Apollo spacecraft with room for three. Once on the Moon, astronauts introduced a variety of gadgets, among them the first four-wheeled lunar roving vehicle and the first live color television broadcasts from the lunar surface, including the remotely controlled transmission of a lunar liftoff.

Automated spacecraft introduced similar variations. Fleeting close-up photographs from spacecraft dispatched on a crash course to the moon were replaced by high-quality photographs from orbiters and the first Surveyor spacecraft to make a soft landing. Brief flybys of Venus and Mars were succeeded by spacecraft that orbited nearby planets and eventually landed on them. Fuzzy photographs from the Pioneer spacecraft that flew by Jupiter and Saturn were followed by Voyager images of incredible detail.

During the early years of the space program, NASA was constantly doing something new, maintaining interest in the effort among the public at large, as well as preventing worker complacency within. Once the space program began to mature, the introduction of new models occurred less frequently. NASA was still flying the space shuttle a quarter century after the first orbital flight in 1981, a long life-span by aircraft standards, to say nothing of the proclivity of consumption-driven Americans for changing their personal transportation vehicles every few years.

NASA's propensity for introducing new gadgets during its formative years was driven by both necessity and culture. The primitive quality of rocket and spacecraft technology prompted model upgrades. In addition, NASA officials embraced an engineering culture that encouraged employees to tinker with new ideas. The preference of the NASA engineer, said one insider, was "invent and build . . . and then go up and invent, build and watch something else work."[25] For a variety of reasons, the civil space agency developed an internal culture that harmonized with consumer expectations. Part of the bargain with technology was the promise of products that at least appeared to be new.

The communication satellite industry provided the basic business model for the commercial space sector. Shortly after the launch of the first Earth-orbiting satellite, scientists and engineers undertook a series of practical experiments designed to test Arthur Clarke's seminal observation that satellites would allow transmission of radio signals between any two points on the globe. The effort began with a supportive government policy and the expectation of substantial business profit. A critical test occurred in 1963, when the experimental communication satellite *Syncom 2* began operating from an orbit twenty-two thousand miles above the surface of the Earth. The satellite, built by the Hughes Aircraft Company, was financed with both corporate and

Space travel advocates envisioned a wide range of commercial activities taking place beyond the Earth, from mining and manufacturing to tourism. By the close of the twentieth century, entrepreneurship had supplanted government work as the dominant means for financing space activities. Visionaries imagined how undertakings like lunar mining could elicit the investments needed to foster the settlement of space. (Artwork by Pat Rawlings; courtesy of NASA)

government funds. Company leaders had began investing in satellite research in 1959 and two years later had received a NASA contract that helped them launch the device. *Syncom 2* was followed by *Syncom 3*, which demonstrated not only the technical feasibility of Clarke's concept but also the ability of corporate and government officials to support each other's work.

In 1962 the U.S. Congress enacted a charter creating a jointly managed corporation called the Communication Satellite Corporation, or Comsat, for the purpose of developing a communication satellite industry. Control of Comsat was vested in a board of directors consisting of six public stockholders, six representatives from the telecommunications industry, and three presidential appointees. The legislation authorized corporate officers to raise operating capital by borrowing money and selling stock. To make the corporation attractive to investors, Congress granted it monopoly status and continued to finance basic research on communication satellite technologies with government funds. As a corporate body, Comsat represented the United States in the formation of the worldwide system of communication satellites known as Intelsat and in 1964 became the managing arm of that system. One year later, participating officials launched the first commercially operational geosynchronous communication device, the seventy-six-pound "Early Bird" satellite.

The business proved enormously profitable. In 1976 Comsat officials reported revenues of $154 million. By 1998 revenues had grown to $616 million annually, with net income of $26 million. Other business firms entered the market, and by 1998 worldwide revenues for all sectors of the satellite communication business topped $56 billion. Nineteen-ninety-eight was a watershed year. Industry analysts estimate that the total level of commercial revenues from privately financed space activities exceeded aggregate worldwide government spending on space that year. For the first time in the history of the space age, business activities were richer than government work.[26]

The watershed created a flurry of high expectations, including financial journalist Lou Dobbs's famous assessment that space commerce would produce as much new wealth as the Internet.[27] Entrepreneurs promoted new ideas like satellite telephone service, satellite radio, microgravity research and manufacturing, alternative launch vehicles, solar-power generation, space mining, and space tourism. Experience with satellite telephone systems, one of the most promising commercial space technologies, illustrates the challenges faced by entrepreneurs seeking to extend the communication satellite model to new fields.

Some thirty years after the creation of Comsat, executives at the Motorola Corporation announced their intent to create a worldwide system in which individuals using hand-held telephone units could communicate with each other from nearly anyplace on the globe. Customers would not need to be near a land-based cellular telephone tower to do so; they could be in the middle of the sea. Unlike Comsat, which employed geosynchronous machines, Motorola executives planned to use low-Earth-orbit satellites. Such satellites cost less than their geosynchronous counterparts and allow users to dial into the system using devices smaller and less elaborate than those needed to communicate with geosynchronous satellites. Unlike geosynchronous satellites, which remain at one spot relative to the face of the Earth, low-Earth-orbit devices whiz around the globe. A fully operational communications network set close to the Earth requires many satellites and an extensive switching system in order to keep customers connected. Planners of the system predicted that the technical difficulties of constructing it would be compensated by the ease of deployment and customer use.

Executives originally envisioned a constellation of seventy-seven satellites and established a corporation named after the element possessing the seventy-seventh number on the periodic table of elements, iridium, a metallic substance resembling platinum. As business plans progressed, Iridium executives settled on a sixty-six satellite constellation instead of seventy-seven, but kept the original company name. Dysprosium, the sixty-sixth element on the periodic table, does not possess such an attractive label, sounding more like dyspepsia than an icon for the commercial space frontier.[28]

Executives in the aerospace business estimated that manufacturers could produce 125 low-Earth-orbit communication satellites for about $700 million. That would provide the necessary redundancy and spares and a head start on replacements once the satellites began to run out of fuel in about seven years. Technicians prepared to launch Iridium satellites on a combination of McDonnell Douglas Delta 2 and Russian Proton rockets. At 1,600 pounds each, the satellites were small enough into fit in a bundle of five to seven per rocket, reducing launch costs on the Delta 2 to about $10 million per machine. Altogether, the satellites could be built and launched for about $16 million each, for a total start-up cost of slightly more than $1 billion for satellites alone. Iridium executives believed that the demand for satellite telephone service would exceed overall expenses, making themselves and their investors terribly rich.

Reality intervened. The aggregate start-up costs for the Iridium Corporation were $5 billion. Iridium executives spent that much during the 1990s organizing their business and building the satellite system. The total cost included satellites, launches, ground stations, engineering centers, the operations contract, licensing around the world, marketing, and the expense of meeting payrolls for nine years before the first customers appeared. Company leaders had planned to spend about $3.5 billion, including the allocation for satellite construction and launch.

Iridium executives began serving customers in November 1998. By then, they were spending $440 million every three months. The company was saddled with about $2 billion in long-term debt. Company executives hoped to attract five million users. If five million subscribers paid an average of just $30 per month to use the system, the business would break even. With more subscribers or higher monthly fees, the business could still prove profitable. Iridium executives planned to charge higher fees. The business plan could work.

By the end of March 1999, the company had attracted only 7,200 satellite telephone subscribers—far short of their corporate goal. In the hands of a user, the phone felt like a brick, with an eight-inch antenna protruding above the handset. Subscribers had to pay more than $3,000 to purchase the phones and were charged as much as $6 per minute to make calls. Handsets required a direct line of sight to satellites to avoid signal breakup. Designed to work practically anywhere on Earth, the telephones were susceptible to failure when users in urban areas walked behind high-rise office buildings.

Ten months after commencing service, in the fall of 1999, Iridium executives gave up and declared bankruptcy. They had attracted only 63,000 customers. Before the satellites crashed back to Earth, a group of investors bought the $5 billion, 66-satellite system for a liquidation sale price of only $25 million. The new investors retained the company name but none of the old executives. They reduced service charges and pitched the system at users who needed to communicate from remote locations, particularly workers at oil and gas platforms and military officers in the field. Company executives signed a contract with the U.S. Department of Defense that provided them with a steady stream of revenues. With practically no debt, using assets purchased at fire-sale prices, the company could prove profitable with less than 100,000 customers. As one commentator observed, "you don't have to do a whole lot to have success with a $25 million investment."[29] Even so, new obstacles arose. Company executives had to deal with competition from alternative ground and satellite providers like Globalstar. They faced the prospect of having to spend substantial sums of money to replace aging satellites, out of which system engineers hoped to squeeze performance lasting ten years. Still, their hopes for business success remained high.

Disappointment in the satellite business spilled into the commercial launch industry. By the mid-1990s, government officials were looking for a way to encourage the development of a new generation of launch vehicles. The space shuttle orbiters, originally designed to fly for not more than a dozen years, were growing old. The technology supporting Atlas and Titan rocketry was older still. Within NASA, the intractably high cost of operating and upgrading the space shuttle consumed funds that could otherwise be invested in new rocket technology. Following the model of private-public cooperation then popular in government reform circles, NASA executives devised a creative strategy for raising the funds necessary to develop the next generation of launch vehicles. They formed a partnership with a private firm.

As a means of encouraging private investment, NASA executives contributed $912 million toward the development of the X-33 launch vehicle. Their partner, the Lockheed Martin Aeronautics Company, eventually contributed $356 million. The first phase of the partnership was designed to produce a test vehicle capable of suborbital flight—the X-33. Using the technology perfected during the development of the X-33, NASA executives expected that Lockheed Martin would build the much larger VentureStar, a single-stage-to-orbit spacecraft that would replace the fleet of aging space shuttles.[30]

Under the terms of the partnership, Lockheed Martin executives would raise the funds necessary to build VentureStar from private investors. In other words, no government funds beyond the initial technology investment in the X-33 would be needed. NASA officials promised to be "anchor tenants" for the system. They believed that the government's promise to use VentureStar, in combination with the growing market for commercial satellite services, would allow Lockheed Martin to attract sufficient capital from the private sector. For a modest technology investment, plus help from the private sector, the government could encourage the development of a new launch vehicle that would serve both public and commercial needs.

It seemed like a workable plan. Between the conception of the idea and its execution, however, the underlying premise for the system collapsed. The Iridium Corporation declared bankruptcy in 1999. The dot.com bubble burst in 2000. The number of commercial space launches, which had tripled during the 1990s, suddenly declined. Lockheed Martin executives assessed the climate for private investment and concluded that it was not sufficient to permit the development of VentureStar. Faced with the usual cadre of technical problems, NASA withdrew its support for the X-33 in 2001.[31]

With the X-33, NASA officials set out to test the feasibility of commercial partnerships for the purpose of financing large rocket ships. Perhaps a bit more government spending would have resolved the obstacles. NASA and Lockheed Martin spent $1.3 billion on the failed development effort; as a point of comparison, the government had spent $9.9 billion of appropriated funds to complete the development work on the space shuttle—the equivalent of $27 billion at the time when work on the X-33 was under way.[32]

Undeterred, advocates of space commercialization tried again. In 2006 NASA officials announced the formation of a Commercial Orbital Transportation Services program, a $500 million effort to encourage private companies to transport cargo and eventually astronauts to and from the International Space Station. In a manner reminiscent of the fictional *2001: A Space Odyssey*, in which the head of the National Council of Astronautics rides to the international space station on a Pan American Airlines shuttle, private companies would finance and own the spacecraft. More than twenty private companies competed for incentive funds, with the understanding that their executives would need to raise private capital to complete the financing. In 2008 NASA officials selected SpaceX and the Orbital Sciences Corporation as award recipients.

In announcing the program, NASA administrator Michael Griffin explained that the time had come for the government to become "a lesser rather than a greater part of what humans do in space." Completion of the International Space Station, he continued, would provide "for the first time a strong, identifiable market for 'routine' transportation service to and from LEO [low Earth orbit]," a turning point that "will be only the first step in what will be a huge opportunity for truly commercial space enterprise."[33]

President Barack Obama extended the philosophy in 2010. Confronted with continuing cost and technical difficulties in the government-run Ares rocket program to replace the NASA space shuttle, Obama proposed that the Congress cancel the program and relinquish low-earth orbit transportation to the private sector. He further proposed that Congress appropriate $6 billion over five years—a fraction of what NASA was spending to develop the Ares I— to encourage the development of an American commercial human space flight industry. The president defended his decision at the Kennedy Space Center.

NASA has always relied on private industry to help design and build the vehicles that carry astronauts to space. . . . By buying the services of space transportation—rather than the vehicles themselves—we can continue to ensure rigorous safety standards are met. But we will also accelerate the pace of innovations as companies—from young startups to established leaders—compete to design and build and launch new means of carrying people and materials out of our atmosphere.[34]

Some fifty years earlier, NASA officials had introduced the Mercury Seven astronauts at a special press conference in Washington, D.C. The seven astronauts became the embodiment of the nation's hopes for space exploration. Speaking at the National Press Club in the nation's capital in 2010, NASA administrator Charlie Bolden made a similar presentation. He introduced seven corporate executives as the new "space pioneers." Joining him on the stage were executives from SpaceX, Orbital Sciences, Blue Origin, the Boeing Company, Paragon Space Development Corporation, Sierra Nevada Corporation, and United Launch Alliance. The seven companies, big and small, proposed to build crew and cargo modules, new rockets, environmental control systems, and emergency detection devices. NASA awarded the companies government funds to help start their work, a down payment on what the Obama administration hoped would be a multibillion-dollar effort "to spur the development of American commercial human spaceflight vehicles."[35]

Some people doubt the ability of NASA employees to encourage a new commercial space transportation industry. "They want to do the right things," said one NASA administrator with extensive business experience, "but this is an area they don't understand." The partnership requires NASA to pick the most promising firms, a process normally relegated to the marketplace. Many apply; few are chosen. Says Lou Dobbs, "Any expectations that government will lead the way toward the future of the space business are sorely misplaced. Even though NASA makes public overtures to supporting commercialization of space, the fact is that it is inherently incapable of doing this successfully."[36]

Any system of government support for space commerce incurs two requirements. First, government support must be sufficiently generous to bridge the gap between business profitability and the high expense of entering space. Start-up costs are substantial. As with previous transportation services receiving government support, industrialists must build ahead of demand. They must commercialize new technologies and construct new infrastructure in the hope that people will use it once it is done. The cost, risk, and time required to complete this work during the early stages of space exploration proved so substantial as to discourage most private investment, with the result that government spending dominated the unfolding frontier. This leads to the second requirement. Government support must contain incentives that encourage business firms to alter the gap between profitability and expense. To the extent that cost, schedule, and risk factors can be reduced, commercialization becomes more feasible. Many people believe that the private market, particularly the engine of private competition, creates much stronger incentives for improvements in cost, schedule, and performance than any government policy. Many people in the business community fear that government subsidies serve to perpetuate systems that are costly, lengthy, and work to the detriment of real commercialization.

Sometimes the system of private-public partnerships works. It worked for the communication satellite industry and before that, for canal building, railroad expansion, and aviation. It fostered the development of the global positioning system (GPS) industry.

GPS is an essential component of an electronic economy. The system provides not only aids to navigation but also supplies the precise timing information necessary to conduct financial transactions and allow wireless communication. In 1999, when the industry was new, producers of GPS equipment sold about $6 billion worth of equipment, more than half of that sum accruing to U.S. firms. Eight years later, sales had risen to nearly $60 billion annually.[37]

The concept underlying the system is relatively simple. A person can determine his or her position by knowing the distance and relationship to points that are known. A well-instructed high school geometry student can perform the required calculations. The known points need not be on the surface of the Earth. Ancient mariners used the known positions of celestial bodies like the Sun and stars to calculate the location of ships on an otherwise featureless sea.

Two advances in technology permitted the use of space for position determination. One was the advent of satellites in predictable orbits; the other, the development of extremely precise clocks. The distance from the person using the system to any properly equipped satellite can be derived from the time required for a signal from the satellite to reach the user. The distance between the user and four GPS satellites provides sufficient information (three coordinates plus time) to determine the user's latitude, longitude, and altitude to a remarkable degree of accuracy.

The basic satellite system was developed by the U.S. Department of Defense, beginning in the 1960s. The original idea was prompted by the need to precisely determine the position of nuclear submarines running silently at sea. Once funded, the military developed a wide range of uses for the technology, from the location of troops on patrol to the ability to guide missiles to precise locations.

In the 1980s, even before the system became fully operational, public officials in the United States stated their intention to make signals from military GPS satellites available to private users. (To prevent unauthorized use, the signals had been degraded, much in the manner that companies transmitting television programs scramble their broadcasts.) Significantly, federal officials in the Reagan-Bush administration promised that the receivable signals would be provided free of charge. President William Clinton reiterated this position in 1996. In a formal policy statement, the White House announced that the U.S. government would continue to provide GPS signals "on a continuous, worldwide basis, free of direct user fees."[38]

Military use of the evolving GPS system during the 1990-91 Gulf War excited interest in commercial applications. If military pilots and troops could use receivers to fix their position with ultraprecise confidence, so could sailors, surveyors, oil and gas explorers, vehicle drivers, pilots in commercial and private aircraft, even mountain climbers. A large number of private firms began producing receivers for commercial use. The military completed its twenty-four-satellite system in 1993, and by 1996 worldwide revenues for commercial GPS producers topped $3 billion.[39]

Government deregulation was required to make the private market function effectively. Before 1991, U.S. manufacturers who sought to sell receivers to customers in other countries had to comply with government export restrictions and obtain a license for each shipment. Suppliers complained that less stringent restrictions on foreign producers gave foreign firms a competitive advantage in commercializing a U.S. technology. The restrictions were lifted in 1991.

The GPS market works in much the fashion that advocates of privatization propose. A relatively large number of firms producing GPS equipment compete with each other to reduce costs, satisfy customers, and locate new markets. To characterize this commercial activity as a purely private one, however, is most inaccurate for the simple reason that the government provides the GPS signals for free. The signals take the form of a government asset created for purposes unrelated to commerce that has commercial value. In the same manner that nineteenth-century government leaders provided land as a means to spur the construction of railroad lines, twentieth-century government officials provided electronic signals as a method of encouraging a global positioning industry.

The cost of the satellite system is not insignificant. The federal government spent about $5 billion, including launch vehicles, for the initial deployment of the system. Analysts believe that continuing modernization will require additional expenditures of at least twice that amount. These figures do not include the expense of operating the system, which is likewise significant. Taken together, such expenditures constitute a public subsidy. It is doubtful that any entrepreneur could have raised the capital necessary to construct and maintain a privately owned constellation of positioning satellites.

As such, GPS represents a pleasant amalgamation of government support and private activity. Through an annual appropriation process, public officials created an asset provided at no charge to private users. An industry of relatively small competitors arose to provide paying customers with equipment that would allow them to use the system. Beyond the existence of the GPS signal, the industry required little government support in the form of loan guarantees, price supports, tax credits, or direct cash subsidies, although some regulatory reform ensued.

The tendency to decry government interference while accepting government subsidies is a noble tradition in American politics. It is a theme that dominated the settlement of the last great American frontier. Some of the greatest advocates of individual enterprise in the history of the United States made their fortunes profiting from federally subsidized timber, water, and transportation in the American West. Advocates of space commercialization have welcomed the benefits of government investment when it supports the research needed to develop new technologies such as communication or weather satellites or when it creates assets such as the global positioning satellite system. Other have requested government help in adjusting the terms and conditions under which new commercial activities are allowed to develop. Space industries do not arise in a governmental vacuum. Existing subsidies and regulations distort existing markets, including those into which space entrepreneurs wish to go. If corporate or governmental officials decided to promote the development of a space-based solar energy industry, for example, they would encounter existing practices that subsidize oil, gas, biofuel, and coal production. At a minimum, someone would need to give the new technology an equal opportunity to succeed. Officials would then be obliged to address a series of government policies affecting the new industry, including those that provide access to communication channels, define liability for any damage caused, and regulate the sale of sensitive technologies to foreign nations. No such industry would develop within the context of a totally free market.[40]

The vision of space commercialization is grand. The practical challenges of making it happen are formidable. The traditional government-contractor relationship that existed between government and industry at the beginning of the space age has been supplanted by a commercial sector that is larger, more independent, and in some respects more mature than its governmental counterparts. This is not your grandparents' space program. As space activities have expanded, so has commercial participation, in a repetition of the old tradition of commerce following in the wake of exploration.

11

Back on Earth

The most significant achievement of that lunar voyage was not that man set foot on the Moon, but that he set eye on the Earth.

—Norman Cousins, 1976

No vision consistently anticipates all of the consequences to flow from it. Secondary and unanticipated effects are constant features of practically every effort to reshape the world. So it has been in space. Of the most pronounced results of the venture, a few have been unexpected, odd, and occasionally self-defeating.

Back on Earth, where nearly all of the results of early space exploration occurred, two startling ones appeared. The first was social and not so much unexpected as unusual, given the conflicting traditions out of which the space movement rose. The second was environmental, and most surprising. The first dealt with the issue of who would fly into space; the second with the manner in which humans view the Earth.

A central feature of the spacefaring vision is the belief that anyone—and eventually everyone—will fly. This belief justified the development of winged spaceships like NASA's space shuttle, and it supports the anticipation of space colonies. It anticipates a future in which space travel becomes accessible to ordinary men, women, children, journalists, and schoolteachers. The belief drew inspiration from the history of aviation, a technology motivated by a similar promise at an earlier time when atmospheric flight seemed risky and remote. Aviation became a great equalizer, not only allowing everyone the opportunity to fly as passengers but also broadening the opportunities for those who would serve as pilots and crew members. According to the American story, the aviation movement provided women with one of their first major opportunities to rise from domesticity and act as the equals of men. Women appeared to play key roles in promoting aviation technology, especially as pilots in that new realm.

The first women pilots served, in the words of Joseph Corn, as "the most effective evangelists of aviation in the period." During the formative years of aviation, women pilots promoted the notion that flight was both safe and easy. "More than the men who barnstormed around the country or crossed the oceans by plane, women pilots domesticated the sky, purging it of associations with death and terror," Corn observes.[1] Although outnumbered by men, women pilots proved on a number of occasions that they were equally capable of mastering this complex technology. In the 1936 Bendix Trophy air race, women pilots captured three of the first five places, including the coveted first prize. The Bendix race was the aviation Super Bowl of its day, a contest to determine the fastest flight time across the United States. Until 1935 women pilots were not allowed to compete with men. In the second year of their participation, women pilots placed first, second, and fifth. In the fledgling aviation industry, women served as test pilots, aerial photographers, and pilots for business enterprises. Many, including Louise Thaden, the winner with teammate Blanche Noyes of the 1936 Bendix race, demonstrated and sold aircraft for commercial firms. "Nothing impresses the safety of aviation on the public quite so much as to see a woman flying an airplane," she observed. If a woman could handle the controls, "it must be duck soup for men."[2]

Beryl Markham, a British citizen, worked as pilot for the infamous Kenyan safari guide Baron von Blixen, scouting game from the air and landing on a variety of unimaginable terrains. In 1936 Markham garnered enormous public interest when she nearly became the first person (male or female) to fly alone from England to New York, a prodigious feat given the prevailing direction of the North Atlantic winds. (Markham crash-landed in Nova Scotia, a few hundred miles short of her official destination.)[3] Charles Lindbergh's wife, Anne Morrow, was an accomplished pilot who flew with her husband around the world. In 1930 United Airlines hired the first stewardesses, women trained in nursing who were prepared to take charge of the passenger cabin in an emergency. United could have hired male stewards, following the tradition of railroad porters, but chose women instead in order to help allay the public's fear of flying. During World War II, females served as Women Airforce Service Pilots, or WASPs. They tested aircraft, delivered airplanes, towed gunnery targets, and flew cargo.[4] Flying seemed to give women a degree of freedom and mobility unimaginable to previous generations.

No woman had a greater impact on the expectation of equal opportunity through aviation than Amelia Earhart. An accomplished pilot, she became a media celebrity whose fame was exceeded only by that of Charles Lindbergh. Earhart won public acclaim in 1932 by repeating Lindbergh's triumph on the date of its fifth anniversary, becoming the first woman to fly across the Atlantic alone. A carefully orchestrated public relations campaign with flights of ever-increasing difficulty kept her in the public eye throughout the 1930s. She used her status to promote the idea that flying was safe and that women were equal to men in their flying ability. Earhart disappeared in 1937 during a long-distance flight around the world, an event that immortalized her fame.[5] To many Americans of that period, female pilots seemed like aliens from another world. Competent and confident of their abilities, they presented a form of female independence that would not become a common part of culture for another sixty years. As a new technology, aviation seemed to forecast social trends. At least that is how it seemed.

Beyond aviation awaited astronautics. To a considerable extent, space exploration drew its strength from aviation pioneers. Airplane model builders, aircraft test pilots, and workers from aviation laboratories nurtured the march into space. In spite of the close association between the two, the precedent apparently established by female pilots did not spill over into the realm of space exploration. It occurred neither in the mythology of space nor within the first astronauts corps. One searches in vain through the earliest images of space exploration for female advocates, either real or fictional, with stature equal to that of women in aviation. Having advanced the premise that humans everywhere would benefit from space technology and be eligible for space travel, early exploration advocates excluded a substantial group of potential supporters from producing it. This exposed a fundamental contradiction in popular impressions about space, wholly inconsistent with the aviation tradition. Space travel promised access to all. Yet under the culture norms of the period, the humans appointed to begin the march into space were invariably depicted as males. Well through the twentieth century, space experts referred to missions with humans on board as "manned" space flight, and flight controllers and rocket engineers were invariably men. This symbolic distinction served to separate a substantial portion of the U.S. population from direct involvement in a technology that promised opportunities and benefits to all.

Women wrote science fiction during the period preceding the first space flights, not an unusual development given the large number of female authors and editors at that time. Many literary historians credit the origin of modern science fiction to Mary Shelley, who penned *Frankenstein* in 1818. During the 1930s and 1940s C. L. (Catherine) Moore wrote interplanetary adventure stories, and Leigh Brackett produced works of science fiction before turning to Hollywood screenplays.[6] The stories produced by women such as Moore and Brackett, however, generally repeated the formulas established by their more numerous male counterparts. During the 1920s and 1930s, science fiction was viewed within the literary marketplace as a type of adventure story that appealed almost exclusively to pubescent males. Men were thought to be the primary consumers, and even stories penned by women aimed to gratify boyhood fantasies. The women who appear in these tales are generally defined by their relationship to their male heroes, as objects to be rescued or educated, protected or tamed. Even the strongest females characters, such as the lovely princess Dejah Thoris in Edgar Rice Burroughs series on Mars, eventually submit to their dominant male.

Practically nowhere in early science fiction can one find memorable female characters of strength and independence. Wilma Deering served as Buck Rogers's girlfriend, and Dale Arden remained perpetually engaged to Flash Gordon. In both cases, their relationship to the male protagonist defined the status of these fictional heroines, which often served to confirm his masculinity. Concurrent with the rise of the American feminist movement, science fiction writers produced female protagonists of independent means. In the screenplay for the 1977 film *Star Wars*, George Lucas allows young Luke Skywalker to attempt a rescue of the beautiful Princess Leia Organa, but then surprises him with the discovery that the rich and powerful princess is perfectly capable of rescuing herself. Two years later, Hollywood produced *Alien*, in which the only match for a shrewd carnivore set loose upon a merchant cargo ship is Ripley, the tough, pragmatic crewmember played by actress Sigourney Weaver. This was a total reversal from the traditional approach. When

Gene Roddenberry produced the pilot for the highly influential *Star Trek* television series, a morality play with many liberation themes, he promoted the topic of sexual equality by placing a woman second in command of the starship and dressing the crew in unisex uniforms. This proved too radical for the moguls at NBC, so after *Star Trek* began its weekly serialization, the woman returned as a nurse and the other female characters were given miniskirts to wear. Not until 1995 did a woman receive command of the fictional *Enterprise*.[7]

Well through the dawn of the space age, works of imagination reinforced sexual stereotypes. Rockets and spacecraft were presented as expensive toys for men and boys. Women were portrayed as technologically clumsy and incapable of comprehending engineering technology, to say nothing of their presumed inability to command men or pilot spacecraft. When George Pal produced *The Conquest of Space* in 1955, he portrayed the all-male crew of his Earth-orbiting space station as little more than rowdy, sex-starved sailors at sea, a formula drawn from naval wartime films. Based on the *Collier's* series and the works of Ley and von Braun, the movie was the most serious attempt at space realism during the decade. In its treatment of women, however, it was light-years behind the future it sought to portray. A similar formula greeted viewers of *Forbidden Planet*, released one year later. In that movie the all-male crew of an investigating spaceship discovers an obsessive scientist and his virginal daughter Altaira on a distant planet, the sole human survivors of a lost expedition. Altaira is saved.[8]

In spite of their apparent aviation role, the requests of women for participation in the new space program were greeted by a succession of bad jokes. Responding to a question about the possibility of female astronauts, Wernher von Braun replied in 1962 that the men in charge of the rocket program were "reserving 110 pounds of payload for recreational equipment." When women took their case to Congress, the chairman of the investigating subcommittee explained that "the whole purpose of space exploration is to some day colonize these other planets and I don't see how we can do that without women." When asked for his opinions on the issue, astronaut John Glenn announced tongue in cheek that he would welcome females into space "with open arms."[9]

A realistic opportunity for women's involvement occurred in 1960 when the Lovelace Foundation for Medical Education and Research initiated a program to test female astronauts. The foundation, headed by W. Randolph Lovelace, had been commissioned by NASA to conduct the physical screening tests for the first group of male applicants to the astronaut corps, a process vividly portrayed in *The Right Stuff*. NASA officials distanced themselves from the female tests, insisting that Lovelace was acting on his own initiative. Nonetheless, the U.S. space community viewed the tests with considerable interest, given the news that the Soviets were training a female cosmonaut.[10]

Dr. Lovelace arranged for Geraldyn Cobb to undergo the same type of physical tests given the Mercury astronauts. The twenty-nine-year-old Cobb, one of the most accomplished aviators of her day, had logged more than seven thousand flight hours, set a number of flight records, including a world altitude mark, and been named pilot of the year by the National Pilot's Association. Elated at her selection, she dreamed of leading women into space. Quietly, she trained hard for the first battery of tests.

In August 1960 Lovelace publicly announced the results of Cobb's performance. She had passed the Mercury astronaut tests and would, in the opinion of the physicians that examined her, "qualify for special missions." Cobb was inundated with requests for personal appearances, and a special article in *Life* magazine appeared. Encouraged by the results, Lovelace asked Jacqueline Cochran, one of the leaders of the women's aviation movement, to select more female candidates. Thirteen women pilots, including Cobb, completed the first round of physical tests. Cobb, meanwhile, completed more trials, including the difficult jet orientation and centrifuge tests at Pensacola, Florida. The women proved that they were as capable of space flight as men and, in fact, enjoyed certain physiological advantages, such as lower oxygen intake and less susceptibility to radiation poisoning.[11]

NASA administrator James Webb acknowledged the existence of the test program and appointed Cobb as a NASA consultant. All of the women prepared for further tests.[12] Within NASA, however, controversy over the program grew. What had begun as a program to test the capability of women as space travelers was becoming a female astronaut program. NASA engineers were preoccupied with the task of getting men into space and did not want to divert resources to the preparation of a new group. NASA, moreover, had a rule that required astronauts to possess experience as military jet test pilots. This automatically eliminated women, because the military, in turn, had a rule that excluded them from going to test pilot school. NASA officials wanted the issue to go away before it spun out of control. As the women prepared for a new round of tests at the Pensacola naval facility, NASA officials announced that they had no requirement for the program. Without a requirement, the navy could not complete the tests.[13]

The visionaries of space flight promised that everyone would fly. In an odd departure from the aviation experience, where women played key roles, early space flight leaders excluded a substantial group of potential supporters from this tradition. In 1960 Geraldyn Cobb passed the physical screening tests given to Mercury astronauts. NASA officials disassociated themselves from the testing program after more women pilots applied and passed the exams. Not until 1995 did an American woman pilot an American spacecraft. In this photograph, Cobb is shown next to the Mercury space capsule in which she hoped to fly. (National Air and Space Museum)

Cobb and her colleagues fought back. Cobb gave speeches to supportive audiences and lobbied Vice President Lyndon Johnson, asking him to use his influence as head of the National Aeronautics and Space Council to make NASA reopen the program.[14] Jane Hart, one of the thirteen women to pass the physiological tests (and wife of Michigan senator Philip Hart) pressured members of Congress. In July 1962 a special panel of the House Science and Astronautics Committee held hearings on the issue. Said Cobb in her opening remarks: "There were women on the *Mayflower* and on the first wagon trains west, working alongside men to forge new trails to new vistas. We ask that opportunity in the pioneering of space."[15]

NASA officials refused to set up an additional training program for women astronauts and allowed Cobb's position as a special NASA consultant to lapse. In her place, Webb appointed Jacqueline Cochran. Although Cochran had helped finance the Lovelace tests, her position on female astronauts was well known. "There is no present real national need for women in such a role," she wrote Cobb, not so long as NASA could find a sufficient number of qualified men. Pushing too hard, Cochran warned, would "retard rather than speed" the acceptance of women astronauts. NASA retained Cochran as its consultant on this issue through the decade.[16]

The first woman to fly in space, Valentina Vladimirovna Tereshkova, piloted the Soviet *Vostok 6* spacecraft on 15 June 1963. Not until 1995 did an American woman pilot an American spacecraft. (Sally Ride, the first American woman to fly in space in 1983, was a mission specialist, not a pilot.) To commemorate the event, shuttle pilot Eileen Collins invited the women who had passed the secret physical exams nearly a quarter-century earlier to watch the launch as her guests. Seven of the eleven surviving women arrived, including Geraldyn Cobb.[17]

In rejecting the pleas of women for a place in the early astronaut corps, NASA officials advanced a vision of spaceflight as something too dangerous for women to try and women as too fragile to engage in it. This contravened the aviation tradition and violated the vision of space as a universal realm. It did fit a prevailing advertising symbol of the time, the helmsman, an individual (invariably a man) who through skill and personality took charge of technology and bent it to human purposes. Certainly early space travel was dangerous. Yet that alone did not disqualify women from engaging in it. What did disqualify women was the desire of men to show the public how that danger could be overcome. In aviation, air company executives had a special interest in convincing the general public that flight was safe. It helped them sell planes and seats on them. In space, the people who ran it saw an advantage in convincing the public that rocketry was dangerous and that they were "man enough" to master it. Not until NASA geared up the promotional campaign for the space shuttle did space advocates begin to sense the substantial contradiction in this stance. If space travel was too dangerous for women, then not everyone could fly, and the dream of universal space travel could not be realized. The willingness of NASA executives to entertain the admission of women to the astronaut corps coincided with the deadline for commencing the test flights of the new space shuttle, which had been sold as a machine that would make space travel routine, safe, and accessible to everyone.

To a certain degree, the visionaries of space travel continued a tradition quietly present in the early aviation industry. In spite of the publicity accorded women pilots, they were not given as a matter of practice an equal role in aviation. Nor did most claim it. Women pilots at that time rarely used their position to promote equal opportunity in general. Their words and actions scrupulously avoided what would have been viewed as radical sentiments such as these.[18]

By refusing to build upon the image of women in aviation, pacesetters in astronautics lost an important opportunity to build support for the U.S. space program among a substantial portion of the voting public. People who took large steps in advancing space flight technology remained conservative in challenging social trends. The price they paid is well recorded in the annals of American public opinion. During the 1980s, pollsters revealed a large gender gap in public support for space exploration. Responding to a Media General/Associated Press poll in 1988, 56 percent of the men interviewed said that the U.S. spent "too little" on space exploration; only 25 percent of the women responding agreed. Similar gaps were recorded by the Gallup organization. True to its roots, space exploration remained a program that drew its support disproportionately from men.[19]

NASA's record with minorities was not much better. Although agency leaders appreciated the value of technological demonstration, they seemed insusceptible to the value of social display. Not until 1978 did NASA recruit the first class of astronauts to contain minorities (three African Americans and one Asian American), as well as the first to contain women (six). NASA officials had resisted efforts to broaden the criteria for astronaut selection through the mid-1970s and reversed course only under the strongest pressure. Support for the U.S. space program among African Americans, not surprisingly, remained low.[20]

Within NASA, the participation of women as space scientists and engineers was likewise restrained. Women were employed as computers, a carry-over from the old wind tunnel research days. Donna Shirley, the aerospace engineer who managed the Mars *Sojourner* project, remembered the practice. She arrived at NASA in 1966. The women, she recalled, "were actually called computers. They solved the trajectory analysis and aerodynamic calculations by brute force, punching in lengthy equations in a room full of noisy mechanical calculators." Space exploration, in spite of promoting a message that stressed participation by all, remained throughout its early stages an endeavor conducted largely by white males. When this began to change under social pressure, opportunities for women and minorities expanded.[21]

In the same manner that practice contradicted promise in the realm of equal opportunity, experience confounded expectations in shaping another social trend: the environmental movement. Many memorable images emerged from the early space program: portraits of humans standing on the Moon and close-up photographs of planetary neighbors. As impressive as they were, no image had a greater social impact than the picture of the whole Earth in space.

The first high-quality, full-color photograph to show the whole Earth as it appears in full sunlight from distant space was taken by a machine. In 1967 the ATS-III satellite took a portrait of the Earth from geosynchronous orbit, far enough away to capture the whole sphere. South America is clearly visible in the portrait, as are parts of the United States and Africa. One year later the astronauts on *Apollo 8* became the first humans to witness the scene of the Earth rising above the lunar surface. Astronaut Bill Anders quickly snapped a photograph. The three astronauts read from the Bible on that famous Christmas Eve flight, immutably linking the image of the Earth with a sense of the divine.[22]

The final trio of explorers, who returned home in the last month of 1972, captured a frame-filling photograph that became, in the words of astronomer Carl Sagan, an "icon of our age."[23] At the bottom of the photograph rests Antarctica, surrounded by delicate storms. Rising above it sits the continent of Africa, where human life began. The lands bordering the Mediterranean Sea, from North Africa through the Arabian Peninsula, are clearly visible across the

As technological optimism spurred space advocates to look away from the Earth, astronauts captured images that encouraged people to look back at it. On the outbound leg of the last mission to the Moon in 1972, the crew of Apollo 17 took this picture of the whole Earth. The photograph became an icon for the age, encouraging people to view the planet as a small and fragile globe with limited resources suspended in a cosmic void. (NASA)

top. The oceans are blue, the deserts ocher, the forests green. No national boundaries can be seen.

Every perspective of Earth serves as a metaphor for the beliefs of its day. When humans were stuck to the surface of the planet, plowing crops on lands away from which residents rarely ventured far, the world seemed huge. People so situated could imagine themselves at the center of creation, with Gods in the heavens and Dante's hell beneath their feet. As humans traversed the seas and visited strange civilizations, new perspectives arose. The Earth still seemed large, but a patchwork of nations embracing a multitude of beliefs appeared. Not surprisingly, the age of exploration supported the age of nationalism, as common but separate communities unified into nation-states in order to preserve and extend their cultures. The view from space created a new meta-phor. It became harder for intelligent people to stand on hills and imagine a landscape divided. From space, Earth looks small and whole, like a spaceship traveling through the cosmos. Space exploration fosters an image of a whole Earth, united and interconnected.

"To see the earth as it truly is, small and blue and beautiful in that eternal silence where it floats, is to see ourselves as riders on the earth together," poet Archibald MacLeish wrote as the first astronauts circled the Moon. Ten years later, at a ceremony honoring outstanding astronauts, President Jimmy Carter observed that, "of all the things we have learned from our explorations of space, none has been more important than this perception of the essential unity of our world. . . . We saw our own world as a single delicate globe of swirl-ing blue and white, green, brown. From the perspective of space our planet has no national boundaries. It is very beautiful, but it is also very fragile. And it is the special responsibility of the human race to preserve it."[24]

The view from space accelerated acceptance of Earth as a single interdepen-dent system. To champions of this perspective, the image expedited a new way of thinking, one captured in a host of publications and related imagery. The first issue of the *Whole Earth Catalog*, released in the fall of 1968, displayed the ATS-III photograph of the Earth on a black background, without commentary, as if the image spoke for itself. The United Nations placed a NASA image of the whole Earth on the cover of its report from the first Conference on the Human Environment. Environmentalists sewed the 1972 image on a blue background for their unofficial whole Earth flag. Al Gore used images of the whole Earth to open and close his environmental call to arms in the award-winning docu-mentary *An Inconvenient Truth*.[25]

True to the vision, the world began to operate as a single system. Space-based communication networks, created by space-age technology, created what Mar-shall McLuhan called a "global village." People who heretofore had considered themselves members of national communities received news and gossip from networks that respected no boundaries. Industrialists constructed globally in-tegrated economies that bypassed national barriers. Futurist Alvin Toffler char-acterized this development as the "Third Wave" of civilization. (The discovery of agriculture launched the first wave nearly ten thousand years ago; the sec-ond wave began with the Industrial Revolution.) The third wave, Toffler main-tained, would produce global networks governing human affairs but would not promote the development of a single world society. To the contrary, societies would splinter as communication networks linked people with common interests heretofore separated geographically. "As the Second Wave produced a mass society," Toffler explained, "the Third Wave de-massifies us, moving the entire social system to a much higher level of diversity."[26]

Just as information and money bypass national boundaries in the post-industrial society, so does the detritus of technology. Pollution migrates across the Earth, a fact visibly confirmed by photographs from space. As images from space became part of public consciousness, so did a new sensitivity to environmental issues. Many factors encouraged the worldwide rise of environmental concerns during the 1970s, but one of the most important was the panorama from space. The environmental movement required its adherents to view the Earth and its habitats as a single ecosystem; images from space provided visual confirmation.

Engineers like to think in terms of systems, an approach that emphasizes feedback loops and inclusionary analysis. Just as a thermostat uses a simple feedback loop to regulate the temperature of a room, so large systems employ complex feedback mechanisms to adjust their parts. In 1970 a professor from the Massachusetts Institution of Technology presented a model of the entire Earth to a group of concerned individuals known as the Club of Rome. Club members had encouraged the professor and his colleagues to calculate the interaction of five basic factors: population, agricultural production, natural resources, industrial production, and pollution. By treating the Earth as a single system, the team predicted the way in which each intricate part reacted to changes in the rest.

The findings, published in 1972 under the title *The Limits to Growth*, created a firestorm of controversy. Armed with the vision of Earth as a single system (and aided by computer simulation techniques), the investigating team explored the consequences of unchecked consumption and population growth. The results were disturbing. "If the present growth trends . . . continue unchanged," the team reported, "the limits to growth on this planet will be reached sometime within the next one hundred years." At that point, the Earth would no longer be able to support a large, industrially dependent population. A catastrophic decline in both population and industrial productivity would be the most likely result.[27]

Critics of the report lambasted the team for ignoring the effects of technology. The very technologies that had permitted the vision of a single Earth, the critics argued, could rescue it from overconsumption and growth. Giant space-based reflectors could beam solar energy down to Earth, advances in robotics could permit the development of pollution-free industries, and humans could migrate to other worlds. The Club of Rome team belittled these arguments. Advances in technology, team leaders observed, "would only delay rather than avoid crises" and offered computer runs of their model to prove it. Only by severely limiting population growth and economic expansion could the human race produce a stable equilibrium and avoid ultimate collapse.[28] Conclusions such as these allowed radical environmentalists to scare the public with convincing warnings of impending doom. Global warming, ozone holes, and exotic viruses provided what many believed to be the first signs of an Earth on the slippery slope of disequilibrium.

The most extraordinary notion to receive its impetus from the belief in the globe as a single ecosystem was the Gaia hypothesis. The concept garnered its name from the Greek goddess of the Earth, mother of the sky and the sea. In its more respectable form, the Gaia hypothesis suggests that the Earth is a self-regulating system wherein features such as the climate and atmosphere are continually adjusted by the organisms that inhabited it in order to maintain conditions suitable for their existence. The average mean temperature of the Earth, for example, has responded for millennia to the presence of greenhouse gases such as methane and carbon dioxide, which are naturally produced by termites and other living things. In the past, fluctuations in naturally produced greenhouse gases have helped stabilize temperature shifts on Earth and counteract fluctuations in the energy produced by the Sun.[29]

The more radical proponents of the hypothesis make the startling claim that the Earth is actually alive. "The Gaia hypothesis is the first comprehensive scientific expression of the profoundly ancient belief that the planet Earth is a living creature," one wrote. According to supporters of this idea, the planet regulates itself in the same manner that animal bodies reflexively adjust to changes in temperature, body chemistry, or the introduction of foreign substances. Given this view, the Earth contains systems that act like vital organs in the animal realm: "Regions of intense biological activity, such as the tropical rain forests and coastal seas, are seen as vital not only to their geographic regions but to the entire global environment, much as the liver or spleen is necessary to the survival of the body as a whole."[30] The living Earth may also have the capacity to resist invaders, which, like viruses in animal bodies, threaten to overwhelm the planet through uncontrolled multiplication. A living Earth may view the human population explosion as such a threat and, so threatened, may strike back in the manner of a human body fighting off a virus, by raising its temperature.

It is hard to visualize an idea as wild as the Gaia hypothesis by just standing on the ground. From space, the Earth, with its seasonal bands of vegetation and moving clouds, looks like a dynamic body capable of self-regulation. It appears fragile, susceptible to alteration from the combined effects of the species that dominates it. From a distance, the atmosphere looks as thin as the skin on an apple. Astronauts who have traveled more than once into space have watched rain forests disappear. Is it so hard to then imagine that with insistent human tinkering the Earth might be transformed into a less-hospitable place?

Images of the whole Earth reshaped not only public consciousness but also the NASA space program. Given those images, NASA initiatives gradually redirected resources aimed at the heavens so as to examine the Earth. In the mid-1980s, advocates proposed that space technology be redirected toward a more complete understanding of the only home that humans had. In her 1987 report to the NASA administrator, Sally Ride suggested that the space agency undertake a wide-ranging "Mission to Planet Earth." Although the United States had used satellites and spacecraft to study the home planet previously, the emphasis on Earth studies was a major departure from NASA's basic orientation toward the outward bound. Repeating the philosophy that guided the new environmental consciousness, Ride announced that "interactive physical, chemical, and biological processes connect the oceans, continents, atmosphere, and biosphere of Earth in a complex way." A network of orbiting platforms, she suggested, joined with ground-based observations, would allow scientists to predict changes in the global environment before it was too late. After a substantial debate over the scale of the undertaking, NASA officials won approval for a series of Earth Observing System satellites to monitor the planet. The first major phase of the program began with the 1999 launch of the $1.3 billion Terra satellite, designed to "simultaneously study clouds, water vapor, aerosol particles, trace gases, terrestrial and oceanic properties, and interaction between them and their effect on atmospheric radiation and climate."[31]

Satellites played a key role in helping humans visualize ozone depletion in the stratosphere. Scientists measure ozone depletion using a variety of devices, including aircraft and balloons. Beginning in 1980, satellites confirmed this existence of a large "ozone hole" that formed each year as the atmosphere over Antarctica warmed. In 1987 world leaders agreed to limit the use of gases responsible for ozone depletion through what became known as the Montreal

Protocol. Advances in satellite technology permitted daily monitoring of ozone depletion. The Earth Probe TOMS satellite produced some of the most dramatic images. Using a total ozone mapping spectrometer, the satellite captured a succession of daily images that, when pieced together, provided a motion picture of the ozone hole spreading like an atmospheric tsunami over the continent below.[32] Visual effects from space helped to maintain the commitment to solutions that with time were designed to repair the hole.

When the environmental movement began, its advocates often criticized the space program for spending money that, they said, would be better devoted to problems back on Earth. They even rallied against specific space flights. In 1989 a coalition of environmental groups sought to halt the launch of the space shuttle *Atlantis*. The immediate cause of their concern was the realization that the *Atlantis* would carry a spacecraft (the *Galileo* mission to Jupiter) powered by a radioisotope thermoelectric generator (RTG). Remembering the *Challenger*'s demise, environmentalists warned that a catastrophic accident might release radioactive fuel from the electric power generator into the atmosphere. Five previous U.S. and Soviet spacecraft using RTGs had plowed back into the atmosphere after mission malfunctions, and one (perhaps three) had disintegrated. NASA's reassurance that its triple-containment system would prevent any future dispersal did little to assuage protesters' fears. They worried that NASA had transformed space technology into a broader class of "risky systems" that humans concerned about the future of the Earth could better live without.[33]

Environmentalists soon learned to treat the space program as an ally for their concerns. The 2008 remake of the classic science fiction film *The Day the Earth Stood Still* emphasized the shift. In the original version, a flying saucer captain and his robot Gort land in Washington, D.C., and threaten humans with extinction if they fail to control the nuclear arms race. The alien arrives because humans stand on the threshold of space travel and, in moving outward, seem poised to extend their weapons beyond the Earth. This is prohibited, the alien warns. In the retold story, the captain and robot land in New York City for the purpose of preventing humans from destroying their own planet. "There are only a handful of planets in the cosmos that are capable of supporting complex life. This one can't be allowed to perish."[34]

The shift, though subtle in some respects, represents an important byproduct of space travel. An otherworldly concern certainly motivated the original vision of space travel. Visions of trips to the Moon and planets adopted the classic model of explorers and settlers departing imperfect conditions for a fresh start in new places. Futurist Alvin Toffler predicted the advent of a "throw away" society in which people would simply discard the old.[35] Space technology offered solutions to the dangers imposed by dwindling local resources by promising new ones: new sources of energy, space age materials, and products of unimaginable potential. Many saw in space technology a means to escape a planet growing too small. It is no accident that the fascination with Gerard O'Neill's proposal for space colonies appeared in conjunction with the first images of the whole Earth, tiny in the cosmic sky. O'Neill himself admitted that he sought to resolve the desire for growth with a planet of limited size.

Actual space travel gave humans a new perspective on their own planet and a sense of the importance of preserving what they had. Astronomer Carl Sagan was one of the leading proponents of the outward-bound space program, of galactic exploration and planetary colonization. Toward the end of his life, he wrestled with the consequences of ignoring the planet on which he was born. Reflecting on a photograph of the Earth as a tiny pale blue dot taken by the *Voyager 1* spacecraft in 1990 as it passed beyond the outer planets, he wrote: "Look again at that dot. That's here. That's home. That's us. On it everyone you love, everyone you know, everyone you ever heard of, every human being who ever was, lived out their lives. . . . There is nowhere else, at least in the near future, to which our species could migrate. Visit, yes. Settle, not yet. Like it or not, for the moment the Earth is where we make our stand."[36] Sagan's statement gave new meaning to his consistently held belief that no long-lived technological civilization could ever survive without mastering space travel. Not only did space travel provide an outward-looking manifesto; it also allowed the species that undertook it to look back at its home.

Imagination and Culture

I will always love the false image I had of you.

—Author unknown

The rise of the U.S. space program was due in part to a concerted effort by writers of popular science and science fiction, along with other opinion leaders, to prepare the public for what they hoped would be the inevitable expansion of humanity into space. These people formulated an enticing vision in which humans moved off of the Earth, explored the Moon and planets, establish settlements, and eventually departed for other stars. In constructing this vision, advocates took fantastic images, some drawn from science fiction, and laid them upon ideas already rooted in American culture, such as the myth of the frontier. The resulting vision of space exploration had the power to excite and entertain or, as in the case of the Cold War, to frighten. The vision prevailed over lesser alternatives and moved onto the national agenda not so much as a result of its technical superiority but because it aroused the imaginations of people who viewed it.

In the beginning, practical experience with new ideas is typically limited. People are free to imagine wondrous consequences of new endeavors. Communication of those ideas often takes the form of gospel—literally, the "good news" that will change the world. Where practical experience is limited, acceptance of the gospel must be based on faith, intuition, or the attractiveness of the vision. Imaginations are free to soar.[1]

Practical experience with space flight was entirely lacking during the years in which the initial vision was revealed. The promise of space travel took the form of a revelation not yet experienced by people who would be affected by it. The good news of space travel led to great expectations, firmly supported by other aspects of American culture. By being first in space, proponents proclaimed, the United States would control the Earth and win the Cold War. Space exploration would promote trust in government. Scientists would discover life on other planets and resolve other great mysteries of the universe. Humans would build large space stations; space travel would become easy and cheap. Space travel would rekindle the frontier spirit as humans left Earth and colonized the cosmos. Humans would build robots and robotic spacecraft to help them explore, with humans firmly in charge. Space exploration would revolutionize life on Earth, creating a cornucopia of consumer goods based on high technology. As it came to pass, the unfolding of events often confuted expectations. Many prophecies did not come true, certainly not on the timetables proposed. The prophets of space flight, having created a rich vision of their endeavor, had to deal with the unfolding realization that much of the actual space venture did not fulfill first expectations.

For many years, observers of policy have treated culture and technology as a backdrop to the action taking place on the national stage. Like a painted screen hung behind actors performing in a theater, culture and technology were thought to provide a frame of reference for the actual play, not to influence it directly. Culture did not receive as much attention as politics and personality, in part because the former did not fluctuate as often as the action taking place.[2] This is changing. Scholars have begun to pay more attention to the role of imagination and popular culture within society at large, to the manner in which such forces affect what the public believes, and to the way in which the institutions of government and commerce respond.[3]

The experience with space, vividly revealed through decades of wondering and achievement, suggests a number of ways in which imagination and popular culture shape what might be termed objective reality. This chapter summarizes the lessons drawn from that experience and compares those observations to other areas in which imagination and culture play a critical role.

Imagination and culture affect reality. In the modern view, tangible aspects of the real world do not exist as something separate from the lives of people within it. Rather, many aspects of the so-called real world are socially constructed. The institutions and policies that make up everyday experience depend as much upon social construction as do personal beliefs. Consider the airplane, a technology invented by human beings. Invention of the airplane required someone to discover the principles of aerodynamics preexisting in nature. Discovery, in that respect, led to invention. Yet discovery did not lead automatically to invention. Between discovery and invention lay imagination. Someone had to imagine an airplane that could actually fly. Many tried; the Wright brothers succeeded. In addition to an understanding of aerodynamics, invention of the airplane required supporting beliefs and a vision.

We know that policies and institutions cannot long exist within cultures in which popular beliefs and myths fail to anticipate them. For a policy to succeed, people must be able to imagine an activity taking place and possess a view of the world into which this image comfortably fits. For an institution to exist, people must envision its creation. The social welfare state was not possible so long as people imagined poverty to be an act of God, as inevitable as the weather. That early point of view supported little more than individual acts of charity and Poor Laws designed to keep the destitute from becoming a nuisance.[4] Science and the Enlightenment created a new world view in which people came to believe that circumstances such as poverty and hunger could be overcome. The malleability of poverty was promoted through works of imagination, such as Charles Dickens's *Oliver Twist*, as well as through scholarly tracts.[5]

Likewise, in the U.S. space program, people had to envision it taking place before public action created the institutions that made it happen. Activities in space are such a common part of modern life that it is hard to imagine life without them. Proponents of space flight in the 1950s encountered the opposite problem. No one had gone into space. No one had seen the Earth from the Moon. No one had used a communication satellite. Most of the public encountered space travel through comic strips and science fiction and viewed it as that "Buck Rogers stuff." Works of imagination helped to create an image of this new endeavor. Some works were fictional, such as the portrayal of a lunar expedition in the intentionally realistic 1950 movie *Destination Moon* or the Rocket to the Moon attraction that opened at the Disneyland theme park in 1955. Others were nonfictional, such as the various *Collier's* articles on space exploration and atomic war. These works of imagination helped the public visualize a future in which space travel could take place and helped motivate support for pursuits of the dream.

Individuals construct subjective views of the world such as these. Individuals are socialized into the ways they are expected to behave, exposed to the phenomena in which they are expected to believe, and introduced to the roles they are expected to play. Their subjective views in turn shape the institutions and policies that make up objective reality. Imagination is the engine that drives this process. The space exploration experience helped humans appreciate the power of subjective beliefs in motivating changes that at first seem impossible.[6]

Societal change is often preceded by major shifts in subjective reality. The beliefs and cultures that people construct are not static. People are not socialized into roles and beliefs that exist in perpetuity. Their beliefs and behaviors may change slowly, and they may not change often, but they do change, as the large number of diverse cultures in the history of the world affirms. When beliefs change, the structures that constitute objective reality change with them. A change in beliefs—what is otherwise known as a culture shift—can have a profound effect on objective conditions. Such shifts typically prepare people for significant alterations in society, government, and their personal lives.

The power of culture shifts has been demonstrated in a number of convincing ways, including the notable change in American conservation policy.[7] Few voices in early America extolled the beauty of wilderness or proclaimed the necessity of preserving it. Most people at that time viewed wild areas as savage, uncontrollable, and evil. So long as Americans retained an image of nature as something repugnant, little support could be mustered for conservation. The conservation movement that swept America during the early twentieth century required a radically different view. Art and imagination made this possible.

Before the age of television and motion pictures, most Americans experienced distant wonders through the display of landscape art. Landscape paintings were exhibited in great galleries much as movies would be shown in the century that followed. Beginning in the early nineteenth century, American artists painted and displayed large landscapes that romanticized wilderness areas as places of great natural beauty. By exaggerating geological features and emphasizing the luminosity of light, artists inspired a reverence for American natural wonders that was every bit as powerful as the pride of Europeans in monuments of antiquity such as the Parthenon. In 1823 Thomas Cole gave up his career as a portrait artist and began to paint landscapes of the Catskill Mountains in New York, giving rise to the Hudson River school of American art. Following his first journey into the American West in 1858, Albert Bierstadt became professionally successful by painting natural wonders such as the Rocky Mountains and Yosemite Valley. Thomas Moran accompanied the Hayden expedition to the Yellowstone Plateau in 1871, and his paintings of the plateau, the Tetons, and the Sierra Nevadas helped build support for the national park movement.[8] Congress appropriated twenty thousand dollars for Moran's paintings of the Grand Canyon of the Yellowstone and the Grand Canyon of the Colorado and placed them in the U.S. Capitol.[9] American natural wonders became a source of national pride.

The movement to exalt the American wilderness was further advanced by literary naturalists such as Henry Thoreau and James Fenimore Cooper.[10] Through their art and fiction, Frederic Remington and Owen Wister generated an imaginative view of the frontier that glorified the Wild West.[11] With his 1902 novel *The Virginian*, Wister created the American Western, one of the dominant art forms of the early twentieth century. A major cultural shift in the American mind took hold: people began to view wild areas as places to be preserved, not destroyed. Without this cultural shift, official support for the conservation movement could not have taken place.

In a similar fashion, a culture shift preceded the arrival of the space program. At the midpoint of the twentieth century, most people associated space travel with the work of science fiction writers. It was entertaining but not feasible. Within ten years, that view had changed. Historically, most cultural shifts take place over longer periods of time. The prevailing attitude toward poverty took centuries to change; the rise of the conservation ethic in the public mind took place over a period of one hundred years. The culture shift that gave rise to space exploration, during which space pioneers converted science fantasies into social realities, occurred in less than ten years. The access to American popular culture provided by the producers of television programs, amusement parks, and popular magazines allowed this cultural transformation to take place far more rapidly than had previously occurred.

Where experts disagree, culture and imagination may help to resolve underlying disputes. In the same manner that imagination helps to prepare societies for policy change, so may it help to resolve disputes over the exact shape of the new policy. In a society saturated with information, experts frequently appear to disagree. Where expert opinion seems divided, amateurs often turn elsewhere for guidance. Imagination can affect the resolution of policy disputes by putting a personal face on the positions taken by combating parties. The history of the effort to deinstitutionalize the mentally ill shows how this can occur.

Before the 1960s, mentally ill persons were commonly placed in large state institutions, grim facilities resembling prisons more than hospitals. As if to acknowledge the objective reality of these institutions, their occupants were called inmates—people to be housed, not patients needing care. A group of reformers sought to close the facilities and replace them with community mental health centers, and a few reformers went so far as to suggest that the "crazy" behavior of inmates was a normal reaction to the grim conditions of incarceration in large state-run institutions. Being a temperamental lot, artists have often expressed reservations about confinement of the mentally ill.[12] In 1962 Ken Kesey published *One Flew Over the Cuckoo's Nest*, a novel depicting conditions at an Oregon state mental institution from the point of view of one of the inmates. The story pits the freewheeling Randle Patrick McMurphy, who possesses the power to cure the occupants from within, against the representative of institutional authority, Big Nurse Ratched.[13] The novel, which subsequently became a Broadway play and an award-winning Hollywood movie, helped undermine public confidence in state-run mental health institutions and make deinstitutionalization what psychiatrist Paul McHugh has called the "cultural fashion" of the day.[14] At that time, the empirical evidence on the virtues of deinstitutionalization was mixed. Expert opinion was divided. The artistic portrayal of the issue, however, captivated public attention. It fit so well into the philosophy of the counterculture movement then sweeping the American scene that many people accepted it without question.

Culture and imagination alone did not prompt the push toward deinstitutionalization. Medical technology, which produced drugs for the treatment of mental disorders, as well as changes in the methods of financing mental health care, also favored the dismantling of large state-run institutions. Imagination, nonetheless, played a critical role. When imagination started pushing in the same direction as technology and economics, sweeping policy change was unavoidable. Tens of thousands of mentally ill persons were released from state-run institutions, often to roam city streets as the homeless poor.[15]

Observers of space policy encountered a similar controversy. In the beginning, experts disagreed about the relative advantages of human versus robotic flight. Scientists argued that robotic spacecraft could perform most of the functions of investigation carried out on space expeditions without the direct involvement of human beings; advocates of piloted flight insisted on the necessity of keeping humans "in the loop." In the absence of extensive experience, imagination helped to resolve this issue. People could more easily imagine humans traveling through space than they could envision humans staying behind and allowing machines to do all the work. In early robot stories, humans accompanied what were often more intelligent mechanical companions. As experience accumulated, the advantages of robotic flight became more apparent. Yet this did not dissuade the most vocal advocates of space travel from placing humans at the apex of the endeavor. That is the manner in which imagination pictured the adventure, and that is the way the policy emerged. Even if people had known that robots might be superior devices for exploration, the public would have had difficulty imagining a space program organized with machines in control because practically no works of imagination portrayed the venture that way.

Works of imagination enlarge public interest in policy debates. They put a human face on otherwise torpid issues. Fiction personalizes complex issues, allowing the public to reduce complicated controversies to familiar values such as loyalty and survival. By enlarging the audience in a policy debate, works of imagination alter the balance of power between proponents and opponents of change. E. E. Schattschneider noted in his classic work on political strife that losers typically seek to escape their minority status by asking the inattentive to join the debate. "The most important strategy of politics," Schattschneider observed, "is concerned with the scope of conflict."[16] People who can use imagination to involve otherwise inattentive persons possess the power to influence the outcome of contentious debates.

Images and works of imagination not only prepare the public for change but occasionally precipitate it. The conventional interpretation of societal change envisions both a preparatory period of altered attitudes and some motivating catalyst. A period of advent precedes some precipitating event that leads to structural change. The initial period does not result in an immediate alteration of institutions and policies; rather, it repositions public attitudes in anticipation of some event that will. In their influential work on policy agendas, Frank Baumgartener and Bryan Jones depict long periods of apparent policy equilibrium broken by punctuating events that allow public officials to initiate change. During the period of equilibrium, new assumptions begin to undercut old beliefs. Without the period of preparation, in which the new ideas flourish, the anticipated change lacks meaning. Oddly, this interpretation is drawn not from sociology or political science but from paleontology. In his analysis of the Burgess shale, paleontologist Stephen Jay Gould paints a picture of long periods of relatively undisturbed equilibrium punctuated by short periods of disturbing change that leave old structures ill suited for their new environment. The changes may be random and the alterations arising from them hard to predict in advance, but they nonetheless occur.[17]

Writers of science fiction prepared the public for the idea of space travel by human beings. Purveyors of popular science convinced people that it could really happen. The punctuating events provided by *Sputnik 1* and *Sputnik 2*, along with the Yuri Gagarin flight, motivated public leaders to support the cause. None of these things could have happened without the others.

Can the images and visions that promote new attitudes also provide the social catalyst that precipitates structural change? The most dramatic example of such an event occurred during the first decade of the twentieth century, when a single work of fiction altered public attitudes and motivated the U.S. Congress to create a permanent system for regulating food and drugs. For some time, reformers had promoted the virtues of federal regulation of the nation's food supply. Their requests for legislation were easily checked by the resistance of industry lobbyists, who commanded congressional majorities. In 1906 Upton Sinclair published *The Jungle*, a novel portraying the lives of immigrants who labored in Chicago's stockyards and meatpacking plants.[18] Sinclair's work was supported by a socialist weekly, *The Appeal to Reason*, where the novel first appeared in serial form. Through a work of fiction, the publishers hoped to arouse public sympathy for the plight of real workers and advance a program of socialist reform.

The characters in Sinclair's novel suffer unimaginable tragedies arising from disease and broken health and the hopeless search for employment. Readers were not moved by Sinclair's descriptions of human exploitation so much as they were terrified by the background information he presented on the meatpacking industry, especially the loose system of local inspection that allowed diseased cattle and hogs to slip into the nation's food supply. Sinclair related industry folklore, such as tales of workers who slipped from factory planks above boiling vats to emerge somewhat later as Durham's Pure Beef Lard.

Scarcely a dozen pages in the book described such incidents, but they gripped the attention of the public and made the twenty-seven-year-old author an instant celebrity. "I aimed at the public's heart," Sinclair observed, "and by accident I hit it in the stomach."[19]

As interest in the book rose, sales of meat products fell. Industry efforts to reassure the public through forums such as the *Saturday Evening Post* failed. President Theodore Roosevelt invited Sinclair to the White House and dispatched a special government investigating team to Chicago to assess the charges. The Neill-Reynolds report confirmed the worst and intensified calls for reform. With no other means to restore public confidence in their products, industry officials accepted federal regulation. Less than six months after the release of the novel, Roosevelt signed legislation establishing the modern system of food and drug regulation. The novel had had a direct impact on public policy, enlisting a new audience in such as way as to reshape an old debate.

The first phase of space exploration occurred in the conventional way. A period of preparation repositioned public attitudes; a precipitating event motivated policy change. No subsequent work of fiction served to shift public policy. Yet a few photographic images may have. The image of the whole Earth as captured from the Moon startled the people who saw it. The effect was wholly unanticipated; most people had looked forward to close-up pictures of the Moon. The public was not prepared for the transformational effect of seeing the Earth as a small blue and white orb in the cosmic void. The effect proved as powerful in helping to coalesce the emerging conservation movement as Upton Sinclair's *Jungle* had helped to encourage government regulation of food.

A similar yet disappointing image appeared in 1965 when the *Mariner 4* spacecraft produced the first close-up pictures of the planet Mars. Seventy years of anticipation had helped the public visualize a sister planet, inhabitable although more arid and desertlike than the Earth. The real image shocked viewers. In combination with the photographs of the whole Earth that shortly followed, the images helped to push the space program back toward a more Earth-centered position in which humans could contemplate the necessity of preserving the apparently rare and special planet on which they currently lived.

Advocates of change often utilize metaphors to explain their visions. As if to complicate this process, the images that help to motivate and precipitate institutional change are socially constructed too. Advocates frequently use metaphors to explain impending and unfamiliar change. The metaphors have a subjective quality.

The very act of imagining how things work influences their operation. At the beginning of the twentieth century, industrialists attempting to visualize large business organizations utilized the metaphor of the machine. It should come as no surprise, then, that industrialists who viewed the firm as a complex machine treated their employees as cogs, buying and selling labor as they would a fan belt or an electric motor, or sought to engineer and control business practices. Gareth Morgan calls this "imaginization."[20] He identifies a half-dozen such metaphors that business and public executives have used to visualize their work. Such metaphors have a self-fulfilling quality. People use them to organize reality. The image of the corporation as a machine produces an institutional reality that is quite different from that of people who view their organization as a brain.

In American government, the dominant cultural image of the public servant is that of the foolish and frequently self-serving bureaucrat. This too has a machine-like quality, in which adherence to hierarchy and rules seems to overpower common sense. The Ministry of Love exists solely to perpetuate the jobs and power of the people who torture their subjects in George Orwell's *1984.* Colonel Cathcart in *Catch-22* endangers the lives of the pilots under his command so that he can win a promotion. A bureaucrat releases hundreds of demons onto New York streets after learning that a *Ghostbusters* team has captured and stored toxic spirits without the proper government permits. Portraying public servants as mindless ciphers is a well-established literary formula guaranteed to please audiences seeking to be entertained.[21]

The formula works well because it fits so closely with other American ideas. Alexis de Tocqueville noted during his nineteenth-century visits that American pioneers depended more on each other than on public officials to resolve mutual affairs. Community barn raising became an icon of this attitude among Americans so inclined to believe.[22] The writings of Mark Twain, the preeminent American novelist, contain a prominent anti-institutional streak.[23] The American western and the frontier myth that helped to sustain it reinforce the idea that the highest moral virtues emerge from hardy individuals working together in a pregovernmental "state of nature" before officialdom arrives. With so many works of imagination preaching this theme, it is hard to imagine the idea of the good bureaucrat.

Likewise, the early history of space flight is replete with metaphors. Space would be the final frontier. Expeditions into space would be like terrestrial expeditions. Spaceships would be like airplanes. Extraterrestrials would look like us, and robots would be faithful servants.[24]

The necessity of using metaphors arises from both mind and media. The human mind thinks new thoughts by recombining familiar images in creatively different ways. The process is called heuristic decision making, and its understanding has won Nobel Prizes and provided the foundation for the concept of artificial intelligence. The mind does not simply invent new ideas. Rather, it searches through existing memories for metaphors that can be used to make sense of new and unfamiliar events.

In the modern world, the media amplifies this process. From the printing press to the television set, methods of communication are known to shape the content of the messages they transmit.[25] In the modern electronic world, the speed at which televised information must be presented discourages the transmittal of complex ideas. Competition for viewers pushes producers to create "infotainment." Both tendencies encourage the use of simple metaphors that have a known capacity for attracting viewers. By treating space as a frontier, for example, purveyors of electronic information gain the assurance of an adequate audience and the confidence that their ideas will be easily understood. The success of the original *Star Trek* television series was due not so much to its quality as science fiction but to the fact that producer Gene Roddenberry told his writers to treat the series as if it was a Western, for which ready audiences at that time already existed.

The use of metaphors promote gaps between expectation and reality. The new realities created by the pursuit of subjective ideas rarely turn out to be as achievable as their supporting visions prescribe. To help people visualize the winged space shuttle, NASA officials utilized the metaphor of the airplane. In one early treatment of the subject, Wernher von Braun suggested that the government could launch two winged space shuttles from Earth every day for a period of six months.[26] NASA was hard pressed to fly the real space shuttle more than six times per year. Obstacles arise. Nature surprises. The future is hard to visualize. The resulting gap widens when advocates communicate their visions through metaphors. Rooted in the past—often in a past that is itself imagined—metaphors do not fully capture the nuances of each new undertaking. A society that relies upon metaphors to motivate change almost certainly invites episodes in which the unfolding reality surprises the people who anticipated it.

The resulting gaps cause some people to grow disillusioned with the policy or the change. This is especially true for people who were inclined to doubt the vision when it first appeared. Motivation wanes, difficulties appear. As the course of space exploration became more apparent, attacks on the originating vision grew. "The dream of space travel is glorious," sociologist William Sims Bainbridge observed, "the contemporary reality is dismal."[27] Completion of the initial flights to the Moon inaugurated a period in which knowledgeable people began to sense that the United States was not going to complete the grand vision of exploration and settlement, certainly not soon. "It's over," concluded one critic of big space endeavors in 1994. "Turn out the lights, have a nice life, you're out of here."[28] Author Paul Theroux offered a similar glimpse of the real future: "Forget rocket-ships, super-technology, moving sidewalks and all the rubbishy hope in science fiction. No one will ever go to Mars and live. A religion has evolved from the belief that we have a future in outer space; but it is a half-baked religion—it is a little like Mormonism or the Cargo Cult. Our future is this mildly poisoned earth and its smoky air. . . . There will be no star wars or galactic empires and no more money to waste on the loony nationalism in space programs."[29]

Many people blamed public disinterest in the civil space program on the challenges of repetition. The United States landed on the Moon six times; it grew dull. The United States launched the winged space shuttle more than one hundred times. The sins of repetition, however, may have lay in excessive expectations. Had Americans, for a reasonable cost, been able to launch winged spaceships fifty times in a single year, would the resulting accumulation of infrastructure in space excited support or dulled it further? Lacking experience, it is hard to tell. What we do know is that the use of metaphors such as the airline industry to explain space travel created expectations that were hard to achieve.

Works of imagination have become so pervasive in American culture that the ability of the government to satisfy all of the expectations created by them grows narrower as the number of images expands. Politicians are obliged by the nature of their jobs to satisfy public expectations, but the expectations that imagination creates grow more numerous and less attainable. Such a situation undermines institutions and generates public distrust with amazing speed.[30]

The strongest believers resolve gaps with higher levels of commitment. Policy advocates frequently oversell the programs they promote. The head of the Atomic Energy Commission, Lewis L. Strauss, predicted in a 1954 speech to the National Association of Science Writers that atomic energy would create electricity so cheap that the power industry would not have to meter it.[31] A series of aviation advocates promised that airplanes would become so inexpensive and easy to fly that by the second half of the twentieth century the personal airplane would replace the automobile in the family garage.[32] In launching the development program for the nation's space shuttle, President Nixon repeated the oft-heard claim that the new transportation system would reduce the access cost to space "by a factor of ten."[33] Endeavors launched with great expectations breed imagination gaps.

What happens to the bearers of the original vision when gaps such as these appear? For some, disillusionment and retreat occurs. In the case of space exploration, this often means a return to fantasy and science fiction, to Star Trek conventions and alternative worlds, a realm that early advocates of exploration worked hard to overcome. The symbolic watershed in this development was the decision of the Walt Disney Company in the early 1990s to close its Mission to Mars ride. The direct descendant of the original Rocket to the Moon attraction that opened during the first year of park operation, Mission to Mars had helped keep alive the belief that space travel was real. As the reality of space travel set in, interest in the ride waned. At Walt Disney World, Disney executives replaced Mission to Mars with Alien Encounter, a total fantasy featuring an extraterrestrial being accidently beamed into a spaceship taking an imaginary voyage. "One way for an attraction to remain timeless is for it to be based in fantasy, rather than reality," a Disney representative observed.[34]

Other individuals react to imagination gaps by shifting the objects of their attention. Their interests change. As young people, they are exposed to a different set of stories. In a fascinating but somewhat-forgotten study undertaken in the 1950s, David McClelland worked to explain the features of various civilizations by examining the content of the folk tales, popular music, and elementary school readers to which the inhabitants were exposed when they were young.[35]

In this regard, the shifting nature of science fiction may foretell the types of science policies that new generations embrace when they become adults. The first generation of Americans to experience space travel was raised on stories about rocket ships, galactic travel, and alien beings. The high-water mark of this perspective occurred between 1966 and 1968 with the inauguration of the *Star Trek* television series and the release of *2001: A Space Odyssey*. Subsequently, the message began to shift.

Since that time, new generations have been increasingly exposed to stories better suited to an electronic and biotechnological age. The stories feature cyborgs, rogue machines, avatars, and alternative universes. In *Battlestar Galactica*, humans escaping from a group of doomed planets do battle with a race of cybernetic machines called Cylons. The story ran for more than thirty years in various forms beginning in 1978. In 1999 the first of the *Matrix* films appeared. The story takes place entirely on Earth, although the setting is as alien as any portrayed in a science fiction film. Ten years later, James Cameron released *Avatar*. In that movie, which won awards for art direction and visual effects, humans travel to the lush, Earth-like moon Pandora by conventional means—an interstellar spaceship. Owing to the planet's toxic atmosphere, however, the Earthlings are obliged to interact with the local inhabitants through unconventional means. Scientists genetically produce replicas of the ten-foot tall, blue-skinned Na'vi and operate them through telepathic means in the manner of an operator assuming the role of an avatar in a computer game.[36]

For six seasons ending in 2010, occupants of a mysterious South Pacific island encounter time travel, parallel universes, and other exotic phenomena. The television series *Lost* did not involve conventional space travel; its characters found themselves transported to mysterious places without ever leaving the surface of the Earth.[37] The series was oddly reminiscent of the Barsoom novels that Edgar Rice Burroughs began writing in 1911, in which the principal character John Carter transports himself to Mars simply by falling into a trance. In both cases, unknown laws of physics allow humans to visit alternative places by merely extending their imaginations.

For the most devoted advocates, the appropriate response to unfulfilled expectations has been neither retreat nor disillusionment. They recommitted themselves to the original vision, with few modifications. They continued to promote human space flight, space stations, lunar visits, missions to Mars, and the search for extraterrestrial life. Persistence of the spacefaring vision in the face of adversity is one of its most remarkable features.

For the next generation, the shifting nature of recent stories may foretell a substantial modification. That generation retains the traditional fascination with unfamiliar places and visits to them—the core of the space travel vision. Yet their preferred routes of discovery rely less upon mechanical spaceships and space stations than upon physics, electronic intelligence, and the reconstruction of human beings.

The response of the first group is not atypical. For some time, social scientists have observed such behavior among groups of true believers in many causes. Forced to confront emerging realities that do not confirm to their expectations, such groups work hard to maintain their beliefs even as their prophecies fail. They increase their level of proselytizing or search for new ideas that confirm old behavior.[38] They ignore unfulfilled predictions, issue rationalizations, reinterpret reality, and discourage members who want the group to modify its beliefs. Social scientists characterize persistent gaps between information and expectations as cognitive dissonance. The information believers receive conflicts with the preconceptions in their minds. True believers will not abandon their beliefs just because they suffer cognitive dissonance, especially if they remain close to other believers. The phenomenon has been observed within groups both strange and conventional.[39]

As for the second group, the consequences of their visions are hard to anticipate. Their interests may remain within the realm of fantasy, or they might just motivate new policies. Regardless of the outcome, one observation seems certain. Any shift in policy is highly unlikely without a preceding shift in storytelling, and the latter—fed by developments in the sciences—has begun to occur.

Persistence of the underlying space exploration vision suggests that it possesses a power for its advocates that goes well beyond an interest in the specific technologies or originating events. Much attention has been given to the influence of the Cold War in explaining Americans' early enthusiasm for space exploration. Motivation within the United States to engage in the Moon race and other great events of the first dozen years of space flight clearly arose from Cold War competition.[40] Now that the Cold War has ended, a substantial part of original rationale for government-sponsored space flight has disappeared. Yet the vision and the willingness to invest in space remains. Advocates of space travel complain that the country spends too little on space flight, but as of the first part of the twenty-first century the U.S. Congress allocated more funds to the National Aeronautics and Space Administration than it did to the combined expenditures of the Environmental Protection Agency, the Federal Bureau of Investigation, and the U.S. National Park Service.[41] Private-sector investments in space activities such as communication satellites dwarf government investments. Space flight is a huge part of American society, for both practical and inspirational reasons.

The inspirational reasons contain a deeper purpose always present but not so frequently discussed in more practical considerations. The purpose has at its roots a type of faith associated with religious movements, lending to space exploration a "higher purpose" that helps to explain why the vision persists even as motivating factors change and obstacles appear.

Religions call upon followers to practice a "childlike faith" that does not question, but rather accepts the movement toward ultimate goals. Faith is "the substance of things hoped for, the evidence of things not seen."[42] Faith entails a personal commitment that allows the suspension of disbelief, even when that commitment lacks solid evidence such as might be provided by living witnesses or scientific observation of natural phenomena. For a scientific proposition to be susceptible to investigation, it must possess the characteristic of falsifiability. In other words, the proposition must be stated in such terms as to allow investigators to prove it wrong. Faith entails something else. It typically requires the faithful to accept a doctrine on the basis of its desirability. Faith need not survive repeated experimentation; it need only retain a constant appeal even when the realities that emerge from it seem to confound expectations.

The promulgation of faith requires not only an underlying vision but a means of transmitting it. The vision must be transmitted broadly and then, in spite of possible arguments to the contrary, widely accepted. The resulting process creates a community of individuals who share a common set of beliefs, what some have termed a community of imagination.[43]

Utilizing the American vernacular, the communication and entertainment industry brought the space narrative out of the scientific and technological world and into the homes of millions of people. Believers in the spacefaring vision functioned as an imagined community because they shared a common narrative regarding space exploration and the future it could provide. In the case of the spacefaring vision, validation occurred in large measure because the media covered it. The media primed Americans to expect space travel using images to which the public could reliably relate.[44] In scarcely ten years, the mental image of space exploration evolved from a form of collective fantasy, designed largely to entertain, into a nationally funded quest for reality.[45] From Chesley Bonestell's paintings of Mars to the voyages of the Starship

Enterprise, the images seemed strangely exotic yet beckoningly familiar. No prerequisite language or scientific skills were required to understand them.[46] The images, nonetheless, led to a common belief—acceptance in this case that space travel and eventual settlement could actually occur.

When the media gives attention to certain ideas, it serves to validate them. By paying attention to space flight before it occurred, the media helped to create an atmosphere of credibility.[47] With the effort under way, coverage continued. Despite the complaints of the faithful, space travel (especially when accidents occur) receives far more media attention than its ranking on the list of national priorities would otherwise justify.

Through communication and media priming, the spacefaring vision came to occupy a central position in American culture, to the point that no public leader with the power to do so could seriously consider abandoning the effort without severe consequences. Space exploration, in the terms used to visualize it, became part of an American higher agenda. The concept refers to a limited number of American movements, often with religious overtones, that test and define the national character. When John Winthrop addressed the Puritans preparing to establish a new community in what would become present-day Massachusetts, he reminded them of the overarching purpose of their undertaking. The year was 1630. The Puritans had left England because they believed that the British people had abandoned their higher purpose, in that case their covenant with God. "The eyes of all people are upon us," Winthrop said. If they failed to achieve their purpose, they would be made "a story and a byword throughout the world." If they succeeded, they would transform the lives of many people to follow. He recounted the purpose for which they had come and announced, "For we must consider that we shall be as a city upon a hill."[48]

The same text has been applied to the vision of space exploration. Its strongest advocates view it as more than an outgrowth of technology, a government policy, or a commercial opportunity. For them, it is an effort to maintain the most salient features of national life. When faced with adversity, their natural reaction is not abandonment, but persistence.

Notes

INTRODUCTION: Imagination

Epigraph: Michael Crichton, *Sphere* (New York: Alfred A. Knopf, 1987), 348.

1. See, for example, John M. Logsdon, *The Decision to Go to the Moon: Project Apollo and the National Interest* (Cambridge, Mass.: MIT Press, 1970); Walter A. McDougall, *The Heavens and the Earth: A Political History of the Space Age* (New York: Basic Books, 1985). See also the extensive list of histories available through the NASA History Office at history .nasa.gov (accessed 1 June 2009).

2. See, for example, Lewis Thomas, *The Fragile Species* (New York: Scribner's, 1992); Roderick Nash, *Wilderness and the American Mind*, 4th ed. (New Haven, Conn.: Yale University Press, 2001); Joseph J. Corn, *The Winged Gospel: America's Romance with Aviation* (1983; Baltimore: Johns Hopkins University Press, 2002); Benedict Anderson, *Imagined Communities: Reflections on the Origin and Spread of Nationalism* (London: Verso Editions, 1983).

3. Arthur C. Clarke, "Hazards of Prophecy: The Failure of Imagination," in Clarke, *Profiles of the Future: An Inquiry into the Limits of the Possible* (New York: Harper and Row, 1962).

4. See the works of Steven J. Dick, *Plurality of Worlds: The Origins of the Extraterrestrial Life Debate from Democritus to Kant* (New York: Cambridge University Press, 1982); *The Biological Universe: The Twentieth-Century Extraterrestrial Life Debate and the Limits of Science* (New York: Cambridge University Press, 1996); or *Life on Other Worlds: The 20th-Century Extraterrestrial Life Debate* (New York: Cambridge University Press, 1998).

5. See Frederick I. Ordway and Randy Liebermann, eds., *Blueprint for Space: Science Fiction to Science Fact* (Washington, D.C.: Smithsonian Institution Press, 1992).

6. See W. Lance Bennett, *News: The Politics of Illusion*, 7th ed. (New York: Longman, 2006); James Fallows, *Breaking the News: How the Media Undermine American Democracy* (New York: Pantheon Books, 1996).

7. "Now, my suspicion is that the universe is not only queerer than we support, but queerer than we *can* suppose." J. B. S. Haldane, *Possible Worlds and Other Papers* (New York: Harper and Brothers, 1927), 298.

8. See Leon Festinger, Henry W. Riecken, and Stanley Schachter, *When Prophecy Fails* (Minneapolis: University of Minneapolis Press, 1956).

9. See Carl Sagan, *Contact: A Novel* (New York: Simon and Schuster, 1985); Sagan, *Pale Blue Dot: A Vision of the Human Future in Space* (New York: Random House, 1994).

CHAPTER 1: The Vision

Epigraph: Prov. 29:18 King James Version. See also Lyndon B. Johnson, "Remarks upon Viewing New Mariner 4 Pictures from Mars," 29 July 1965, *Public Papers of the Presidents of the United States* (Washington, D.C.: GPO, 1966), 805.

1. NASA, "Minutes of Meeting of Research Steering Committee on Manned Space Flight," 25–26 May 1959, 2, NASA Headquarters, Washington, D.C.

2. NASA Office of Program Planning and Evaluation, "The Long Range Plan of the National Aeronautics and Space Administration," 16 December 1959, 28, NASA History Office, NASA Headquarters, Washington, D.C.

3. A few persons suggested that objects might not travel *into* space, but *through* it, as within wormholes that evaded the familiar constraints of space and time. Early suggestions for human reengineering, as in the creation of cyborgs better suited for space travel, quietly disappeared. One of the more interesting alternatives suggests that any species that engages in technological advancement for a sufficiently long enough time will abandon its corporeal form. See Steven J. Dick, "They Aren't Who You Think," *Mercury* 32 (November–December 2003): 18–26; Olaf Stapledon, *Last and First Men: A Story of the Near and Far Future* (New York: J. Cape and H. Smith, 1931); Carl Sagan, *Contact: A Novel* (New York: Simon and Schuster, 1985); Manfred E. Clynes and Nathan S. Kline, "Cyborgs and Space," *Astronautics*, September 1960, 26–27, 74–75; and Roger D. Launius and Howard E. McCurdy, *Robots in Space: Technology, Evolution, and Interplanetary Travel* (Baltimore: Johns Hopkins University Press, 2008).

4. Willy Ley, *Rockets: The Future of Travel beyond the Stratosphere* (New York: Viking Press, 1944), chap. 1; Wernher von Braun and Frederick I. Ordway, *Space Travel: A History* (New York: Harper and Row, 1985), chap. 1; Johannes Kepler, *Somnium: The Dream* (Madison: University of Wisconsin Press, 1967).

5. Edward Everett Hale, *The Brick Moon* (1869; reprint, New York: Spiral Press, 1971); also see Frederick I. Ordway, "Dreams of Space Travel from Antiquity to Verne," in *Blueprint for Space: Science Fiction to Science Fact*, ed. Frederick I. Ordway and Randy Liebermann (Washington, D.C.: Smithsonian Institution Press, 1992), 35–48; and Ley, *Rockets*, chaps. 1 and 2.

6. Edgar Rice Burroughs, *A Princess of Mars* (Garden City, N.Y.: Doubleday, 1917).

7. William Shakespeare, *The Tempest* (New York: Oxford University Press, 1987); Nicholas Nayfack, *Forbidden Planet* (MGM, 1956).

8. Daniel Defoe, *Robinson Crusoe* (1719; New York: Knopf, 1992); C. S. Lewis, *The Lion, the Witch and the Wardrobe* (New York: Collier Books, 1950); Lewis Carroll, *Alice's Adventures in Wonderland and Through the Looking Glass* (New York: New American Library, 1960); L. Frank Baum, *The Wizard of Oz* (New York: Ballantine Books, 1979).

9. K. E. Tsiolkovsky, "Exploration of the Universe with Reaction Machines" (1926), in *Collected Works of K. E. Tsiolkovskiy*, vol. 2, *Reactive Flying Machines*, ed. A. A. Blagonravov, NASA technical translation TT F-237 (Washington, D.C.: National Aeronautics and Space Administration, 1965), 212.

10. Jules Verne, *From the Earth to the Moon; All Around the Moon* (New York: Dover Publications, 1962); Ron Miller, "The Origin of the Rocket-Propelled Spaceship," *Quest* 4 (Winter 1995): 4–7.

11. "The Autobiography of K. E. Tsiolkovskii," in *Interplanetary Flight and Communi-*

cation, vol. 3, no. 7, ed. N. A. Rynin (Jerusalem: Israel Program for Scientific Translations, 1971), 3; K. E. Tsiolkovsky, "Investigation of Universal Space by Reactive Devices" (1926), in *Works on Rocket Technology,* ed. M. K. Tikhonravov, NASA technical translation TT F-243 (Washington, D.C.: National Aeronautics and Space Administration, 1965), 208–15; K. E. Tsiolkovsky, *Beyond the Planet Earth,* trans. Kenneth Syers (New York: Pergamon Press, 1960). See also Michael Stoiko, *Pioneers of Rocketry* (New York: Hawthorne Books, 1974).

12. Hermann Oberth, "From My Life," *Astronautics* 4 (June 1959): 39.

13. Fritz Lang, *Frau im Mond* (1929; available through Foothill Video, Tujunga, Calif.).

14. Willy Ley, *Rockets, Missiles, and Men in Space* (New York: Viking Press, 1968), 114.

15. Taken from Frank H. Winter, "Man, Rockets and Space Travel," *Mankind* 2 (December 1969): 23.

16. Esther C. Goddard and G. Edward Pendray, eds., *The Papers of Robert H. Goddard* (New York: McGraw-Hill, 1970), 1:7. See also von Braun and Ordway, *Space Travel,* 44; and R. H. Goddard to H. G. Wells, 20 April 1932, in J. D. Hunley, "Robert H. Goddard, Enigmatic Space Pioneer," unpublished paper, 1993, NASA History Office.

17. Quoted by Tom Crouch, " 'To Fly to the Moon': Cosmic Voyaging in Fact and Fiction from Lucian to Sputnik," in *Science Fiction and Space Futures, Past and Present,* ed. Eugene Emme, AAS History series, vol. 5 (San Diego: American Astronautical Society, 1982), 8.

18. Goddard and Pendray, *Papers of Robert H. Goddard* 1:117, 419–20; 3:1612.

19. Ibid., 1:117, 121; Robert H. Goddard, *A Method of Reaching Extreme Altitudes,* Smithsonian Miscellaneous Collections, vol. 71, no. 2 (Washington, D.C.: Smithsonian Institution, 1919). See also John M. Logsdon, ed., *Exploring the Unknown: Selected Documents in the History of the U.S. Civil Space Program,* NASA SP-4218 (Washington, D.C.: National Aeronautics and Space Administration, 1995), 1:86–133.

20. "Topics of the Times," *New York Times,* 13 January 1920; Frank H. Winter, *Rockets into Space* (Cambridge, Mass.: Harvard University Press, 1990), 19, 29.

21. Frank H. Winter, *Prelude to the Space Age: The Rocket Societies, 1924–1940* (Washington, D.C.: Smithsonian Institution Press, 1983).

22. *Bulletin,* American Interplanetary Society, June 1930, reproduced in Ordway and Liebermann, *Blueprint for Space,* 111.

23. G. Edward Pendray, "32 Years of ARS History," *Astronautics and Aerospace Engineering* 1 (February 1963): 124.

24. Willy Ley, "How It All Began," *Space World* 1 (June 1961): 48, 50.

25. Woodford A. Helfin, "Who Said It First? 'Astronautics,' " *Aerospace Historian* 16 (Summer 1969): 44–47. See also Winter, *Prelude to the Space Age,* 25.

26. H. E. Ross, "The B.I.S. Space-ship," *Journal of the British Interplanetary Society* 5 (January 1939): 4–9; R. A. Smith, "The B.I.S. Coelostat," *Journal of the British Interplanetary Society* 5 (July 1939): 22–27; Arthur C. Clarke, "An Elementary Mathematical Approach to Astronautics," *Journal of the British Interplanetary Society* 5 (January 1939): 26–28; and other articles in the January 1939 and July 1939 issues of the journal. See also H. E. Ross, "The British Interplanetary Society's Astronautical Studies, 1937–39," in *First Steps Toward Space,* ed. Frederick C. Durant and George S. James, Smithsonian Annals of Flight no. 10 (Washington, D.C.: Smithsonian Institution Press, 1974), 209–16;

Arthur C. Clarke, "We Can Rocket to the Moon—Now!" *Tales of Wonder* 7 (Summer 1939): 84–88.

27. Winter, *Prelude to the Space Age*; Frederick I. Ordway and Mitchell R. Sharpe, *The Rocket Team* (New York: Thomas Y. Crowell, 1979).

28. Ley, "How It All Began," 23–25, 48–52; *Current Biography, 1953* (New York: H. W. Wilson, 1954), s.v. "Willy Ley," 356–59.

29. Walter Dornberger, quoted in Ordway and Sharpe, *Rocket Team*, 18.

30. Quoted in Daniel Lang, "A Reporter at Large: A Romantic Urge," *New Yorker* 27 (21 April 1951): 74.

31. Ordway and Sharpe, *Rocket Team*, 19.

32. Michael J. Neufeld, *The Rocket and the Reich: Peenemünde and the Coming of the Ballistic Missile Era* (New York: Free Press, 1995); Neufeld, *Von Braun: Dreamer of Space, Engineer of War* (New York: Alfred A. Knopf, 2007); Ordway and Sharpe, *Rocket Team*, 1–11, 79, and 251–52.

33. Ernst Stuhlinger and Frederick I. Ordway, *Wernher von Braun: Crusader for Space* (Malabar, Fla.: Krieger Publishing, 1994), chap. 2.

34. Quoted in *Current Biography, 1953*, s.v. "Willy Ley."

35. G. Edward Pendray, "The First Quarter Century of the American Rocket Society," *Jet Propulsion* 25 (November 1955): 587.

36. Ibid.

37. The appendix to the novel, a technical treatise, was published in Germany and the United States. See Wernher von Braun, *The Mars Project* (Urbana: University of Illinois Press, 1953). The novel was later issued as von Braun, *Project Mars: A Technical Tale* (Wheaton, Ill.: CG Publishing, 2006). Von Braun later published a fictional account of a trip to the Moon. See Wernher von Braun, *First Men to the Moon* (New York: Holt, Rinehart & Winston, 1958) and "First Men to the Moon," *Reader's Digest*, January 1961, 175–92.

38. See Stephen J. Pyne, "Seeking Newer Worlds: An Historical Context for Space Exploration," in *Critical Issues in the History of Spaceflight*, ed. Steven J. Dick and Roger D. Launius, SP-2006-4702 (Washington, D.C.: National Aeronautics and Space Administration, 2006), 7–36.

39. See Steven Pyne, *The Ice* (Iowa City: University of Iowa Press, 1986); Henry M. Stanley, *How I Found Livingston* (Montreal: Dawson, 1872); David Mountfield, *A History of African Exploration* (New York: Hamlyn, 1976); Richard A. Van Orman, *The Explorers: Nineteenth Century Expeditions in Africa and the American West* (Albuquerque: University of New Mexico Press, 1984); Alfred Runte, *National Parks: The American Experience*, 2nd ed. (Lincoln: University of Nebraska Press, 1987); George Kennan, "Announcement," National Geographic Society, Washington, D.C., October 1888; Sam Moskowitz, "The Growth of Science Fiction from 1900 to the Early 1950s," in Ordway and Liebermann, *Blueprint for Space*, 69–82.

40. Ley, *Rockets*, 3.

41. Philip E. Cleator, *Rockets through Space: The Dawn of Interplanetary Travel* (New York: Simon and Schuster, 1936), 203.

42. Maurice K. Hanson, "The Payload on the Lunar Trip," *Journal of the British Interplanetary Society* 5 (January 1939): 16; "Report of the Technical Committee," *Journal of the British Interplanetary Society* 5 (July 1939): 19.

43. See, for example, George Pal, *Destination Moon* (Eagle Lion, 1950); and George Pal, *The Conquest of Space* (Paramount, 1955).

44. Daniel J. Boorstin, *The Discoverers* (New York: Random House, 1983), 278.

45. Lonora Foerstel and Angela Gilliam, *Confronting the Margaret Mead Legacy* (Philadelphia: Temple University Press, 1992); Joseph Conrad, *Heart of Darkness* (New York: Penguin Books, 1995). See also, James Hilton, *Lost Horizon* (New York: M. Morrow, 1933).

CHAPTER 2: **Making Space Flight Seem Real**

Epigraph: "Man Will Conquer Space Soon," *Collier's*, 22 March 1952, 22–23.

1. George Gallup, ed., *The Gallup Poll: Public Opinion, 1935–1971* (New York: Random House, 1972), 2:875.

2. Walter A. McDougall, *The Heavens and the Earth: A Political History of the Space Age* (New York: Basic Books, 1985), 12; Daniel Bell, "Technology, Nature, and Society," in Bell, *The Winding Passage* (Cambridge: Abt Books, 1980).

3. Richard Adams Locke, *The Moon Hoax; or, A Discovery that the Moon Has a Vast Population of Human Beings* (Boston: Gregg Press, 1975); Roger Lancelyn Green, *Into Other Worlds: Space-Flight in Fiction, from Lucian to Lewis* (New York: Arno Press, 1975), chap. 7.

4. Edward Everett Hale, *The Brick Moon and Other Stories* (Freeport, N.Y.: Books for Libraries Press, 1970).

5. Rem Koolhaas, *Delirious New York* (New York: Oxford University Press, 1978).

6. Brian Horrigan, "Popular Culture and Visions of the Future in Space, 1901–2001," in *New Perspectives on Technology and American Culture*, ed. Bruce Sinclair (Philadelphia: American Philosophical Society, 1986), 52.

7. Marshall B. Tymn and Mike Ashley, *Science Fiction, Fantasy, and Weird Fiction Magazines* (Westport, Conn.: Greenwood Press, 1985).

8. Sam Moskowitz, *Science Fiction by Gaslight: A History and Anthology of Science Fiction in the Popular Magazines, 1891–1911* (Westport, Conn.: Hyperion Press, 1968); Paul A. Carter, *The Creation of Tomorrow: Fifty Years of Magazine Science Fiction* (New York: Columbia University Press, 1977).

9. Robert C. Dille, ed., *The Collected Works of Buck Rogers in the 25th Century* (New York: A & W Publishers, 1977).

10. John F. Kasson, *Amusing the Million: Coney Island at the Turn of the Century* (New York: Hill and Wang, 1978), 61: Frank H. Winter and Randy Liebermann, "A Trip to the Moon," *Air & Space* 9 (November 1994): 62–76.

11. Gary Grossman, *Saturday Morning T.V.* (New York: Dell Publishing, 1981), 138.

12. E. E. "Doc" Smith, *The Skylark of Space* (New York: Pyramid Books, 1928).

13. The quotation is taken from the "Writers Guide" to the *Star Trek* television series, 17 April 1967, lent by Gregory Jein to the National Air and Space Museum. The entire quotation follows: "Let's go back to the days when some of us were working on the first television westerns. We did *not* recreate the Old West as it actually existed; instead we created a new Western form, actually a vast colorful backdrop *against which any kind of story could be told.*"

14. David Lasser, *The Conquest of Space* (New York: Penguin Press, 1931), 5.

15. *Bulletin*, American Interplanetary Society, June 1930, reproduced in *Blueprint for Space: Science Fiction to Science Fact*, ed. Frederick I. Ordway and Randy Liebermann (Washington, D.C.: Smithsonian Institution Press, 1992), 111.

16. Jo Ranson, "Radio Dial-Log," *Brooklyn Daily Eagle*, 1 March 1932; Frank H. Winter, *Prelude to the Space Age: The Rocket Societies, 1924–1940* (Washington, D.C.: Smithsonian Institution Press, 1983), 76.

17. Philip E. Cleator, *Rockets through Space: The Dawn of Interplanetary Travel* (New York: Simon and Schuster, 1936), 163.

18. Willy Ley, *Rockets and Space Travel* (New York: Viking Press, 1947); Chesley Bonestell and Willy Ley, *The Conquest of Space* (New York: Viking Press, 1949).

19. Arthur C. Clarke, *The Exploration of Space* (New York: Harper and Brothers, 1951), 183–84. See also Neil McAleer, *Odyssey: The Authorised Biography of Arthur C. Clarke* (London: Victor Gollancz, 1992); and *Current Biography, 1966*, ed. Charles Moritz (New York: H. W. Wilson, 1967), s.v. "Arthur C. Clarke," 49–52.

20. Clarke, *Exploration of Space*, 61–62.

21. Ibid., 182.

22. Willy Ley, letter of invitation, First Annual Symposium on Space Travel, 1951, American Museum of Natural History, Hayden Planetarium Library, New York.

23. Willy Ley, "Thirty Years of Space Travel Research," and American Museum of Natural History, "Space Travel Symposium Held at Hayden Planetarium," 12 October 1951, American Museum of Natural History.

24. Robert R. Coles, "The Conquest of Space," First Annual Symposium on Space Travel, 12 October 1951, American Museum of Natural History.

25. Robert R. Coles, "The Role of the Planetarium in Space Travel," Second Symposium on Space Travel, 13 October 1952, American Museum of Natural History.

26. Frederick I. Ordway and Mitchell R. Sharpe, *Rocket Team* (New York: Thomas Y. Crowell, 1979).

27. Wernher von Braun, "The Early Steps in the Realization of the Space Station," Second Symposium on Space Travel, 13 October 1952, American Museum of Natural History.

28. Joseph M. Chamberlain, "Introductory Remarks," 3, Third Symposium on Space Travel, 4 May 1954, American Museum of Natural History.

29. American Museum of Natural History, "Space-Travel Specialists Confer at Planetarium," 4 May 1954, American Museum of Natural History.

30. R. C. Truax, "A National Space Flight Program," Third Symposium on Space Travel, 4 May 1954, American Museum of Natural History.

31. Harry Hansen, ed., *The World Almanac and Book of Facts for 1952* (New York: New York World-Telegram and the Sun, 1952).

32. Fred L. Whipple, "Recollections of Pre-Sputnik Days," in Ordway and Liebermann, *Blueprint for Space*, 129.

33. "Man Will Conquer Space Soon," *Collier's*, 22 March 1952, cover and 22–23.

34. Wernher von Braun, "Man on the Moon: The Journey," *Collier's*, 18 October 1952, 51.

35. Wernher von Braun and Cornelius Ryan, "Baby Space Station," *Collier's*, 27 June 1953, 34.

36. Fred L. Whipple, "Is There Life on Mars?" *Collier's*, 30 April 1954, 21.

37. "Journey into Space," *Time*, 8 December 1952, 62.

38. "The Seer of Space: Lifetime of Rocket Work Gives Army's Von Braun Special Insight into the Future," *Life*, 18 November 1957, cover, 133–39.

39. Walt Disney, JustDisney.com, "Disneyland's History," 2005, justdisney.com/disneyland/history.html (accessed 16 May 2009).

40. David R. Smith, "They're Following Our Script: Walt Disney's Trip to Tomorrowland," *Future*, May 1978, 55.

41. Ward Kimball, "Man in Space" (Walt Disney, 1955).

42. Ibid.

43. Ward Kimball, "Man and the Moon" (Walt Disney, 1955); also titled "Tomorrow the Moon."

44. Ward Kimball, "Mars and Beyond" (Walt Disney, 1957).

45. Ryan A. Harmon, "Yesterday, Today, and Tomorrowland," *Disney News* 26 (Spring 1991): 18–23; Bruce Gordon and David Mumford, "Tomorrowland, 1986: The Comet Returns," unpublished manuscript, Walt Disney Archives, Burbank, California. Disney chose 1986 because it was the year that Halley's Comet would return.

46. "Blast Off!" *Vacationland* 5 (Spring 1961): 13, published by Disneyland, Inc., Anaheim History Room, Anaheim Public Library, Anaheim, California.

47. Kimball, "Tomorrow the Moon."

48. "Space Travel: The Trip to the Moon Disneyland Style," *Disneyland News*, 10 March 1956, 10–11; "Feeling of Space Journey Accompanies Trip to Moon," *Disneyland News*, January 1957.

49. Smith, "They're Following Our Script," 56.

50. Harmon, "Yesterday, Today, and Tomorrowland," 19.

51. Walt Disney Productions, *Disneyland Diary: 1955–Today*, 1982; Jon C. A. DeKeles, "Disneyland Concordance," 1982; both in Anaheim History Room.

52. "Tomorrowland Built with Aid of Experts," *Disneyland News*, September 1955, 6.

53. David A. Hardy, *Visions of Space* (New York: Gallery Books, 1989).

54. James Nasmyth and James Carpenter, *The Moon* (London: J. Murray, 1874).

55. T. E. R. Phillips, *Splendour of the Heavens* (London: Hutchinson and Company, 1923).

56. *Astounding Science-Fiction*, June 1938 and April 1939.

57. Arthur C. Clarke, "Space Future: Visions of Space," *Spaceflight* (May 1986): 201.

58. F. Barrows Colton, "News of the Universe," *National Geographic* 76 (July 1939): 1–32.

59. "Interview: Chesley Bonestell," *Space World* V (December 1985): 9–12. Also see Frederick C. Durant and Ron Miller, *Worlds Beyond: The Art of Chesley Bonestell* (Norfolk: Donning Company, 1983); Frederick I. Ordway, *Visions of Spaceflight: Images from the Ordway Collection* (New York: Four Walls Eight Windows, 2001).

60. See Carl Sagan, *Pale Blue Dot: A Vision of the Human Future in Space* (New York: Random House, 1994), 115.

61. "Solar System," *Life*, 29 May 1944, 78–86; quotation from 80.

62. "Trip to the Moon: Artist Paints Journey by Rocket," *Life*, 4 March 1946, 73–76.

63. Quoted from Durant and Miller, *Worlds Beyond*, 9. For one of the working drawings, see 7.

64. Willy Ley and Wernher von Braun, *The Exploration of Mars* (New York: Viking Press, 1956).

65. Cornelius Ryan, ed., *Across the Space Frontier* (New York: Viking Press, 1952); Cornelius Ryan, ed., *Conquest of the Moon* (New York: Viking Press, 1953).

66. Mike McIntyre, "Celestial Visions," *Air and Space* 1 (August–September 1986): 86–92.

67. See Ron Miller, *Space Art* (New York: Starlog Magazine, 1978).

68. Clarke, *Exploration of Space*; R. A. Smith (text by Arthur C. Clarke), *The Exploration of the Moon* (London: Frederick Muller, 1954).

69. Jack Coggins and Fletcher Pratt, *Rockets, Jets, Guided Missiles and Space Ships* (New York: Random House, 1951); Jack Coggins and Fletcher Pratt, *By Space Ship to the Moon* (New York: Random House, 1952). For later artists, see Ben Bova, *Vision of the Future: The Art of Robert McCall* (New York: Harry N. Abrams, 1982); Robert McCall, *The Art of Robert McCall* (New York: Bantam Books, 1992); Ron Miller and William K. Hartmann, *The Grand Tour: A Traveler's Guide to the Solar System* (New York: Workman, 1981); and William K. Hartmann, Ron Miller, and Pamela Lee, *Out of the Cradle: Exploring the Frontiers beyond Earth* (New York: Workman, 1984); Pat Rawlings Web site, www.patrawlings.com (accessed 17 June 2009).

70. Megan Prelinger, *Another Science Fiction: Advertising the Space Race, 1957–1962* (New York: Blast Books, 2010).

71. Gail Morgan Hickman, *The Films of George Pal* (New York: A. S. Barnes, 1977), 36–46; Robert A. Heinlein, *Rocket Ship Galileo* (New York: Charles Scribner's Sons, 1947).

72. Robert Heinlein, "Shooting Destination Moon," in *Requiem: New Collected Works by Robert Heinlein*, ed. Yoji Kondo (New York: Tom Doherty Associates, 1992), 121.

73. Cobbett A. Steinberg, *Reel Facts* (New York: Vintage Books, 1978), 344–45.

74. David Wingrove, *The Science Fiction Film Source Book* (Harlow, U.K.: Longman, 1985), 303.

75. George Pal, *Conquest of Space* (Paramount, 1955).

76. Wingrove, *Science Fiction Film Source Book*, 304.

77. Gallup, *Gallup Poll: Public Opinion, 1935–1971*, 2:875, 1306.

78. Ibid., 2:1521–22. See also Roger D. Launius, "Public Opinion Polls and Perceptions of U.S. Human Spaceflight," *Space Policy* 19 (August 2003): 163–75.

79. Robert A. Devine, *The Sputnik Challenge* (New York: Oxford University Press, 1993), 102–5.

80. Hugh L. Dryden, "Space Technology and the NACA," *Aeronautical Engineering Review* 17 (March 1958): 32–44. Dryden could not attend the institute's annual meeting and his statement was delivered by John Victory, executive secretary of the NACA. See also Michael H. Gorn, ed., *Hugh L. Dyden's Career in Aviation and Space*, Monographs in Aerospace History no. 5 (Washington, D.C.: NASA History Office, 1996).

81. National Advisory Committee for Aeronautics, "On the Subject of Space Flight," a resolution, 16 January 1958, NASA History Office, NASA Headquarters, Washington, D.C.

82. Dryden, "Space Technology and the NACA," 33.

83. NASA, "Minutes of Meeting of Research Steering Committee on Manned Space Flight," 2, NASA Headquarters, Washington, D.C., 25–26 May 1959.

84. Ibid., 9.

85. Ibid., 10.

86. NASA, "Minutes of Meeting of Research Steering Committee on Manned Space Flight," Ames Research Center, Moffett Field, California, 25–26 June 1959, 6.

87. NASA Office of Program Planning and Evaluation, "The Long Range Plan of the

National Aeronautics and Space Administration," 16 December 1959, 28, NASA History Office, table 1.

88. Daniel Herman, Workshop on Automated Space Station, Washington, D.C., 18 March 1984, in U.S. Senate Committee on Appropriations, a subcommittee, *Department of Housing and Urban Development, and Certain Independent Agencies Appropriations for Fiscal Year 1985,* 98th Cong., 2nd sess., 1984, 1266.

89. NASA, "NASA Long Range Planning Conference," concluding remarks by Dr. Paine, transcript of proceedings, Wallops Island, Virginia, 14 June 1970, 32, NASA History Office.

90. NASA, "Post-Apollo Space Program: Directions for the Future," summary of National Aeronautics and Space Administration's Report to the President's Space Task Group, September 1969, i; NASA, "America's Next Decades in Space," a report for the Space Task Group, September 1969; Space Task Group, *Post-Apollo Space Program;* all in NASA History Office.

91. National Commission on Space, *Pioneering the Space Frontier* (New York: Bantam Books, 1986), frontispiece.

92. Ibid., opposite contents page.

93. NASA, "NASA Establishes Office of Exploration," release 87-87, *NASA News,* 1 June 1987.

94. The White House, Office of the Press Secretary, "Fact Sheet: Presidential Directive on National Space Policy," 11 February 1988. See also Ronald Reagan, "1988 Legislative and Administrative Message: A Union of Individuals," *Weekly Compilation of Presidential Documents* 24, no. 4 (25 January 1988): 119.

95. NASA, "Civil Space Exploration Initiative," undated, in the possession of Howard E. McCurdy; "The Decision to Send Humans Back to the Moon and On to Mars," Space Exploration Initiative History Project, March 1992, NASA History Office. Also see Edward McNally interview, 7 August 1992, Anchorage, Alaska.

96. George Bush, "Remarks on the 20th Anniversary of the *Apollo 11* Moon Landing," 20 July 1989, *Public Papers of the Presidents of the United States, 1989* (Washington, D.C.: GPO, 1990), bk. 2, p. 991.

97. Ibid., 993.

98. NASA, NASA Facts, "President Bush Delivers Remarks on U.S. Space Policy," 14 January 2004; Frank Sietzen and Keith L. Cowing, *New Moon Rising: The Making of America's New Space Vision and the Remaking of NASA* (Burlington, Ontario: Apogee Books, 2004); Dwayne Day, "Doomed to Fail: the Birth and Death of the Space Exploration Initiative," *Spaceflight* 37 (March 1995): 79–83.

CHAPTER 3: The Cold War

Epigraphs: Statement of Democratic Leader Lyndon B. Johnson to the Meeting of the Democratic Conference on 7 January 1958, statements of LBJ collection, box 23, Lyndon Baines Johnson Library, Austin, Texas; Clark R. Chapman and David Morrison, "Chicken Little Was Right," *Discover* 12 (May 1991): 40–43.

1. The White House, *Introduction to Outer Space* (Washington, D.C.: GPO, 1958), 10.

2. Aerospace Industries Association of America, *Aerospace Facts and Figures, 1962* (Washington, D.C.: American Aviation Publications, 1962), 20. Funds available for missile development and production, fiscal years 1954 through 1961.

3. U.S. Bureau of the Budget, *The Budget of the United States Government,* fiscal year 1958 (Washington, D.C.: GPO, 1957), message of the president, sec. M, p. 45.

4. "Address to the Nation by President Eisenhower on the Cost of Government," 14 May 1957, reproduced in Robert L. Branyan and Lawrence H. Larsen, *The Eisenhower Administration, 1953–1961: A Documentary History* (New York: Random House, 1971), 2:845.

5. "What Are We Waiting For?" *Collier's,* 22 March 1952, 23.

6. President's Science Advisory Committee, "Report of Ad Hoc Panel on Man-in-Space," December 1960, 8, NASA History Office, NASA Headquarters, Washington, D.C. See also John Logsdon, *The Decision to Go to the Moon* (Cambridge, Mass.: MIT Press, 1970), 34–35.

7. "What Are We Waiting For?" 23.

8. Wernher Von Braun, "Early Steps in the Realization of the Space Station," 5, Second Symposium on Space Travel, 13 October 1952, American Museum of Natural History, Hayden Planetarium Library, New York.

9. Milton W. Rosen, *The Viking Rocket Story* (New York: Harper and Brothers, 1955).

10. Milton W. Rosen, "A Down-to-Earth View of Space Flight," 1, Second Symposium on Space Travel, 13 October 1952, American Museum of Natural History.

11. Ibid., 5.

12. William L. Laurence, "2 Rocket Experts Argue 'Moon' Plan," *New York Times,* 14 October 1952; Robert C. Boardman, "Space Rockets with Floating Base Predicted," *New York Herald Tribune,* 14 October 1952.

13. Laurence, "2 Rocket Experts Argue"; Boardman, "Space Rockets with Floating Base Predicted."

14. "Journey into Space," *Time,* 8 December 1952, 62, 68.

15. Ibid., 70.

16. See "Chief Whip in Scientific Race," *Business Week,* 16 November 1957, 42–43; *Current Biography, 1959,* ed. Charles Moritz (New York: H. W. Wilson, 1960), s.v. "James R. Killian," 229–31.

17. White House, *Introduction to Outer Space.*

18. Ibid., 6.

19. Ibid., 10.

20. Ad Hoc Committee on Space (Jerome B. Wiesner, chair), "Report to the President Elect," 10 January 1961, 15–16, NASA History Office.

21. President's Science Advisory Committee. "The Next Decade in Space," February 1970, 2–3, NASA History Office. See also NASA, *Outlook for Space: A Synopsis* (Washington, D.C.: National Aeronautics and Space Administration, 1976).

22. See Karlyn Keene and Everett Ladd, "Government as Villain," *Government Executive* 20 (January 1988): 11–16.

23. George Gallup, ed., *The Gallup Poll: Public Opinion, 1935–1971* (New York: Random House, 1972), 3:1720.

24. Media General/Associated Press Public Opinion Poll, June and July 1988, Media General Research, Richmond, Virginia; George Gallup, *The Gallup Poll: Public Opinion, 1989* (Wilmington, Del.: Scholarly Resources, 1990), 169–73. See also Herbert E. Krugman, "Public Attitudes toward the Apollo Space Program, 1965–1975," *Journal of Communication* 27 (Autumn 1977): 87–93; Michael A. G. Michaud, "The New Demographics

of Space," *Aviation Space* 2 (Fall 1984): 46–47; and Roger D. Launius, "Public Opinion Polls and Perceptions of US Human Spaceflight," *Space Policy* 19 (2003): 163–75.

25. Dwight D. Eisenhower, "Farewell Radio and Television Address to the American People, January 17, 1961," in *Public Papers of the Presidents of the United States, 1960–61* (Washington, D.C.: GPO, 1961), 1038.

26. Stephen E. Ambrose, *Eisenhower*, vol. 2, *The President* (New York: Simon and Schuster, 1984), 257.

27. Dwight D. Eisenhower, "Why I Am a Republican," *Saturday Evening Post*, 11 April 1964, 19.

28. Technological Capabilities Panel (James Killian, chair), "Meeting the Threat of Surprise Attack," 14 February 1955, Jet Propulsion Laboratory, Pasadena, California.

29. R. Cargill Hall, "Origins of U.S. Space Policy: Eisenhower, Open Skies, and Freedom of Space," in *Exploring the Unknown: Selected Documents in the History of the U.S. Civil Space Program*, vol. 1, ed. John M. Logsdon, NASA SP-4407 (Washington, D.C.: National Aeronautics and Space Administration, 1995). Also see Constance McLaughlin Green and Milton Lomask, *Vanguard: A History* (Washington, D.C.: Smithsonian Institution Press, 1971).

30. See Hall, "Origins of U.S. Space Policy," 1:213–29.

31. Hugh L. Dryden, memorandum for Dr. James R. Killian, 18 July 1958, NASA History Office; NASA, *Mercury Project Summary*, NASA SP-45 (Houston: Manned Spacecraft Center, 1963), 2; Loyd S. Swenson, James M. Grimwood, and Charles C. Alexander, *This New Ocean: A History of Project Mercury*, NASA SP-4201 (Washington, D.C.: GPO, 1966), esp. 110–11.

32. Dwight D. Eisenhower, "Annual Budget Message to Congress: Fiscal Year 1962," 16 January 1961, *Public Papers of the Presidents of the United States, 1960–61*, 414.

33. President's Science Advisory Committee, "Report of Ad Hoc Panel on Man-in-Space," 14 November 1960, 5, NASA History Office.

34. White House, *Introduction to Outer Space*, 15.

35. Dwight D. Eisenhower, "Are We Headed in the Wrong Direction?" *Saturday Evening Post*, 11–18 August 1962, 24. See also Dwight R. Eisenhower, "Spending into Trouble," *Saturday Evening Post*, 18 May 1963, 19.

36. Eisenhower, "Farewell Radio and Television Address," 1038.

37. Eisenhower, "Why I Am a Republican," 19.

38. Robert L. Rosholt, *An Administrative History of NASA, 1958–1963*, NASA SP-4101 (Washington, D.C.: GPO, 1966), 79–81.

39. Ibid., 213–14.

40. Henry C. Dethloff, *Suddenly, Tomorrow Came: A History of the Johnson Space Center*, NASA SP-4307 (Houston: Johnson Space Center, 1993), chap. 3.

41. William J. Jorden, "Soviet Fires Earth Satellite into Space; It Is Circling the Globe at 18,000 M.P.H.; Sphere Traced in 4 Crossings over U.S.; 560 Miles High," *New York Times*, 5 October 1957; "Satellite Announcement Brings Mixed Reaction," *New York Times*, 5 October 1957; W. H. Lawrence, "Eisenhower Gets Missile Briefing," *New York Times*, 9 October 1957. See also "Satellites and Our Safety," *Newsweek*, 21 October 1957, 29–39; and E. Nelson Hayes, "Tracking Sputnik I," in *The Coming of the Space Age*, ed. Arthur C. Clarke (New York: Meredith Press, 1967), 11–12.

42. Harry Schwartz, "A Propaganda Triumph," *New York Times*, 6 October 1957,

sec. A, p. 43. See also "Satellite: The World Takes a Second Look," *U.S. News & World Report*, 18 October 1957, 110.

43. "A Time of Danger," *Time*, 11 November 1957, 23.

44. Mr. Hagerty's News Conference, Saturday, 5 October 1957, Dwight D. Eisenhower Library, Abilene, Kansas. See also "Soviet Fires Earth Satellite into Space: Device Is 8 Times Heavier Than One Planned by U.S.," *New York Times*, 5 October 1957, A1.

45. "Soviet Fires Earth Satellite into Space," A1; W. H. Lawrence, "Eisenhower Gets Missile Briefing," *New York Times*, 9 October 1957, A1.

46. Dwight D. Eisenhower, "The President's News Conference of October 9, 1957," *Public Papers of the Presidents of the United States, 1957* (Washington, D.C.: GPO, 1958), 73.

47. Dwight D. Eisenhower, "Radio and Television Address to the American People on Science in National Security," November 7, 1957, *Public Papers of the Presidents of the United States, 1957*, 794.

48. "Into Space: Man's Awesome Adventure," *Newsweek*, 14 October 1957, 38.

49. "Defense: The Race to Come," *Time*, 21 October 1957, 21. See also "Russia's Satellite, a Dazzling New Sight in the Heavens," *Time*, 21 October 1957, 19–35.

50. "Space Travel, Not in Five or Ten Years," *U.S. News & World Report*, 18 October 1957, 106; "Sputniks and Budgets," *New Republic*, 14 October 1957, 3.

51. Gallup, *Gallup Poll: Public Opinion, 1935–1971*, 2:1467, 1522.

52. National Advisory Committee for Aeronautics to Dr. Killian's Office, 6 August 1958, NASA History Office.

53. Swenson, Grimwood, and Alexander, *This New Ocean*, 91–101; Hugh L. Dryden, "Space Technology and the NACA," *Aeronautical Engineering Review* 17 (March 1958): 33.

54. U.S. Senate Armed Services Committee, Preparedness Investigating Subcommittee, *Inquiry into Satellite and Missile Programs*, 85th Cong., 1st and 2nd sess., 1957 and 1958; Lyndon B. Johnson and Styles Bridges, Statement of the Senate Preparedness Subcommittee, 23 January 1958, from Bryce Harlow Papers, Dwight D. Eisenhower Library, Abilene, Kansas.

55. Robert A. Divine, *The Sputnik Challenge: Eisenhower's Response to the Soviet Satellite* (New York: Oxford University Press, 1993).

56. Gallup, *Gallup Poll: Public Opinion, 1935–1971*, 3:1720.

57. George Reedy, memo to Lyndon B. Johnson, 17 October 1957, LBJ Library, Senate Papers, box 421.

58. G. Edward Pendray, "Next Stop the Moon," *Collier's*, 7 September 1946, 12 (emphasis added).

59. Ibid., 77.

60. Robert S. Richardson, "Rocket Blitz from the Moon," *Collier's*, 23 October 1948, 24–25, 44–46.

61. Ibid., 25.

62. Caleb B. Laning and Robert A. Heinlein, "Flight into the Future," *Collier's*, 30 August 1947, 36.

63. Jack Coggins and Fletcher Pratt, *By Space Ship to the Moon* (New York: Random House, 1952), 1–2.

64. Wernher von Braun, "Crossing the Last Frontier," *Collier's*, 22 March 1952, 74.

65. Ibid., 26.

66. "What Are We Waiting For?" 23. See also "Hello, Down There," *Collier's*, 16 Sep-

tember 1955, 82, and Cornelius Ryan, *Across the Space Frontier* (New York: Viking Press, 1952), xiii–xiv.

67. "What Are We Waiting For?" 23.

68. George Pal, *Destination Moon* (Eagle Lion, 1950).

69. "Treaty on Principles Governing the Activities of States in the Exploration and Use of Outer Space, Including the Moon and Other Celestial Bodies," 10 October 1967, in U.S. Arms Control and Disarmament Agency, *Arms Control and Disarmament Agreements* (Washington, D.C.: Arms Control and Disarmament Agency, 1975), 46–55; Philip D. O'Neill, "The Development of International Law Governing the Military Use of Outer Space," in *National Interests and the Military Use of Space*, ed. William J. Durch (Cambridge, Mass.: Ballinger Publishing, 1984), 169–99.

70. William E. Burrows, *Deep Black: Space Espionage and National Security* (New York: Random House, 1987).

71. Paul B. Stares, *The Militarization of Space: U.S. Policy, 1945–1984* (Ithaca, N.Y.: Cornell University Press, 1985).

72. White House, *Introduction to Outer Space*, 12.

73. Burrows, *Deep Black*.

74. "Journey into Space," 67, 73.

75. John 6. *The New American Bible* (New York: Benziger, 1970).

76. J. Gordon Melton, *The Encyclopedia of American Religions*, vol. 79 (Wilmington, N.C.: McGrath Publishing, 1978), 459–61; Charles H. Lippy and Peter W. Williams, eds., *Encyclopedia of the American Religious Experience*, vol. 2 (New York: Charles Scribner's Sons, 1988), 834–35.

77. Wm. H. Barton, "The End of the World," *Sky* 1 (October 1937): 3–14, 19; "The Talk of the Town," *New Yorker*, 9 July 1949, 11–13.

78. Max Wilhelm Meyer, *The End of the World* (Chicago: C. H. Kerr, 1905); Joseph McCabe, *The End of the World* (London: G. Routledge, 1920); George Gamow, *Biography of the Earth* (New York: Viking Press, 1941); Geoffrey Dennis, *The End of the World* (London: Eyre and Spottiswoode, 1930).

79. "Hayden Planetarium Shows Four Ways in Which the World May End," *Life*, 1 November 1937, 54–58. See also "Picture Show," *Coronet*, July 1947, 27–34; Barton, "End of the World," 19.

80. Lincoln Barnett, "The Earth Is Born," *Life*, 8 December 1952, 87, 99–100.

81. Richard F. Dempewolff, "Five Roads to Doomsday," *Popular Mechanics*, February 1950, 83.

82. Albert Einstein, as told to Raymond Swing, "Atomic War or Peace," *Atlantic*, November 1947, 29.

83. Quoted from Kenneth Heuer, *The End of the World* (New York: Rinehart, 1953), 144–45.

84. *Life's Picture History of World War II* (New York: Time, 1950).

85. William L. Laurence, "How Hellish Is the H Bomb?" *Look*, 21 April 1953, 31–35.

86. John Lear, "Hiroshima, U.S.A.: Can Anything Be Done about It?" *Collier's*, 5 August 1950, 11–15.

87. Robert E. Sherwood, "The Third World War," *Collier's*, 27 October 1951, 18–31ff.

88. W. Warren Wager, *Terminal Visions: The Literature of Last Things* (Bloomington: Indiana University Press, 1982); Carl B. Yoke, ed., *Phoenix from the Ashes: The Literature*

of the Remade World (New York: Greenwood Press, 1987); Eric S. Rabkin, Martin H. Greenberg, and Joseph D. Olander, *The End of the World* (Carbondale: Southern Illinois University Press, 1983).

89. Ray Bradbury, *The Martian Chronicles* (New York: Bantam Books, 1950), 143–45.

90. Nevil Shute, *On the Beach* (New York: Ballantine, 1957); Stanley Kramer, *On the Beach* (United Artists, 1959); Cobbett S. Steinberg, *Film Facts* (New York: Facts on File, 1980).

91. David Weishart, *Them!* (Warner Brothers, 1954).

92. Gallup, *Gallup Poll: Public Opinion, 1935–1971*, 2:907, 1018, 1225, 1365, 1434–43, 1460, 1523; 3:1726.

93. Curtis Peebles, *Watch the Skies! A Chronicle of the Flying Saucer Myth* (Washington, D.C.: Smithsonian Institution Press, 1994).

94. Donald A. Keyhoe, "The Flying Saucers Are Real," *True*, January 1950, 11–13, 83–87. See also H. B. Darrach and Robert Ginna, "Have We Visitors from Space?" *Life*, 7 April 1952, 80–96.

95. Richard Kyle, *The Religious Fringe* (Downers Grove, Ill.: InterVarsity Press, 1993), 282–84; H. Taylor Buckner, "Flying Saucers Are for People," *Trans-action* 3 (May–June 1966): 10–13; Robert S. Ellwood, *Religious and Spiritual Groups in Modern America* (Englewood Cliffs, N.J.: Prentice-Hall, 1973), chap. 4; Peebles, *Watch the Skies!*, chap. 7.

96. Quote from Julian Blaustein, *The Day the Earth Stood Still* (Twentieth Century Fox, 1951). See also Scott Derrickson, *The Day the Earth Stood Still* (Twentieth Century Fox, 2008).

97. Lawrence J. Tacker, *Flying Saucers and the U.S. Air Force* (New York: D. Van Nostrand, 1960).

98. David Michael Jacobs, *The UFO Controversy in America* (Bloomington: Indiana University Press, 1975).

99. Carl Sagan, "What's Really Going On?" *Parade*, 7 March 1993, 3–7; Joseph Klaits, *Servants of Satan: The Age of Witch Hunts* (Bloomington: Indiana University Press, 1985).

100. Edward U. Condon, *Scientific Study of Unidentified Flying Objects* (New York: E. P. Dutton, 1969), 514; "If You're Seeing Things in the Sky," *U.S. News & World Report*, 15 November 1957, 122–26.

101. Statement of Democratic Leader Lyndon B. Johnson, 3–4. See also Johnson, "The Vision of a Greater America," *General Electric Forum*, July–September 1962, 7–9.

102. Johnson, "Vision of a Greater America," 3.

103. Ibid.

104. "Man on the Moon: The Epic Journey of Apollo 11," CBS interview with Lyndon B. Johnson by Walter Cronkite, 21 July 1969.

105. W. Stuart Symington, Address, Veterans Day, Jefferson City, Missouri, 11 November 1957, from Lee Saegesser, "High-Ground Advantage," NASA History Office.

106. John F. Kennedy, "If the Soviets Control Space," *Missiles and Rockets*, 10 October 1960, 12.

107. Peter J. Roman, *Eisenhower and the Missile Gap* (Ithaca, N.Y.: Cornell University Press, 1995).

108. Homer A. Boushey, "Who Controls the Moon Control the Earth," excerpts from a speech delivered before the Aero Club of Washington, D.C., 28 January 1958, *U.S. News & World Report*, 7 February 1958, 54. Also see Boushey's testimony in Select Committee

on Astronautics and Space Exploration, *Astronautics and Space Exploration*, 85th Cong., 2nd sess., 1958, 521–26.

109. See the remarks of Thomas B. White, U.S. Air Force Chief of Staff, to the National Press Club of Washington, D.C., 29 November 1957; General James H. Doolittle, 1959; Eugene M. Zuckert, 22 April 1961; and General Thomas D. White, 30 June 1961; all in Saegesser, "High-Ground Advantage."

110. Paul Palmer, "Soviet Union vs. U.S.A.—What Are the Facts?" *Reader's Digest*, April 1958, 44.

111. White House, *Introduction to Outer Space*, 12; Palmer, "Soviet Union vs. U.S.A.," 44.

112. T. R. B., "Washington Wire," *New Republic*, 25 November 1957, 2.

113. George R. Price, "Arguing the Case for Being Panicky," *Life*, 18 November 1957, 126. See also "Stepping Up the Pace," *Newsweek*, 21 October 1957, 30–34; "The Feat That Shook the Earth," *Life*, 21 October 1957, 19–35; and "World Will Be Ruled from Skies Above," *Life*, 17 May 1963, 4.

114. James R. Killian, *Sputnik, Scientists, and Eisenhower* (Cambridge, Mass.: MIT Press, 1977), 7.

115. Logsdon, *Decision to Go to the Moon*, 95–99, 126–29.

116. James R. Hansen, *Engineer in Charge: A History of the Langley Aeronautical Laboratory, 1917–1958*, NASA SP-4305 (Washington, D.C.: GPO, 1987), 376.

117. McCurdy interview with Max Faget, 9 November 1987, Houston, Texas.

118. Gallup, *Gallup Poll: Public Opinion, 1935–1971*, 2:907, 1018, 1225, 1240–41, 1345, 1451, 1523, 1539; 3:1595, 1632, 1674, 1812, 1842, 1881, 1934, 1944, 1973. See also Vernon Van Dyke, *Pride and Power: The Rationale of the Space Program* (Urbana: University of Illinois Press, 1964).

119. See Leon Festinger, *When Prophecy Fails* (Minneapolis: University of Minneapolis Press, 1956).

120. Jeffrey Klein and Dan Stober, "The American Empire in Space," *San Jose Mercury News*, 2 August 1992; Jack Manno, *Arming the Heavens* (New York: Dodd, Mead, 1984); U.S. Space Command, Office of History, "The Role of Space Forces: Quotes from Desert Shield/Desert Storm," Peterson Air Force Base, Colorado, May 1993; Thomas Karas, *The New High Ground* (New York: Simon and Schuster, 1983); Jeff Kueter, "The War in Space Has Already Begun," George C. Marshall Institute *Policy Outlook*, October 2006, marshall .org/pdf/materials/459.pdf (accessed 26 June 2010).

121. Jim Yardley and William J. Broad, "Heading for the Stars, and Wondering If China Might Reach Them First," *New York Times* (22 January 2004).

122. Reuters, "China Eyes 2017 Moon Landing," 4 November 2005; Ian O'Neill, "Griffin: China Could Beat US in Moon Race," Universetoday.com (15 July 2008); "Chinese Space Program," Wikipedia, en.wikipedia.org/wiki/Chinese_space_program (accessed 5 March 2009); "Asia Could Win Next 'Space Race,' US Scientists Fear," *Moon Daily*, 30 September 2007; James Oberg, "Will China's Space Plan Skip the Moon?" *Space News*, 24 May 2004.

123. Dwayne A. Day, "Exploding Moon Myths: Or Why There's No Race to Our Nearest Neighbor," *Space Review*, 12 November 2007, thespacereview.com/article/999/1 (accessed 5 March 2009); Day, "Mysterious Dragon: Myth and Reality of the Chinese Space Program," *Space Review*, 7 November 2005, thespacereview.com/article/492/1 (accessed

5 March 2009); Dean B. Cheng, "The Long March Upward: A Review of China's Space Program," in *Harnessing the Heavens: National Defense through Space*, ed. Paul G. Gillespie and Grant T. Weller (Chicago: Imprint Publications, 2008), 151–63.

124. Commission to Assess United States National Security Space Management and Organization (Donald Rumsfeld, Chair), *Report of the Commission*, 11 January 2001.

125. Chapman and Morrison, "Chicken Little Was Right," 40; Clark R. Chapman, "Statement on the Threat of Impact by Near-Earth Asteroids," U.S. Congressional Hearings on Near-Earth Objects and Planetary Defense, 21 May 1998, NASA Ames Research Center, impact.arc.nasa.gov/gov_asteroidperils_2.cfm (accessed 17 June 2009).

126. Chapman and Morrison, "Chicken Little Was Right," 40. See also David Morrison "Target Earth!" *Astronomy* 23 (October 1995): 34–41; David Morrison, ed., "The Spaceguard Survey: Report of the NASA International Near-Earth-Object Detection Workshop," 11–12, Jet Propulsion Laboratory, Pasadena, California; Duncan Steel, *Rogue Asteroids and Doomsday Comets* (New York: John Wiley and Sons, 1995); Robert Roy Britt, "The Odds of Dying," *Live Science*, 6 January 2005, livescience.com/environment/050106 _odds_of_dying.html (accessed 15 June 2010).

127. George Pal, *When Worlds Collide* (Paramount, 1951).

128. Larry Niven and Jerry Pournelle, *Lucifer's Hammer* (Chicago: Playboy Press, 1977); Arthur C. Clarke, *The Hammer of God* (New York: Bantam Books, 1993); Michael Bay, *Armageddon* (Touchstone Pictures, 1998); David Brown and Richard D. Zanuck, *Deep Impact* (Paramount Pictures, 1998).

129. Luis W. Alvarez, Walter Alvarez, Frank Asaro, and Helen V. Michel, "Extraterrestrial Cause for the Cretaceous-Tertiary Extinction," *Science* 208 (1980): 1095–1108; John Noble Wilford, *The Riddle of the Dinosaur* (New York: Alfred A. Knopf, 1985), chaps. 14 and 15; R. Ganapathy, "Evidence for a Major Meteorite Impact on the Earth 34 Million Years Ago: Implication for Eocene Extinctions," *Science* 216 (May 1982): 885–86. See also Morrison, "Spaceguard Survey," 8–11.

130. John R. Spencer and Jacqueline Mitton, *The Great Comet Crash: The Impact of Comet Shoemaker-Levy 9 on Jupiter* (New York: Cambridge University Press, 1995); "Close Encounter: Earth-Asteroid Near Miss," *Sky News*, 4 March 2009.

131. NASA Solar System Exploration Division, "Report of the Near-Earth Objects Survey Working Group," NASA Office of Space Science, Washington, D.C., June 1995; NASA, "Near Earth Asteroid Rendezvous Press Kit," February 1996, 5, NASA Headquarters, Washington, D.C.; Carl Sagan, *Pale Blue Dot: A Vision of the Human Future in Space* (New York: Random House, 1994), 327.

132. H.R. 4489, "An Act to Authorize Appropriations to the National Aeronautics and Space Administration," 103rd Cong., 2nd sess., sec. 225. See also U.S. House Committee on Science, Space, and Technology, *Report to Accompany H.R. 5649: National Aeronautics and Space Administration Multiyear Authorization Act of 1990*, report 101-763, 101st Cong., 2nd sess., 26 September 1990, 29–30; Morrison, "Spaceguard Survey," 49.

CHAPTER 4: Apollo: The Aura of Competence

Epigraph: Tom Horton, "On Environment," *Baltimore Sun*, 22 July 1984.

1. John Logsdon, *The Decision to Go to the Moon* (Cambridge, Mass.: MIT Press, 1970).

2. Total NASA appropriations increased from $524 million in fiscal year 1960 to

$5.3 billion in fiscal year 1965. NASA, *Pocket Statistics* (Washington, D.C.: National Aeronautics and Space Administration, 1993), sec. C, p. 16.

3. John F. Kennedy, "Special Message to Congress on Urgent National Needs," 25 May 1961, *Public Papers of the Presidents of the United States, 1961* (Washington, D.C.: GPO, 1962), 404.

4. Charles L. Schultze, memorandum for the President, 24 January 1966, 3, WHCF EX\FI 4, box 22, Lyndon Baines Johnson Library, Austin, Texas; Robert A. Divine, "Lyndon B. Johnson and the Politics of Space," in *The Johnson Years: Vietnam, the Environment, and Science*, ed. Robert A. Divine (Lawrence: University Press of Kansas, 1987), 238–39. See also Glen P. Wilson, "The Legislative Origins of NASA: The Role of Lyndon B. Johnson," *Prologue: Quarterly of the National Archives* 25 (Winter 1993): 363–73; Robert Dallek, "Johnson, Project Apollo, and the Politics of Space Program Planning," in *Spaceflight and the Myth of Presidential Leadership*, ed. Roger Launius and Howard E. McCurdy (Urbana: University of Illinois Press, 1997), 68–91.

5. See Roger D. Launius, "Compelling Rationales for Spaceflight? History and the Search for Relevance," in Steven J. Dick and Roger D. Launius, *Critical Issues in the History of Spaceflight*, NASA SP-4702 (Washington, D.C.: National Aeronautics and Space Administration, 2006): 37–70; George Gallup, ed., *The Gallup Poll: Public Opinion, 1935–1971* (New York: Random House, 1972), 3:1720, 1952.

6. W. Henry Lambright, *Powering Apollo: James E. Webb of NASA* (Baltimore: Johns Hopkins University Press, 1995).

7. Paul. R. Abramson, *Political Attitudes in America* (San Francisco: W. H. Freeman, 1983), 12. See also Seymour Martin Lipset and William Schneider, *The Confidence Gap* (New York: Free Press, 1983).

8. NASA History Division, *Pocket Statistics*, 1997, sec. B, NASA Major Launch Record, history.nasa.gov/pocketstats/ (accessed 18 June 2009).

9. See Richard P. Hallion, "The Development of American Launch Vehicles," in *Space Science Comes of Age: Perspectives in the History of the Space Sciences*, ed. Paul A. Hanle and Von del Chamberlain (Washington, D.C.: Smithsonian Institution Press, 1981), 115–34.

10. R. Cargill Hall, *Lunar Impact: A History of Project Ranger*, NASA SP-4210 (Washington, D.C.: National Aeronautics and Space Administration, 1977); House Committee on Science and Astronautics, Subcommittee on NASA Oversight, *Investigation of Project Ranger*, 88th Cong., 2nd sess., 1964.

11. Loyd S. Swenson, James M. Grimwood, and Charles C. Alexander, *This New Ocean: A History of Project Mercury*, NASA SP-4201 (Washington, D.C.: National Aeronautics and Space Administration, 1966), 372–77.

12. Tom Wolfe, *The Right Stuff* (New York: Farrar, Straus, Giroux, 1979), 280–96. See also William J. Perkinson, "Grissom's Flight: Questions," *Baltimore Sun*, 22 July 1961.

13. "Month's Delay for Glenn Seen," *Washington Star*, 31 January 1962; Art Woodstone, "Television's $1,000,000 (When & If) Manshoot; Lotsa Prestige & Intrigue," *Variety*, 24 January 1962; Swenson, Grimwood, and Alexander, *This New Ocean*, 419–22.

14. Nate Haseltine, "Vanguard Fails, Burns in Test Firing; Hill Critics See Blow to U.S. Prestige," *Washington Post*, 7 December 1957; Joseph Alsop, "Making the Worst of It," *Washington Post*, 9 December 1957; Milton Bracker, "Vanguard Rocket Burns on Beach; Failure to Launch Satellite Assailed as Blow to U.S. Prestige," *New York Times*, 7 December 1957; "Sputternik," *New York Times*, 10 December 1957.

15. Ad Hoc Committee on Space (Jerome B. Wiesner, chair), "Report to the President Elect," 10 January 1961, 16–17, NASA History Office, NASA Headquarters, Washington, D.C.

16. Swenson, Grimwood, and Alexander, *This New Ocean*, 272–79, 291–97, 326–28, 335–38.

17. James Barr, "Is Mercury Program Headed for Disaster?" *Missiles and Rockets*, 15 August 1960, 12.

18. Martin J. Collins and Sylvia K. Kraemer, *Space: Discovery and Exploration* (Washington, D.C.: Smithsonian Institution, 1993), 277.

19. See "Americans Jubilant over Shepard's Flight," *Washington Post*, 6 May 1961; Robert Conley, "Nation Exults over Space Feat," *New York Times*, 6 May 1961.

20. "The High Price of History," *Television*, April 1962, 65.

21. "8,000 Eyes in Orbit," *Life*, 2 March 1962, 2–3.

22. Joseph Arthur Angotti, "A Descriptive Analysis of NBC's Radio and Television Coverage of the First Manned Orbital Flight" (master's thesis, Indiana University, June 1965), 117–19; "Space: The New Ocean," *Time*, 2 March 1962, 14; Robert B. Voas, "John Glenn's Three Orbits in *Friendship 7*," *National Geographic*, June 1962, 792–827.

23. See T. Keith Glennan, *The Birth of NASA: The Diary of T. Keith Glennan*, ed. J. D. Hunley, NASA SP-4105 (Washington, D.C.: National Aeronautics and Space Administration, 1993), 20–21.

24. See James L. Kauffman, *Selling Outer Space: Kennedy, the Media, and Funding for Project Apollo, 1961–1963* (Tuscaloosa: University of Alabama Press, 1994).

25. Wolfe, *Right Stuff*, 24.

26. Ibid., 120.

27. "Space Voyagers Rarin' to Orbit," *Life*, 20 April 1959, 22.

28. Richard Slotkin, *Gunfighter Nation: The Myth of the Frontier in Twentieth-Century America* (New York: Atheneum, 1992).

29. Tom Wolfe, "The Last American Hero," in Wolfe, *The Kandy-Kolored Tangerine-Flake Streamline Baby* (New York: Farrar, Straus, and Giroux, 1965).

30. Chuck Yeager and Leo Janos, *Yeager: An Autobiography* (New York: Bantam Books, 1985).

31. *The Adventures of Ozzie and Harriet*, ABC, 3 October 1952 to 3 September 1966; *Father Knows Best*, CBS, 3 October 1954 to 27 March 1955, 22 September 1958 to 17 September 1962, NBC, 31 August 1955 to 17 September 1958, ABC, 30 September 1962 to 3 February 1967; *Leave It to Beaver*, CBS, 11 October 1957 to 26 September 1958, ABC, 3 October 1958 to 12 September 1963. Also see Steven Mintz and Susan Kellogg, *Domestic Revolutions: A Social History of American Family Life* (New York: Free Press, 1988), chap. 9.

32. John H. Glenn, "A New Era: May God Grant Us the Wisdom and Guidance to Use It Wisely," *Vital Speeches of the Day*, 15 March 1962, 324–26; Dora Jane Hamblin, "Applause, Tears and Laughter and the Emotions of a Long-Ago Fourth of July," *Life*, 9 March 1962, 34.

33. Agreement among Malcolm Scott Carpenter et al. and C. Leo DeOrsey, 28 May 1959; agreement between Leo DeOrsey and Time Incorporated, 5 August 1959; both in NASA History Office.

34. *Life* magazine: 20 April, 14, 21 September, and 14 December 1959; 29 February, 21 March, 11 April, 9 May, 1 August, and 3 October 1960; 27 January, 3 March, 12, 19

May, 28 July, 4 August, and 8 December 1961; 2 February, 2, 9 March, 18 May, 1, 8 June, and 12, 26 October 1962; 24, 31 May, and 7 June 1963.

35. See Larry J. Sabato, *Feeding Frenzy: How Attack Journalism Has Transformed American Politics* (New York: Free Press, 1991).

36. Don A. Schanche to P. Michael Whye, 28 December 1976, NASA History Office.

37. See Robert Sherrod, "The Selling of the Astronauts," *Columbia Journalism Review* 12 (May–June 1973): 16–25.

38. Dora Jane Hamblin to P. Michael Whye, 18 January 1977, NASA History Office.

39. "Backing up the Men, Brave Wives and Bright Children," *Life*, 20 April 1959, 24–25. See also "Seven Brave Women behind the Astronauts," *Life*, 21 September 1959, 142–63.

40. Loudon S. Wainwright, "New Astronaut Team, Varied Men with One Goal, Poise for the Violent Journey," *Life*, 3 March 1961, 24–25.

41. "High Dreams for a Man and His Sons," *Life*, 8 June 1962, 38.

42. Jack Mann, "Rene Carpenter's Own Orbit," *Washington Post/Potomac*, 16 June 1974, 8–19.

43. Hamblin to Whye, 18 January 1977.

44. See David Halberstam, *The Best and the Brightest* (New York: Random House, 1969), esp. 41.

45. U.S. Department of Commerce, *United States Science Exhibit: Seattle World's Fair*, final report (Washington, D.C.: GPO, 1963), 3.

46. Murray Morgan, *Century 21: The Story of the Seattle World's Fair, 1962* (Seattle: Acme Press, distributed by the University of Washington Press, 1963), 81–86.

47. U.S. Department of Commerce, *United States Science Exhibit: World's Fair in Seattle, 1962* (Seattle: Craftsman Press, 1962).

48. Ibid., 7.

49. Ibid.

50. NASA, "NASA Technical Exhibit: Space for the Benefit of Mankind," Century 21 Exposition, Seattle, Washington, 1962, from the private collection of Norman P. Bolotin, Redmond, Washington; Washington State Department of Commerce and Economic Development, "Seattle World's Fair, 1962: Official Souvenir Program," 15, Seattle Center Foundation, Seattle, Washington.

51. Morgan, *Century 21*, 25.

52. "National Air and Space Museum—Historical Perspective," *Aerospace* 14 (June 1976): 2–3; "Freedom Seven Rests, at Last," *National Aeronautics*, April 1963, 2; Philip Hopkins, "The National Air Museum," *National Aeronautics*, June 1964, 4–5.

53. Flip Schulke, Debra Schulke, Penelope McPhee, and Raymond McPhee, *Your Future in Space: U.S. Space Camp Training Program* (New York: Crown Publishers, 1986). See also Edward O. Buckbee and Charles Walker, "Spaceflight and the Public Mind," in *Blueprint for Space: Science Fiction to Science Fact*, ed. Frederick I. Ordway and Randy Liebermann (Washington, D.C.: Smithsonian Institution Press, 1992), 189–97.

54. See, for example, "Space," *Time*, 2 March 1962, 11–18; Voas, "John Glenn's Three Orbits," 792–827; Jerry E. Bishop, "Gemini Team Maneuvers in Three Orbits of Earth While Probing for Knowledge to Aid U.S. Moon Flight," *Wall Street Journal*, 24 March 1965.

55. See Walter Sullivan, ed., *America's Race for the Moon* (New York: Random House,

1962); M. Scott Carpenter, L. Gordon Cooper Jr., John H. Glenn Jr., Virgil I. Grissom, Walter M. Schirra Jr., Alan B. Shepard Jr., and Donald K. Slayton, *We Seven* (New York: Simon and Schuster, 1962); Edgar M. Cortright, ed., *Apollo Expeditions to the Moon*, NASA SP-350 (Washington, D.C.: GPO, 1975).

56. See David Sanford, "Admen in Orbit," *New Republic*, 17 December 1966, 13–15.

57. Courtney G. Brooks, James M. Grimwood, and Loyd S. Swenson, *Chariots for Apollo: A History of Manned Lunar Spacecraft*, NASA SP-4205 (Washington, D.C.: National Aeronautics and Space Administration, 1979), 266.

58. See CBS Television, *February 14 15 16 17 18 19 20* (New York: Columbia Broadcasting System, n.d.).

59. "High Price of History," 65.

60. See, for example, "Networks in High Gear for G-T 6," *Broadcasting*, 25 October 1965, 78; Edwin Diamond, "Perfect Match: TV and Space," *Columbia Journalism Review* 4 (Summer 1965): 18–20.

61. James Barron, "Jules Bergman," *New York Times*, 13 February 1987, sec. D, p. 20; Jules Bergman, "The Reluctant Astronaut, the Suborbital Water-Skier, Carpenter's Snafu," *TV Guide*, 2 March 1974, 8–13; "Bergman Gives Views on Science, Technology and Medicine," *Langley Researcher*, 20 May 1983, 4–5.

62. Diamond, "Perfect Match: TV and Space," 20.

63. Art Buchwald, "Countdown 1966," *Washington Post*, 26 August 1965.

64. See Blake Clark, "A Job for the Next Congress: Stop the Race to the Moon," *Reader's Digest*, January 1964, 75–79; Joe Alex Morris, "How Haste in Space Makes Waste," *Reader's Digest*, July 1964, 82–87; and Joe Alex Morris, "The Pork Barrel Goes into Orbit," *Reader's Digest*, August 1964, 87–92; or Amitai Etzioni, *The Moon-Doggle* (Garden City, N.Y.: Doubleday, 1964).

65. "The Cosmic Circus," *New York Herald Tribune*, 24 March 1965, 28.

66. "One of Our Finest Hours," *New York Times*, 21 February 1962, 44.

67. "Go!" *Washington Post*, 21 February 1962, sec. A, p. 24; "Jumping Gemini," *Washington Post*, 24 March 1965, sec. A, p. 20.

68. Brooks, Grimwood, and Swenson, *Chariots for Apollo*, chap. 3.

69. John F. Kennedy, "Address at Rice University in Houston on the Nation's Space Effort," 12 September 1962, *Public Papers of the Presidents of the United States, 1962* (Washington, D.C.: GPO, 1963), 669.

70. James E. Webb and Robert S. McNamara, "Recommendations for Our National Space Program: Changes, Policies, Goals," report to the Vice President, 8 May 1961, in *Exploring the Unknown: Selected Documents in the History of the U.S. Civil Space Program*, vol. 1, ed. John M. Logsdon, NASA SP-4407 (Washington, D.C.: National Aeronautics and Space Administration, 1995), 1:427, 444.

71. Kennedy, "Special Message to Congress on Urgent National Needs," 404.

72. Logsdon, *Decision to Go to the Moon*, 118; Webb and McNamara, "Recommendations for Our National Space Program," 443.

73. Arnold W. Frutkin Oral History, 4 April 1974, by Eugene M. Emme and Alex Roland, 28–29, and Arnold W. Frutkin Oral History, 30 July 1970, by John M. Logsdon, 17–18, both in NASA Historical Reference Collection, NASA History Office. See also Arnold W. Frutkin, *International Cooperation in Space* (Englewood Cliffs, N.J.: Prentice-Hall, 1965), chap. 3. There were numerous meetings between U.S. and Soviet representatives

toward this end. See Arnold W. Frutkin, "Record of US-USSR Talks on Space Cooperation," 27 March–1 May 1962; Hugh L. Dryden, "Bilateral Meeting with Soviets on Outer Space," 30 May 1962, 1, 4, 6, 7 June 1962; James E. Webb to the President, 31 January 1964, all in NASA Historical Reference Collection, NASA History Office.

74. Dodd L. Harvey and Linda C. Ciccoritti, *U.S.-Soviet Cooperation in Space* (Miami: Center for Advanced International Studies, University of Miami, 1974), 78–79.

75. "Text of President Kennedy's Address on Peace Issues at U.N. General Assembly," *New York Times*, 21 September 1963, sec. C, p. 6. See Yuri Karash, "The Price of Rivalry in Space," *Baltimore Sun*, 19 July 1994, 11A; Walter A. McDougall, *The Heavens and the Earth: A Political History of the Space Age* (New York: Basic Books, 1985), 394–96.

76. "Major Legislation—Appropriations," *Congressional Quarterly Almanac, 1963* (Washington, D.C.: Congressional Quarterly, 1964), 170.

77. Charles L. Schultze, Memorandum for the President, 20 September 1966, with attached memorandum, subject: NASA's budget and Mr. Webb's letter, 1 September 1966, WHCF EX\OS, box 2, LBJ Library.

78. Schultze, Memorandum for the President, 24 January 1966; Charles L. Schultze, Memorandum for the President, 11 August 1967, WHCF EX\FI 4, box 30 (emphasis removed), LBJ Library.

79. Lyndon B. Johnson to James Webb, 29 September 1967, WHCF confidential file, box 43; James E. Webb, Memorandum for the President, 10 August 1967, WHCF EX\FI 4, box 29; both in LBJ Library.

80. Lyndon B. Johnson, "President's News Conference at the LBJ Ranch," 29 August 1965, *Public Papers of the Presidents of the United States, 1965* (Washington, D.C.: GPO, 1966), 2:945. See also Lyndon B. Johnson, "Remarks Following an Inspection of NASA's Michoud Assembly Facility near New Orleans," 12 December 1967, *Public Papers of the Presidents of the United States, 1967* (Washington, D.C.: GPO, 1968), 2:1123.

81. Lyndon B. Johnson, "Remarks upon Presenting the NASA Distinguished Service Medal to the Apollo 8 Astronauts," 9 January 1969, *Public Papers of the Presidents of the United States, 1969* (Washington, D.C.: GPO, 1970), 2:1247.

82. NASA, "Post-Apollo Space Program: Directions for the Future," summary of National Aeronautics and Space Administration's Report to the President's Space Task Group, September 1969; Space Task Group, *The Post-Apollo Space Program: Directions for the Future*, September, 1969, 22; both in NASA History Office.

83. NASA, "Program Review: Manned Space Science and Advanced Manned Missions," Office of Programming, NASA Headquarters, 7 October 1965, 36; NASA, "Apollo Applications Program: Program Review Document," 15 November 1966, NASA History Office; Arnold S. Levine, *Managing NASA in the Apollo Era*, NASA SP-4102 (Washington, D.C.: GPO, 1982), 242–61. See also NASA, "Lunar Studies," 25 June 1964; and U.S. House Committee on Science and Astronautics, *Summary Report: Future Programs Task Group*, 89th Cong., 1st sess., 1965; all in NASA History Office.

84. Paul S. Boyer, *By the Bomb's Early Light* (New York: Pantheon, 1985).

85. See Norman Mailer, *The Armies of the Night* (New York: Signet, 1968), 135–51.

86. Rachel Carson, *Silent Spring* (Boston: Houghton Mifflin, 1962).

87. Paul Ehrlich, *The Population Bomb* (New York: Ballantine Books, 1968). See also Donella Meadows, Dennis L. Meadows, Jorgen Randers, and William W. Behrens III, *The Limits to Growth* (New Hyde Park, N.Y.: University Books, 1972); and Kirkpatrick Sale, *The*

Green Revolution: The American Environmental Movement, 1962–1992 (New York: Hill and Wang, 1993).

88. Joseph Heller, *Catch-22* (New York: Dell, 1961); Kurt Vonnegut, *Slaughterhouse Five* (New York: Delta Books, 1969).

89. Alvin Weinberg, "Impact of Large-Scale Science in the United States," *Science* 134 (21 July 1961): 161–64.

90. Robert J. Samuelson, *The Good Life and Its Discontents* (New York: Times Books, 1995); Robert McNamara, *In Retrospect: The Tragedy and Lessons of Vietnam* (New York: Times Books, 1995).

91. John F. Kennedy, "Letter to the President of the Senate and to the Speaker of the House on Development of a Civil Supersonic Air Transport," 14 June 1963, *Public Papers of the Presidents of the United States, 1963* (Washington, D.C.: GPO, 1964), 476.

92. See Mel Horwich, *Clipped Wings: The American SST Conflict* (Cambridge, Mass.: MIT Press, 1982).

93. Homer A. Neal, Tobin L. Smith, and Jennifer B. McCormick, *Beyond* Sputnik: *U.S. Science Policy in the Twenty-First Century* (Ann Arbor: University of Michigan Press, 2008), chap. 12.

94. Brooks, Grimwood, and Swenson, *Chariots for Apollo*, chap. 9.

95. Herbert E. Krugman, "Public Attitudes toward the Apollo Space Program, 1965–1975," *Journal of Communication* 27 (Autumn 1977): 87–93; George Gallup, ed., *Gallup Poll: Public Opinion, 1935–1971* (New York: Random House, 1972), 3:1952.

96. Krugman, "Public Attitudes toward the Apollo Space Program"; Chesley Bonestell and Willy Ley, *Conquest of Space* (New York: Viking Press, 1949); Brooks, Grimwood, and Swenson, *Chariots for Apollo*, 281.

97. "Columbuses of Space," *New York Times*, 22 December 1968; "Two Faces of Science," *New York Times*, 29 December 1968; Robert J. Donovan, "Moon Voyage Turns Men's Thoughts Inward," *Los Angeles Times*, 29 December 1968; "Apollo 8," *Philadelphia Sunday Bulletin*, 22 December 1968; "Footprints in the Dirty Sand," *Washington Post*, 28 December 1968.

98. "Apollo 8: 'Millennial' Event," *Los Angeles Times*, 29 December 1968.

99. Stanley Kubrick and Victor Lyndon, *Dr. Strangelove; or, How I Learned to Stop Worrying and Love the Bomb* (Columbia Pictures, 1963); Michael Gruskoff and Douglas Trumbull, *Silent Running* (Universal Pictures, 1971); Mort Abrahams, *Planet of the Apes* (APJAC/Twentieth Century Fox, 1968). See also Dino DeLaurentis, *Barbarella* (Paramount Pictures, 1967); Francis Ford Coppola and Lawrence Sturhahn, *THX 1138* (Warner Brothers, 1970); and Byron Kennedy, *Mad Max* (Orion Pictures, 1979).

100. Stanley Kubrick, *2001: A Space Odyssey* (MGM, 1968); Arthur C. Clarke, *2001: A Space Odyssey* (New York: New American Library, 1968).

101. Gene Roddenberry, *Star Trek*, NBC, 8 September 1966 to 9 September 1969; Irwin Allen, *Lost in Space*, CBS, 15 September 1965 to 11 September 1968; Stephen E. Whitfield and Gene Roddenberry, *The Making of Star Trek* (New York: Ballantine Books, 1968).

102. Michael Crichton, *The Andromeda Strain* (New York: Dell, 1969); Robert Wise, *The Andromeda Strain* (Universal Studios, 1971).

103. Gordon Carroll, David Giler, and Walter Hill, *Alien* (Brandywine-Shussell/Twentieth Century Fox, 1979). See also Michael Deeley and Ridley Scott, *Blade Runner* (Warner Brothers, 1982).

104. Julia Phillips and Michael Phillips, *Close Encounters of the Third Kind* (Columbia Pictures, 1977); Steven Spielberg and Kathleen Kennedy, *E.T.: The Extra-Terrestrial* (Universal Studios, 1982); Lili Zanuck, *Cocoon* (TCF, 1985).

105. See Charles Murray and Catherine Bly Cox, *Apollo: The Race to the Moon* (New York: Simon and Schuster, 1989).

106. See Jim Lovell and Jeffrey Kluger, *Lost Moon: The Perilous Voyage of Apollo 13* (Boston: Houghton Mifflin, 1994).

107. Brian Grazer, *Apollo 13* (Universal, 1995); "Peerless Voyage: NASA's 'Finest Hour' Comes to the Silver Screen," *Spaceflight*, September 1995, 319–21. See also Norman Mailer, *A Fire on the Moon* (Boston: Little, Brown, 1969).

108. "The U.S. Space Program," Media General/Associated Press Public Opinion Poll, 22 June to 2 July 1988.

109. Columbia Accident Investigation Board, *Report*, vol. 1 (Washington, D.C.: GPO, August 2003); Matthew L. Ward, "Expert: Shuttle Mistakes Repeated," *Oakland Tribune*, 13 April 2003.

110. See Howard E. McCurdy, *Inside NASA: High Technology and Organizational Change in the U.S. Space Program* (Baltimore: Johns Hopkins University Press, 1993), 146–55.

111. Michael A. G. Michaud, *Reaching for the High Frontier* (New York: Praeger, 1986), chap. 6; Michaud, "The New Demographics of Space," *Aviation Space*, Fall 1984, 46–47; Roger D. Launius, "Public Opinion Polls and Perceptions of US Human Spaceflight," *Space Policy* 19 (August 2003): 163–75; Krugman, "Public Attitudes toward the Apollo Space Program." See also Gallup, *Gallup Poll: Public Opinion, 1935–1971*, 3:1952, 2184; Elizabeth H. Hastings and Philip K. Hastings, eds., *Index to International Public Opinion, 1979–1980* (Westport, Conn.: Greenwood Press, 1981), 73; Elizabeth H. Hastings and Philip K. Hastings, eds., *Index to International Public Opinion, 1982–1983* (Westport, Conn.: Greenwood Press, 1984), 183; "The Public's Agenda," *Time*, 30 March 1987, 37; "The U.S. Space Program," Media General/Associated Press Public Opinion Poll 21, 22 June–2 July 1988; George Gallup, *The Gallup Poll: Public Opinion, 1988* (Wilmington, Del.: Scholarly Resources, 1989), 106–8.

112. Robert P. Mayo to Thomas O. Paine, 28 July 1969, NASA History Office; Space Task Group, *Post-Apollo Space Program*; Richard Witkin, "Agnew Proposes a Mars Landing," *New York Times*, 17 July 1969.

113. Caspar Weinberger, memo to the President, "Future of NASA," 12 August 1971, NASA History Office. See also Roger D. Launius, "NASA and the Decision to Build the Space Shuttle, 1969–72," *Historian* 57 (Autumn 1994): 17–34.

114. Weinberger, memo to the President, 12 August 1971. See also John Logsdon, "The Decision to Develop the Space Shuttle," *Space Policy* 2 (May 1986): 103–19; and Logsdon, "The Space Shuttle Decision: Technological and Political Choice," *Journal of Contemporary Business* 7, no. 3 (1978): 13–30.

115. Richard M. Nixon, "Space Shuttle Program," statement by the president announcing the decision to proceed with the development of the new Space Transportation System, *Weekly Compilation of Presidential Documents* 8, no. 2 (5 January 1972): 27–28. See also Nixon, "The Future of the United States Space Program," *Weekly Compilation of Presidential Documents* 6, no. 10 (7 March 1970): 328–31.

116. W. Henry Lambright, "Apollo: Critical Factors in Success and Implications for

Climate Change," a paper delivered at the Solutions Summit for Climate Change, Nashville, Tennessee, 14 May 2008.

CHAPTER 5: Mysteries of Life

Epigraph: Gordon Foreman, *A Brief History of Time* (Paramount Home Video, 1993).

1. Joseph Campbell, *The Inner Reaches of Outer Space* (New York: Harper and Row, 1986). See also Joseph Campbell, *The Power of Myth* (New York: Doubleday, 1988).

2. Stephen Hawking, *A Brief History of Time* (New York: Bantam, 1988), 7–11, 174–75.

3. Pierre Simon Laplace, *The System of the World*, trans. J. Pond (London: R. Phillips, 1809). See also Agnes M. Clerke, *A Popular History of Astronomy during the Nineteenth Century* (London: Adam and Charles Black, 1902), chap. 9.

4. Isabel M. Lewis, "Life on Venus and Mars?" *Nature Magazine*, September 1934, 134.

5. Willy Ley, *Rockets and Space Travel* (New York: Viking Press, 1947), 34. See also A. Pannekoek, *A History of Astronomy* (New York: Barnes and Noble, 1969), chap. 35.

6. Ward Kimball, "Man and the Moon" (Walt Disney, 1955).

7. Arthur C. Clarke, *2001: A Space Odyssey* (New York: New American Library, 1968).

8. Ley, *Rockets*, 37–39; Pannekoek, *History of Astronomy*, 381; Camille Flammarion, *Popular Astronomy* (London: Chatto and Windus, 1907), 370.

9. See John Noble Wilford, *Mars Beckons* (New York: Alfred A. Knopf, 1990), 22; Samuel Glasstone, *The Book of Mars*, NASA SP-179 (Washington, D.C.: GPO, 1968).

10. A. Lawrence Lowell, *Biography of Percival Lowell* (New York: Macmillan, 1935). See also William Graves Hoyt, *Lowell and Mars* (Tucson: University of Arizona Press, 1976).

11. Percival Lowell, *Mars* (New York: Houghton Mifflin, 1895); *Mars and Its Canals* (New York: Macmillan, 1907); and *Mars as the Abode of Life* (New York: Macmillan, 1908).

12. Lowell, *Mars*, 207–8.

13. Lowell, *Mars and Its Canals*, 377, 382.

14. Lowell, *Mars as the Abode of Life*, 214–15. See also Samuel Phelps Leland, *World Making: A Scientific Explanation of the Birth, Growth and Death of Worlds* (Chicago: Leland, 1906), 62–67.

15. Alfred Russel Wallace, *Is Mars Habitable?* (London: Macmillan, 1907). See also Wallace, *Man's Place in the Universe* (New York: McClure, Phillips, 1903), and William H. Pickering, "Recent Studies of the Martian and Lunar Canals," *Popular Astronomy*, February 1904.

16. H. G. Wells, *The Time Machine and War of the Worlds*, ed. Frank D. McConnell (New York: Oxford University Press, 1977), chap. 1, 125.

17. Frank D. McConnell, introduction to Wells, *Time Machine and the War of the Worlds*, 3–10.

18. Wells, *Time Machine and War of the Worlds*, 124.

19. Lowell, *Mars as the Abode of Life*, 216.

20. See "Mars," in *The Encyclopedia of Science Fiction*, by John Clute and Peter Nicholls (New York: St. Martin's Press, 1993), 777–79.

21. Burroughs, *Princess of Mars*. See also Richard A. Lupoff, *Barsoom: Edgar Rice Burroughs and the Martian Vision* (Baltimore: Mirage Press, 1976); Erling B. Holtsmark, *Edgar Rice Burroughs* (Boston: Twayne, 1986), chap. 2.

22. Hadley Cantril, *The Invasion from Mars: A Study of the Sociology of Panic* (Princeton, N.J.: Princeton University Press, 1940).

23. Kurt Neumann, *Rocketship X-M* (Lippert, 1950).

24. George Pal, *War of the Worlds* (Paramount, 1953); Steven Spielberg, *War of the Worlds* (Paramount, 2008).

25. See Arthur C. Clarke, *Childhood's End* (New York: Harcourt, Brace, and World, 1953).

26. Ray Bradbury, *The Martian Chronicles* (New York: Bantam Books, 1950), 32, 49–66, 102–10.

27. Ward Kimball, "Mars and Beyond" (Walt Disney, 1957).

28. Ray Bradbury, Arthur C. Clarke, B. Murray, Carl Sagan, and W. Sullivan, *Mars and the Mind of Man* (New York: Harper and Row, 1973), 22.

29. Daniel J. Boorstin, *Discoverers* (New York: Random House, 1983), pt. 12.

30. T. H. White, ed., *The Bestiary: A Book of Beasts; Being a Translation from a Latin Bestiary of the Twelfth Century* (New York: Perigree, 1980), 24–25; Willy Ley, *Dawn of Zoology* (Englewood Cliffs, N.J.: Prentice-Hall, 1968), 97. For the origin of the goose story, see Robert Bartlett, *Gerald of Wales* (Oxford: Clarendon Press, 1982), 136–37.

31. See Sully Zuckerman, *Great Zoos of the World: Their Origins and Significance* (Boulder, Colo.: Westview Press, 1980), and Marian Murray, *Circus! From Rome to Ringling* (Westport, Conn.: Greenwood Press, 1956).

32. White, *Bestiary*, 22–24.

33. John Ashton, *Curious Creatures in Zoology* (Detroit: Singing Tree Press, 1890), vi.

34. Josephine Waters Bennett, *The Rediscovery of Sir John Mandeville* (New York: Kraus Reprint, 1954).

35. Antonio Pigafetta, *Magellan's Voyage: A Narrative Account of the First Circumnavigation* (New Haven, Conn.: Yale University Press, 1969); Richard C. Temple, *The World Encompassed and Analogous Contemporary Documents Concerning Sir Francis Drake's Circumnavigation of the World* (New York: Cooper Square, 1969). No major narrative of the English discovery of North America exists. See James A. Williamson, *The Cabot Voyages and Bristol Discovery under Henry VII* (Cambridge: Cambridge University Press, 1962). See also Alexander de Humboldt and Aime Bonpland, *Personal Narrative of Travels to the Equinoctial Regions of the New Continent during the Years 1799–1804* (New York: AMS Press, 1966); John Bakeless, *The Eyes of Discovery: America as Seen by the First Explorers* (New York: Dover, 1950); Walter Raleigh, *The Discovery of the Large, Rich, and Beautiful Empire of Guiana* (Cleveland: World, 1966). Descriptions of races far exceed descriptions of beasts.

36. See Stephen E. Ambrose, *Meriwether Lewis, Thomas Jefferson and the Opening of the American West* (New York: Simon and Schuster, 1996).

37. Charles Darwin, *The Voyage of the Beagle* (New York: P. F. Collier and Son, 1909), chap. 17; Charles Darwin, *The Zoology of the Voyage of the Beagle* (New York: New York University Press, 1987).

38. Darwin, *Voyage of the Beagle*, 218 (December 25th).

39. See Steven J. Dick, *Plurality of Worlds: The Origins of the Extraterrestrial Life Debate from Democritus to Kant* (New York: Cambridge University Press, 1982), and Michael J. Crowe, *The Extraterrestrial Life Debate, 1750–1900* (New York: Cambridge University Press, 1986).

40. Aristotle, *On the Heavens*, trans. W. K. C. Guthrie (Cambridge, Mass.: Harvard University Press, 1939), chap. 8. Also see Friedrich Solmsen, *Aristotle's System of the Physical World* (Ithaca, N.Y.: Cornell University Press, 1960).

41. St. Thomas Aquinas, *Summa Theologica* (New York: Benziger Brothers, 1947), pt. 1, question 47, art. 3.

42. See "The Solar System: Exception or Rule?" in Walter Sullivan, *We Are Not Alone: The Search for Intelligent Life on Other Worlds*, rev. ed. (New York: McGraw-Hill, 1964), chap. 5; S. F. Dermott, *The Origin of the Solar System* (New York: John Wiley and Sons, 1978).

43. Immanuel Velikovsky, *Worlds in Collision* (Garden City, N.Y.: Doubleday, 1950). See also Immanuel Velikovsky, *Ages in Chaos* (Garden City, N.Y.: Doubleday, 1952) and *Earth in Upheaval* (Garden City, N.Y.: Doubleday, 1955); Ronald N. Bracewell, *The Galactic Club: Intelligent Life in Outer Space* (San Francisco: W. H. Freeman, 1974).

44. Genesis 1:27.

45. Bernard de Fontenelle, *Entretiens sur la pluralité des mondes* (1686; reprint, Berkeley and Los Angeles: University of California Press, 1990).

46. H. G. Wells, "The Things That Live on Mars," *Cosmopolitan* 44 (March 1908): 335–42.

47. Kenneth Heuer, *Men of Other Planets* (New York: Pellegrini and Cudahy, 1951), 150. See also Steven J. Dick, *The Biological Universe: The Twentieth Century Extraterrestrial Life Debate and the Limits of Science* (New York: Cambridge University Press, 1996); Gene Bylinsky, *Life in Darwin's Universe: Evolution and the Cosmos* (Garden City, N.Y.: Doubleday, 1981); David Milne, David Raup, John Billingham, Karl Niklaus, and Kevin Padian, eds., *The Evolution of Complex and Higher Organisms*, NASA SP-478 (Washington, D.C.: National Aeronautics and Space Administration, 1985); Willy Ley, "What Will 'Space People' Look Like," *This Week Magazine*, 10 November 1957; Alan D. Foster, *Alien Omnibus* (Warner Books, 1987).

48. I. S. Shklovskii and Carl Sagan, *Intelligent Life in the Universe* (San Francisco: Holden-Day, 1966), 329.

49. Bonnie Dalzell, "Exotic Bestiary for Vicarious Space Voyagers," *Smithsonian*, October 1974, 84–91.

50. See Wayne Douglas Barlowe and Ian Summers, *Barlowe's Guide to Extra-Terrestrials* (New York: Workman, 1979), and Isaac Asimov, "Anatomy of a Man from Mars," *Esquire*, September 1965, 113–17, 200.

51. Howard Hawks, *The Thing* (RKO Radio Pictures, 1951).

52. Gordon Carroll, David Giller, and Walter Hill, *Alien* (Fox, 1979).

53. Gary Kurtz, *Star Wars* (Fox, 1977). Also see Shane Johnson, *Star Wars Technical Journal* (New York: Ballantine Books, 1995).

54. *E.T.: The Extra-Terrestrial, Star Wars, Return of the Jedi*, and *The Empire Strikes Back*. Mark S. Hoffman, *The World Almanac and Book of Facts* (New York: Pharos Books, 1991), 308.

55. Henry Norris Russell, "Anthropcentrism's Demise," *Scientific American* 169 (July 1943): 18–19.

56. H. Spencer Jones, *Life on Other Worlds* (New York: Macmillan, 1940), 244. See also "Life Beyond Earth?" *Time*, 7 October 1940, 62; and Bruce Bliven, "Is There Life on Other Planets," *Reader's Digest*, February 1955, 103–7.

57. See Crowe, *Extraterrestrial Life Debate*, chap. 9, 378–86.

58. Camille Flammarion, "Are the Planets Inhabited?" *Harper's*, November 1904, 844.

59. Harlow Shapley, "Coming to Terms with the Cosmos," *Saturday Review*, 6 September 1958, 54. See also Harlow Shapley, *Of Stars and Men: The Human Response to an Expanding Universe* (Boston: Beacon Press, 1958).

60. See "Space Theology," *Time*, 19 September 1955, 81.

61. See Loren C. Eiseley, "Little Men and Flying Saucers," *Harper's*, March 1953, 86–91; Wyn Wachhorst, "Seeking the Center at the Edge: Perspectives on the Meaning of Man in Space," *Virginia Quarterly Review* 69 (Winter 1993): 1–23; William Williams, *The Universe No Desert, the Earth No Monopoly* (Boston: James Munroe, 1855).

62. "Pictures Show Evidence of Life on Planet Mars," *Life*, 29 May 1944, 83. See also David Todd, "Professor Todd's Own Story of the Mars Expedition," *Cosmopolitan* 44 (March 1908): 343–51.

63. "Buck Rogers Baedeker," *Newsweek*, 28 February 1949, 48–49.

64. NASA, *Mariner-Mars 1964: Final Project Report*, NASA SP-139 (Washington, D.C.: GPO, 1967), 273–89; Glasstone, *Book of Mars*. Turn-of-the-century astronomers may have mapped these features. See William Sheehan, "Did Barnard & Mellish Really See Craters on Mars?" *Sky & Telescope*, July 1992, 23–25.

65. "An End to the Myths about Men on Mars," *U.S. New & World Report*, 9 August 1965, 4.

66. Lyndon B. Johnson, "Remarks upon Viewing New Mariner 4 Pictures from Mars," 29 July 1965, *Public Papers of the Presidents of the United States, 1965* (Washington, D.C.: GPO, 1966), 2:806.

67. NASA, *Mariner-Mars 1964: Final Project Report*, 318–22.

68. "Is There Life on Mars—or Earth?" *Time*, 7 January 1966, 44; Steven D. Kilston, Robert R. Drummond, and Carl Sagan, "A Search for Life on Earth at Kilometer Resolution," *Icarus* 5 (January 1966): 79–98.

69. Stuart Auerbach, "Mariner 7 Photographs Mysterious Mars Canals," *Washington Post*, 5 August 1969, sec. A, p. 1.

70. Senate Aeronautical and Space Sciences Committee, *Future NASA Space Programs*, 91st Cong., 1st sess., 1969, 6–7, 68.

71. Stuart Auerbach, "Mars Pocked by Moon-Like Depressions," *Washington Post*, 1 August 1969, sec. A, p. 1. See also "Surface of Mars Similar to Moon," *Baltimore Sun*, 1 August 1969, 1, 4.

72. William K. Hartmann and Odell Raper, *The New Mars: The Discoveries of Mariner 9*, NASA SP-337 (Washington, D.C.: GPO, 1974).

73. Quoted from Henry S. F. Cooper, *The Search for Life on Mars* (New York: Holt, Rinehart and Winston, 1976), 78–79.

74. See Norman Horowitz, *To Utopia and Back* (New York: W. H. Freeman, 1986), chaps. 5 and 6; Wilford, *Mars Beckons*.

75. Kathy Sawyer, "NASA Prepares Craft for a Deep Encounter of the Martian Kind," *Washington Post*, 21 September 1992, sec. A, p. 3.

76. Horowitz, *To Utopia and Back*, xi.

77. NASA, Phoenix Mars Mission, "Bright Chunks at Phoenix Lander's Mars Site Must Have Been Ice," 19 June, 2008, phoenix.lpl.arizona.edu/06_19_pr.php (accessed 24

January 2009); Marc Kaufman, "Mars Vents Methane in What Could Be Sign of Life," *Washington Post*, 16 January 2009, sec. A, p. 4.

78. See Richard Corliss, "The Invasion Has Begun," *Time*, 8 July 1996, 58–64.

79. "Meteorite Find Incites Speculation on Mars Life," *Space News*, 5–11 August 1966, 2; David S. McKay, Everett K. Gibson Jr., Kathie L. Thomas-Keprta, Hojatollah Vali, Christopher S. Romanek, Simon J. Clemett, Xavier D. F. Chillier, Claude R. Maechling, and Richard N. Zare, "Search for Past Life on Mars: Possible Relic Biogenic Activity in Martian Meteorite ALH84001," *Science* 273 (16 August 1996): 924–30.

80. David Colton, "Discovery Would Equal Finding the New World," *USA Today*, 8 August 1996; NASA, *NASA News*, "Statement of Daniel S. Goldin, NASA Administrator," release 96-159, 6 August 1996; NASA, *NASA News*, "NASA Briefing Wednesday on Discovery of Possible Martian Life," note to editors N96-53, 6 August 1966; William J. Clinton, "NASA Discovery of Possible Life on Mars," *Weekly Compilation of Presidential Documents* 32, no. 32 (7 August 1996): 1417–18.

81. John Noble Wilford, "On Mars, Life's Getting Tougher (If Not Impossible)," *New York Times*, 22 December 1996.

82. NASA News Release, "Jupiter's Europa Harbors Possible 'Warm Ice' or Liquid Water," release 96-164, 13 August 1996, NASA Headquarters, Washington, D.C.

83. See Carl Sagan, *Cosmos* (New York: Random House, 1980), 299–302; Carl Sagan and Frank Drake, "The Search for Extraterrestrial Intelligence," *Scientific American* 232 (May 1975): 80–89.

84. Robert Jastrow and Malcolm H. Thompson, *Astronomy: Fundamentals and Frontiers*, 4th ed. (New York: John Wiley and Sons, 1984), 71; Sagan, *Cosmos*, 299.

85. See "Life on a Billion Planets," *Time*, 3 March 1958, 42–42; "Anybody Out There?" *Time*, 23 November 1959, 84–84; "Advice from Space," *Time*, 29 December 1961, 26. Also see Frank D. Drake, *Intelligent Life in Space* (New York: Macmillan, 1962).

86. Isaac Asimov, *Foundation* (Garden City, N.Y.: Doubleday, 1951), *Foundation and Empire* (Garden City, N.Y.: Doubleday, 1952), and *Second Foundation* (Garden City, N.Y.: Doubleday, 1953).

87. Gene Roddenberry, *Star Trek*, NBC, September 1966 to September 1969, seventy-nine episodes.

88. Asimov, *Foundation*, 3–6.

89. Frank D. Drake, *Intelligent Life in Space* (New York: Macmillan, 1962).

90. Quoted from Carl Sagan, *The Cosmic Connection* (New York: Doubleday, 1973), 25–26.

91. House Science and Technology Committee, *The Possibility of Intelligent Life Elsewhere in the Universe*, 94th Cong., 1st sess., 1975, 24–27; Stanford University and NASA Ames Research Center, "Project Cyclops: A Design Study of a System for Detecting Extraterrestrial Intelligent Life," CR 114445, July 1973, Moffett Field, California, NASA/Ames Research Center.

92. House Science and Technology Committee, *The Possibility of Intelligent Life Elsewhere in the Universe*, 32–53. See also Sagan and Drake, "Search for Extraterrestrial Intelligence," 80–89, and Frank Drake and Dava Sobel, *Is Anyone Out There? The Scientific Search for Extraterrestrial Intelligence* (New York: Delacorte Press, 1992), 182–84.

93. James C. Fletcher, "NASA and the 'Now' Syndrome," from an address to the National Academy of Engineering, Washington, D.C., November 1975, NASA brochure,

7. See also Fletcher, "Space: 30 Years into the Future," *Acta Astronautica* 19, no. 11 (1989): 855–57.

94. Joseph Smith, *The Holy Scriptures*, The Reorganized Church of Jesus Christ of Latter Day Saints (Independence, Mo.: Herald Publishing House, 1944), a revelation given to Joseph the Seer, para. 21; Roger D. Launius, "A Western Mormon in Washington, D.C.: James C. Fletcher, NASA, and the Final Frontier," *Pacific Historical Review* 64 (May 1995): 217–41.

95. See Philip Morrison, John Billingham, and John Wolfe, eds., *The Search for Extraterrestrial Intelligence (SETI)*, NASA SP-419 (Washington, D.C.: GPO, 1977).

96. NASA, "SETI," National Aeronautics and Space Administration publication NP-114, June 1990.

97. Lance Frazer, "Small Change, High Gain," *Ad Astra*, September 1989, 19.

98. Rob Meckel, "Proxmire 'Fleeces' NASA over Communications," Proxmire biography file, NASA History Office, NASA Headquarters, Washington, D.C.; "Senate Rejects NASA Space Signal Plan," *Newport News Times-Herald*, 8 August 1978.

99. William Triplett, "SETI Takes the Hill," *Air and Space*, November 1992, 80–86; Lance Frazer, "Listening for Life," *Ad Astra*, September 1989, 16–22.

100. Dava Sobel, "Is Anybody Out There?" *Life*, September 1992, 14. See also Peter Bond, "Extra-Terrestrials Search Stepped Up," *Spaceflight*, January 1993, 6–7.

101. *Congressional Record*, 22 September 1993, S12151.

102. Sebastian von Hoerner, "Where Is Everybody?" in *The Quest for Extraterrestrial Life*, ed. Donald Goldsmith (Mill Valley, Calif.: University Science Books, 1980), 252; Michael H. Hart, "An Explanation of the Absence of Extraterrestrials on Earth," *Quarterly Journal of the Royal Astronomical Society* 16 (June 1975): 128–35; Eric M. Jones, "Colonization of the Galaxy," *Icarus* 28 (1976): 421–22; J. Freeman and M. Lampton, "Interstellar Archeology and the Prevalence of Intelligence," *Icarus* 25 (1975): 368–69; Sagan, *Cosmos*, chap. 12.

103. Michael H. Hart, "An Explanation for the Absence of Extraterrestrials on Earth," *Quarterly Journal of the Royal Astronomical Society* 16 (1975): 128–35. See also Michael H. Hart and Ben Zuckerman, eds., *Extraterrestrials: Where Are They?* (New York: Pergamon Press, 1982); Steven J. Dick, *Life on Other Worlds: The 20th-Century Extraterrestrial Life Debate* (Cambridge: Cambridge University Press, 1998; Dick, *The Biological Universe: The Twentieth-Century Extraterrestrial Life Debate and the Limits of Science* (Cambridge: Cambridge University Press, 1996).

104. Steven Spielberg, *Men in Black* (Columbia Pictures, 1997).

105. See Sagan, *Cosmos*, 301.

106. Steve Starkey and Robert Zemeckis, *Contact* (Warner Brothers, 1997).

107. Peter D. Ward and Donald Brownlee, *Rare Earth: Why Complex Life Is Uncommon in the Universe* (New York: Copernicus, 2000), xiv. See also Stephen Webb, *Where Is Everybody? If the Universe Is Teeming with Aliens, Where Is Everybody? Fifty Solutions to the Fermi Paradox and the Problem of Extraterrestrial Life* (New York: Copernicus Books, 2002).

108. See Stephen H. Dole, *Habitable Planets for Man* (ca. 1964, Santa Monica, Calif.: RAND, 2007), chap. 8; Neil F. Comins, *What If the Moon Didn't Exist?* (New York: HarperCollins, 1993).

109. Von Hoerner, "Where Is Everybody?" 251–53.

110. Dick, *Life on Other Worlds*, 254.

111. Hawking, *A Brief History of Time*, 124–25.

112. *Congressional Record*, 22 September 1993, S12152. See also Mariane K. Meuse, "Space Explodes! Alien Media Invades Earth," *Ad Astra*, January–February 1992, 42–46, 55.

113. See Michio Kaku, *Hyperspace: A Scientific Odyssey through Parallel Universes, Time Warps, and the Tenth Dimension* (New York: Oxford University Press, 1994), 196–201; George Smoot and Keay Davidson, *Wrinkles in Time* (New York: William Morrow, 1993), 83–86; NASA, *NASA Facts*, "COBE Observes Primeval Explosion," Goddard Space Flight Center, Greenbelt, Maryland, n.d.

114. Craig Covault, "Cosmic Background Explorer to Observe Big Bang Radiation," *Aviation Week & Space Technology*, 6 November 1989, 36–41.

115. Quoted from Paul Hoversten, "Relics of Universe's Birth Found," *USA Today*, 24 April 1992; and Thomas H. Maugh, "'Holy Grail' of the Cosmos," *Los Angeles Times*, 24 April 1992, Washington ed.

116. Malcolm W. Browne, "Despite New Data, Mysteries of Creation Persist," *New York Times*, 12 May 1992, sec. C, pp. 1, 10; NASA, "Cosmic Background Explorer Observes the Primeval Explosion," Goddard Space Flight Center, Greenbelt, Maryland, n.d.

117. Kathy Sawyer, "Big Bang "Ripples' Have Universal Impact," *Washington Post*, 3 May 1992, sec. A, p. 1.

118. Hoversten, "Relics of Universe's Birth Found." Also see Billy Goodman, "Ancient Whisper," *Air & Space*, April–May 1992, 55–61.

119. Campbell, *The Power of Myth*.

120. Genesis 1:3–5.

121. Sawyer, "Big Bang 'Ripples' Have Universal Impact," sec. A, p. 20.

122. H. G. Wells, *The Time Machine* (New York: Berkley, 1963). George Pal released the motion picture version in 1960. Pal, *The Time Machine* (MGM, 1960).

123. See Paul J. Nahin, *Time Machines: Time Travel in Physics, Metaphysics, and Science Fiction* (New York: American Institute of Physics, 1993); Peter Nicholls, ed., *The Science in Science Fiction* (New York: Knopf, 1983), chap. 5.

124. NASA, "Frontiers in Cosmology," Hubble Space Telescope Fact Sheet, Space Telescope Science Institute, Baltimore Maryland, n.d., NASA History Office.

125. Lockheed Missiles and Space Company, "Hubble: A Window into the Universe," 1986. See also NASA, "Hubble Space Telescope Media Reference Guide," published for NASA by Lockheed Missiles and Space Company, Sunnyvale, California, 1990; Joseph J. McRoberts, *Space Telescope*, NASA EP-166 (Washington, D.C.: GPO, n.d.). All in NASA History Office.

126. Steven Spielberg, Bob Gale, and Neil Canton, *Back to the Future* (Universal Studios, 1985).

127. Kaku, *Hyperspace*, x (emphasis removed). Also see Kaku and Jennifer Trainer, *Beyond Einstein: The Cosmic Quest for the Theory of the Universe* (New York: Bantam Books, 1987); and Hawking, *Brief History of Time*.

128. Kip S. Thorne, *Black Holes and Time Warps* (New York: W. W. Norton, 1994); Barry Parker, *Cosmic Time Travel: A Scientific Odyssey* (New York: Plenum Press, 1991).

129. Michael S. Morris, Kip S. Thorne, and Ulvi Yurtsever, "Wormholes, Time Machines, and the Weak Energy Condition," *Physical Review Letters* 61 (26 September 1988): 1446–49; Michael S. Morris and Kip S. Thorne, "Wormholes in Spacetime and Their Use

for Interstellar Travel," *American Journal of Physics* 56 (May 1988): 395–412. See also, Nahin, *Time Machines*, tech note 9.

130. Lewis Carroll, *Through the Looking Glass* (New York: Grosset and Dunlap, 1946); C. S. Lewis, *Lion, the Witch and the Wardrobe* (New York: Collier Books, 1950).

131. NASA, News Release 95-216, "Hubble Finds New Black Hole and Unexpected Mysteries," 4 December 1995.

132. Kaku, *Hyperspace*, chap. 10.

133. Patrick J. Kiger, "The New Galileo," *Baltimore Magazine*, February 1990, 107.

134. *Congressional Record*, 24 June 1993, H4057. See also *Congressional Record*, 23 June 1993, H3974–78, and 19 October 1993, H8114–24; and *Congressional Record*, 10 July 1991, S9430–43.

135. See David C. Black, ed., *Project Orion: A Design Study of a System for Detecting Extrasolar Planets*, NASA SP-436 (Washington, D.C.: National Aeronautics and Space Administration, 1980); NASA, "TOPS: Toward Other Planetary Systems," a report by the Solar System Exploration Division, n.d.; D. DeFrees, *Exobiology in Earth Orbit*, NASA SP-500 (Washington, D.C.: GPO, 1989).

136. See Milne et al., *Evolution of Complex and Higher Organisms*, 153–54.

137. John Noble Wilford, "New Discoveries Turn Astronomers toward Hunt for New Planets," *New York Times*, 23 January 1996; Wilford, "The Search for Solar Systems Accelerates amid New Clues," *New York Times*, 21 April 1987; Malcolm Brown, "Clues Point to Young Planet Systems Nearby," *New York Times*, 12 June 1992; NASA, *NASA News*, release 92-226, "Hubble Discovers Proloplanetary Disks around New Stars," 16 December 1992.

138. The Planetary Society, press release "Planetary Society Charges Administration with Blurring its Vision for Space Exploration," 6 February 2006.

CHAPTER 6: The Extraterrestrial Frontier

Epigraph: Herbert F. Solow and Robert H. Justman, *Inside Star Trek* (New York: Pocket Books, 1996), 149.

1. See Robert McCall, *The Art of Robert McCall* (New York: Bantam Books, 1992), 28–29, for a view of the mural. For the full quotation, see Ayn Rand, "The Soul of an Individualist," 1961, aynrandlexicon.com/lexicon/creators.html (accessed 27 June 2010). See also Stephen M. Fjellman, *Vinyl Leaves: Walt Disney World and America* (Boulder, Colo.: Westview Press, 1992), chap. 5; Richard R. Beard, *Walt Disney's Epcot* (New York: Harry N. Abrams, 1982), 136–63.

2. See Patricia Nelson Limerick, *The Legacy of Conquest: The Unbroken Past of the American West* (New York: W. W. Norton, 1987); Richard White, *"It's Your Misfortune and None of My Own": A New History of the American West* (Norman: University of Oklahoma Press, 1991); Donald Worster, *Rivers of Empire: Water, Aridity, and the Growth of the American West* (New York: Pantheon Books, 1985); William Cronon, George Miles, and Jay Gitlin, eds., *Under the Open Sky: Rethinking America's Western Past* (New York: W. W. Norton, 1992); and Richard Slotkin, *Gunfighter Nation: The Myth of the Frontier in Twentieth-Century America* (New York: Atheneum, 1992).

3. "Day at Tranquility," *Washington Daily News*, 21 July 1969; "One Small Step—One Giant Leap," *Washington Post*, 21 July 1969.

4. Manfred van Ehrenfried, *Adventures on Santa Maria and Future Ships Sailing the Oceans of Space* (Glenside, Pa.: Custom Comic Services, 1991), 1.

5. National Commission on Space, *Pioneering the Space Frontier* (New York: Bantam Books, 1986), 8. See also Joseph F. Shea, "Manned Space Flight Program," address at the third national conference on the peaceful uses of space, Chicago, 6 May 1963, NASA News release, NASA History Office, NASA Headquarters, Washington, D.C.

6. Buzz Aldrin, "The Mars Transit System," *Air & Space*, October–November 1990, 41, 42. See also Buzz Aldrin and John Barnes, *Encounter with Tiber* (New York: Warner Books, 1996).

7. "Excerpts of Remarks by Governor Ronald Reagan," 4, America Legion State Convention, Sacramento, California, 26 June 1970, Hoover Institution on War, Revolution, and Peace, Stanford, California.

8. John Logsdon, *Decision to Go to the Moon* (Cambridge, Mass.: MIT Press, 1970), 35. See also Ira C. Eaker, "Columbus and the Moon: Debates on Voyages Are Similar," *San Diego Union*, 22 September 1963; Daniel Goldin, "Celebrating the Spirit of Columbus," *National Forum* 72 (Summer 1992): 8–9.

9. Howard E. McCurdy, *The Space Station Decision: Incremental Politics and Technological Choice* (Baltimore: Johns Hopkins University Press, 1990), 184. See also Lyndon B. Johnson, "Remarks Following an Inspection of NASA's Michoud Assembly Facility near New Orleans," 12 December 1967, *Public Papers of the Presidents of the United States, 1967* (Washington, D.C.: GPO, 1968), 2:1123.

10. James M. Beggs, "Why the United States Needs a Space Station," 2–3, remarks prepared for delivery at the Detroit Economic Club and Detroit Engineering Society, 23 June 1982, NASA History Office; reprinted under the same title in *Vital Speeches*, 1 August 1982, 615–17.

11. James M. Beggs, "The Wilbur and Orville Wright Memorial Lecture," Royal Aeronautical Society, London, England, 13 December 1984, NASA History Office; Michael Ryan, "Why They Come to the Ice," *Parade*, 11 July 1993, 5. See also James Beggs, "Space Tomorrow: The Antarctica Model," *IEEE Spectrum* 20 (September 1983): 89–90; and "A Terrestrial Testing Ground for Space Exploration," in NASA, *HQ Bulletin*, 19 February 1991, 1, 3.

12. Ronald Reagan, "United States Space Policy," remarks on the completion of the fourth mission of the space shuttle *Columbia*, *Weekly Complication of Presidential Documents* 18, no. 27 (4 July 1982): 870. See also NASA Grant Nsg-253-62; Bruce Mazlish, ed., *The Railroad and the Space Program: An Exploration in Historical Analogy* (Cambridge, Mass.: MIT Press, 1965).

13. George H. W. Bush, "Remarks at the Texas A&I University Commencement Ceremony in Kingsville, Texas," *Weekly Compilation of Presidential Documents* 26, no. 20 (11 May 1990): 749.

14. Robert Zubrin, *The Case for Mars: The Plan to Settle the Red Planet and Why We Must* (New York: Free Press, 1996), 297.

15. Patricia Nelson Limerick, "Imagined Frontiers: Westward Expansion and the Future of the Space Program," in *Space Policy Alternatives*, ed. Radford Byerly (Boulder, Colo.: Westview Press, 1992), 249–61.

16. Frederick Jackson Turner, "The Significance of the Frontier in American History," in John M. Farager, *Rereading Frederick Jackson Turner* (New York: Henry Holt, 1994),

31–60. See also Louis Hartz, *The Liberal Tradition in America* (New York: Harcourt, Brace, 1955); Clinton Rossiter, *Conservatism in America* (New York: Knopf, 1955); Alexis de Tocqueville, *Democracy in America* (New York: Vintage Books, 1954); Aaron Wildavsky, *The Rise of Radical Egalitarianism* (Washington, D.C.: American University Press, 1991); Seymour Martin Lipset, *American Exceptionalism: A Double-Edged Sword* (New York: W. W. Norton, 1996).

17. Turner, "The Significance of the Frontier in American History," 32.

18. See George Rogers Taylor, ed., *The Turner Thesis: Concerning the Role of the Frontier in American History*, 3rd ed. (Lexington, Mass.: D. C. Heath, 1972).

19. See, for example, Eric M. Jones, ed., "The Space Settlement Papers," *Journal of the British Interplanetary Society* 39 (July 1986): 291–311.

20. Brad Darrach and Steve Petranek, "Mars: Our Next Home," *Life*, May 1991, 34.

21. "Go!" *Washington Post*, 21 February 1962, sec. A, p. 24. See also Robert J. Donovan, "Moon Voyage Turns Men's Thoughts Inward," *Los Angeles Times*, 29 December 1968; "One Small Step—One Giant Leap," *Washington Post*, 21 July 1969; and "At Path, Not an End," *Washington Post*, 21 July 1969.

22. James M. Beggs, "The Wilbur and Orville Wright Memorial Lecture," Royal Aeronautical Society, London, England, 13 December 1984, NASA History Office; Daniel Goldin, "Celebrating the Spirit of Columbus," *Phi Kappa Phi Journal*, Summer 1992, 8. See also Synthesis Group on America's Space Exploration Initiative, *America at the Threshold* (Washington, D.C.: GPO, 1991), iv, and Carl Sagan, *The Demon-Haunted World: Science as a Candle in the Dark* (New York: Ballantine, 1996).

23. Beggs, "Wilbur and Orville Wright Memorial Lecture," 2. Also see Thomas O. Paine, "1969: A Space Odyssey," 9–11, address at the American Institute of Aeronautics and Astronautics, Washington, D.C., 7 November 1968; and Paine, "Thomas A. Edison Memorial Lecture," 11–12, address at the Naval Research Laboratory, Washington, D.C., 11 March 1969; all in NASA History Office.

24. John F. Kennedy, "Address at Rice University in Houston on the Nation's Space Effort," 12 September 1962, *Papers of the Presidents of the United States, 1962* (Washington, D.C.: GPO, 1963), 373.

25. Carl Sagan, *Pale Blue Dot: A Vision of the Human Future in Space* (New York: Random House, 1994), 371, 382; James C. Fletcher, "NASA and the 'Now' Syndrome," address to the National Academy of Engineering, Washington, D.C., November 1975, 3.

26. Walter J. Hickel, "In Space: One World United," in *Lunar Bases and Space Activities of the 21st Century*, ed. W. W. Mendell (Houston: Lunar and Planetary Institute, 1985), 15, 17; Beggs, "Wilbur and Orville Wright Memorial Lecture," 2.

27. Paine, "Thomas A. Edison Memorial Lecture," 11.

28. Ibid.

29. Ibid., 10.

30. Paul D. Lowman, "Lunar Bases and Post-Apollo Lunar Exploration: An Annotated Bibliography of Federally-Funded American Studies, 1960–82," Geophysics Branch, Goddard Space Flight Center, Greenbelt, Maryland, October 1984; NASA, "NASA Initiates Lunar Base Study Program," news release no. 63-91, 6 May 1963; both in NASA History Office, NASA Headquarters, Washington, D.C.

31. Boeing Company, "Initial Concept of Lunar Exploration Systems for Apollo

(LESA)," D2-1000057, NASW 792, Boeing Company Aero-Space Division. See also House Subcommittee on Manned Space Flight, *1965 NASA Authorization*, 88th Cong., 2nd sess., 1964, pt. 2, 587–626; and House Committee on Science and Astronautics, *Future National Space Objectives*, 89th Cong., 2nd sess., 1966; NASA, "Apollo Applications Program: Program Review Document," 15 November 1966, 3–4, NASA History Office.

32. Lockheed, "Study of Mission Modes and System Analysis for Lunar Exploration (MIMOSA), Final Report," Lockheed Missiles and Space Company, NAS 8-20262, Sunnyvale, California. See also NASA, Future Programs Task Group, "Summary Report," 47–48, NASA Headquarters, January 1965, NASA History Office.

33. Arthur C. Clarke, *2001: A Space Odyssey* (New York: New American Library, 1968), 61–65. See also Paul D. Lowman, "Lunar Bases: A Post-Apollo Evaluation," in Mendell, *Lunar Bases and Space Activities*, 38, 42–43.

34. Robert Zubrin, *The Case for Mars: The Plan to Settle the Red Planet and Why We Must* (New York: Free Press, 1996), 297.

35. Ray Bradbury, "The Million Year Picnic," in *The Martian Chronicles* (Garden City, N.Y.: Doubleday, 1958), 214.

36. Thomas More, *Utopia* (Boston: Adamant Media Corporation, 2005); Alan Heimert and Andrew Delbanco, eds., *The Puritans in America: A Narrative Anthology* (Cambridge, Mass.: Harvard University Press, 1985); Stephen J. Stein, *The Shaker Experience in America* (New Haven, Conn.: Yale University Press, 1994); James Hilton, *Lost Horizon* (New York: M. Morrow, 1933).

37. See Brian Alexander, *Rapture: How Biotech Became the New Religion* (New York: Basic Books, 2003); Ray Kurzweil, *The Singularity Is Near: When Humans Transcend Biology* (New York: Penguin Books, 2005).

38. Zubrin, *The Case for Mars*, 296–98.

39. Michael A. G. Michaud, *Reaching for the High Frontier* (New York: Praeger, 1986), 65; Gerard K. O'Neill, "The Colonization of Space," *Physics Today* 27 (September 1974): 32–40. See also Gerard K. O'Neill, *The High Frontier: Human Colonies in Space* (New York: William Morrow, 1976), app. 1.

40. O'Neill, "Colonization of Space," 36; Gerard K. O'Neill, "A Lagrangian Community?" *Nature* 250 (23 August 1974): 636.

41. Ibid., 37, 39; O'Neill, *High Frontier*, 64.

42. O'Neill, "Colonization of Space," 37.

43. Ibid., 37.

44. In 1969 a Saturn 5 launch vehicle cost a total of $255 million in hardware and operational costs to place a lunar module weighing fifteen hundred kilograms (thirty-three hundred pounds) on the Moon. Howard E. McCurdy, "The Cost of Space Flight," *Space Policy* 10 (November 1994): 277–89.

45. O'Neill, *High Frontier*, 138–41.

46. Stewart Brand, *Space Colonies* (New York: Penguin Books, 1977), 15. See also O'Neill, "Colonization of Space," 37–38; Jerry Grey, ed., *Space Manufacturing Facilities (Space Colonies)* (New York: American Institute of Aeronautics and Astronautics, 1977).

47. Tsiolkovsky, *Beyond the Planet Earth*, trans. Kenneth Syers (New York: Pergamon Press, 1960), chaps. 12 and 13; J. D. Bernal, *The World, the Flesh, and the Devil* (London: Methuen, 1929; Bloomington: University of Indiana Press, 1969); J. N. Leonard, *Flight into Space* (New York: Signet, 1954), chap. 22; Dandridge M. Cole and Donald W. Cox,

Notes to Pages 170–173   359

Islands in Space: The Challenge of the Planetoids (New York: Chilton, 1964); Dandridge M. Cole, "Extraterrestrial Colonies," *Navigation* 7 (Summer–Fall 1960): 83–98; Arthur C. Clarke, *Islands in the Sky* (New York: Holt, Rinehart, and Winston, 1954); Krafft A. Ehricke, "Extraterrestrial Imperative," *Bulletin of the Atomic Scientists* 27 (November 1971): 18–26. See also Michaud, *Reaching for the High Frontier*, chap. 4.

48. See Brand, *Space Colonies*; Michaud, *Reaching for the High Frontier*, chap. 5; and O'Neill, *High Frontier*, 253–61. Richard D. Johnson, *Space Settlements: A Design Study*, NASA SP-413 (Washington, D.C.: GPO, 1977).

49. Quoted from Michaud, *Reaching for the High Frontier*, 86.

50. Frederick Ordway and Randy Liebermann, eds., *Blueprint for Space* (Washington, D.C.: Smithsonian Institution Press, 1992), 193–96; National Space Society vision statement, 20 March 2009, www.nss.org (accessed 20 March 2009).

51. Michaud, *Reaching for the High Frontier*, chap. 7.

52. "The Mars Declaration," in "The Way to Mars" (Pasadena, Calif.: Planetary Society, 1987); *Washington Post*, 26 May 1988: A7.

53. Trudy E. Bell, "Space Activists on Rise," *Insight* (National Space Institute), August–September 1980: 1, 3, 13–16.

54. Michaud, *Reaching for the High Frontier*, chaps. 8 and 9; Space Frontier Foundation Web site, space-frontier.org/; Mars Society Web site, www.marssociety.org/portal (both accessed 20 March 2009).

55. National Commission on Space, *Pioneering the Space Frontier*, 141.

56. Ibid., 86.

57. Thomas O. Paine, "Mars Colonization," *Phi Kappa Phi Journal* 72 (Summer 1992): 26. See also National Commission on Space, *Pioneering the Space Frontier*, 89; Benton Clark, "Chemistry of the Martian Surface: Resources for the Manned Exploration of Mars," in *The Case for Mars*, ed. Penelope U. Boston, Science and Technology Series, vol. 57 (San Diego: American Astronautical Society, 1984), 197–208; and Christopher P. McKay, *The Case for Mars II*, ed. Christopher P. McKay, Science and Technology Series, vol. 62 (San Diego: American Astronautical Society, 1985), chap. 7 ("Utilizing Martian Resources").

58. National Commission on Space, *Pioneering the Space Frontier*, 72. See also *Planetary Explorer*, a publication of the General Dynamics Space Systems Division, San Diego, California, 1988; Rick N. Tumlinson, "Why Space? Personal Freedom," the 1995 Frontier Files, 1995, space-frontier.org/History/frontierfiles.html (accessed 20 March 2009).

59. James Fletcher, "Excerpts from Remarks Prepared for Delivery, National Space Symposium," Colorado Springs, Colorado, 14 April 1988.

60. Leonard David, "Moon Fever," *Space World*, August 1988, 6–8.

61. Mendell, *Lunar Bases and Space Activities*; Craig Covault, "Manned U.S. Lunar Station Wins Support," *Aviation Week & Space Technology*, 19 November 1984, 73–86; Darren L. Burnham, "Back to the Moon with Robots?" *Spaceflight* 35 (February 1993): 54–57; David, "Moon Fever," 7.

62. Boston, *Case for Mars*, x.

63. McKay, *Case for Mars II*, ix–x; Carol R. Stoker, ed., *The Case for Mars III: Strategies for Exploration—General Interest and Overview*, Science and Technology series, vol. 74 (San Diego: American Astronautical Society, 1989). See also Zubrin, *The Case for Mars*.

64. Duke B. Reiber, *The NASA Mars Conference*, Science and Technology series, vol. 71

(San Diego: American Astronautical Society, 1988); NASA, "NASA Establishes Office of Exploration," release 87-87, *NASA News*, 1 June 1987.

65. Sally K. Ride, *Leadership and America's Future in Space*, Report to the Administrator (Washington, D.C.: National Aeronautics and Space Administration, 1987), 32–35; "Journey to Mars," transparency in NASA, Office of Exploration, "Exploration Initiative: A Long-Range, Continuing Commitment," January 1990; NASA, "Report of the 90–Day Study on Human Exploration of the Moon and Mars," NASA Headquarters, November 1989; Synthesis Group, *America at the Threshold*, 21.

66. Alcestis Oberg and James Oberg, "The Future of Mars," in Boston, *Case for Mars*, 311–13.

67. Ross Rocklynne, "Water for Mars," *Astounding Stories*, April 1937, 10–46.

68. Will Stewart, "Collision Orbit," in *Seventy-Five: The Diamond Anniversary of a Science Fiction Pioneer—Jack Williamson*, ed. Stephen Haffner and Richard A. Hauptmann (1942; Royal Oak, Mich.: Haffner Press, 2004), 80–108. The story originally appeared in the July 1942 issue of *Astounding Science-Fiction*.

69. Robert A. Heinlein, *Farmer in the Sky* (New York: Ballantine, 1950).

70. Carl Sagan, "The Planet Venus," *Science* 133 (24 March 1961): 857–58.

71. M. M. Averner and R. D. MacElroy, eds., *On the Habitability of Mars: An Approach to Planetary Ecosynthesis*, NASA SP-414 (Springfield, Va.: National Technical Information Service, 1976). Also see Christopher P. McKay, "Living and Working on Mars," in Reiber, *NASA Mars Conference*, 522; "Terraforming: Making an Earth of Mars," *Planetary Report* 6 (November–December 1987): 26–27.

72. Alcestis R. Oberg, "The Grass Roots of the Mars Conference," in Boston, *Case for Mars*, ix–xii. See also James Edward Oberg, *New Earths: Transforming Other Planets for Humanity* (Harrisburg, Pa.: Stackpole Books, 1981).

73. Brad Darrach and Steve Petranek, "Mars: Our Next Home," *Life*, May 1991, 32–35.

74. Carl Sagan, *Pale Blue Dot: A Vision of the Human Future in Space* (New York: Random House, 1994), 339. See also James B. Pollack and Carl Sagan, "Planetary Engineering," in *Near-Earth Resources*, ed. J. Lewis and M. Matthews (Tucson: University of Arizona Press, 1992).

75. George H. W. Bush, "Remarks on the 20th Anniversary of the *Apollo 11* Moon Landing," 20 July 1989, *Public Papers of the Presidents of the United States, 1989* (Washington, D.C.: GPO, 1990), 2:992; George H. W. Bush, "Remarks at the Texas A&I University Commencement Ceremony in Kingsville, Texas," 11 May 1990, *Public Papers of the Presidents of the United States, 1990* (Washington, D.C.: GPO, 1991), 644.

76. Dwayne A. Day, "Aiming for Mars, Grounded on Earth: Part Two," *Space Review*, 23 February 2004; Andrew Lawler, "Newsmaker Forum: Barbara Mikulski," *Space News*, 13 November 1989; George Bush, "Remarks to Employees of the George C. Marshall Space Flight Center in Huntsville, Alabama," *Weekly Compilation of Presidential Documents* 26, no. 25 (20 June 1990): 981–83. See also "Bush Goes on Counterattack against Mars Mission Critics," *Congressional Quarterly Weekly Report*, 23 June 1990, 1958; Gregg Easterbrook, "The Case against Mars," in Stoker, *Case for Mars III*, 49–54.

77. "Transcript of President Bush's Speech at NASA Headquarters," Space.com, 14 January 2004, www.space.com/news/bush_transcript_040114.html (accessed 24 March 2009). See also The White House, President George W. Bush, "President Bush Announces

New Vision for Space Exploration Program Fact Sheet: A Renewed Spirit of Discovery," 14 January 2004.

78. See Patricia Nelson Limerick, *The Legacy of Conquest: The Unbroken Past of the American West* (New York: W. W. Norton, 1987), 323–24; White, *"It's Your Misfortune and None of My Own,"* chap. 21; James R. Grossman, ed., *The Frontier in American Culture* (Berkeley and Los Angeles: University of California Press, 1994).

79. Quoted from Brian W. Dippie, "The Winning of the West Reconsidered," *Wilson Quarterly* 14 (Summer 1990): 73.

80. Thomas O. Paine, "Head of NASA Has New Vision of 1984," *New York Times,* 17 July 1969; Konstantin Tsiolkovsky, quoted by Nicholas Daniloff, *The Kremlin and the Cosmos* (New York: Knopf, 1972), 20; O'Neill, *High Frontier,* 219.

81. Arthur E. Bestor, *Backwoods Utopias,* 2nd ed. (Philadelphia: University of Pennsylvania Press, 1970).

82. Roland Emmerich, *Stargate* (Metro-Goldwyn-Mayer, 1994); George Lucas, *Star Wars Episode IV: A New Hope* (Twentieth Century Fox, 1977). See also Karl Wittfogel, *Oriental Despotism: A Comparative Study of Total Power* (New Haven, Conn.: Yale University Press, 1957).

83. George Orwell, *1984: A Novel* (London: Secker & Warburg, 1949); Aldous Huxley, *Brave New World* (New York: Harper and Brothers, 1946); David Sivier, "The Development of Politics in Extraterrestrial Colonies," *Journal of the British Interplanetary Society* 53 (September–October, 2000): 290–96.

84. See Limerick, *Legacy of Conquest,* chap. 3.

85. Sagan, *Pale Blue Dot,* xiv.

86. Ibid., xii. See also James A. Michener, "Space and the Human Quest," *National Forum* 72 (Summer 1992): 3–5.

87. Sagan, *Pale Blue Dot,* 50. See Bryan Appleyard, *Understanding the Present: Science and the Soul of Modern Man* (London: Picador/Pan Books, 1992).

88. Sagan, *Pale Blue Dot,* 53, 377.

CHAPTER 7: Stations in Space

Epigraph: House Science and Technology Committee, Subcommittee on Space Science and Applications, *NASA's Space Station Activities,* 98th Cong., 1st. sess., 1983, 4.

1. Ronald Reagan, "Address before a Joint Session of the Congress on the State of the Union," 25 January 1984, *Public Papers of the Presidents of the United States, 1984* (Washington, D.C.: GPO, 1986), 1:90.

2. Willard B. Robinson, *American Forts: Architectural Form and Function* (Urbana: University of Illinois Press, 1977); Robert B. Roberts, *Encyclopedia of Historic Forts: The Military, Pioneer, and Trading Posts of the United States* (New York: Macmillan, 1988).

3. Roland Huntford, *The Last Place on Earth* (New York: Atheneum, 1983).

4. John Hunt, *The Ascent of Everest* (Seattle: Mountaineers, 1993); Micheline Morin, *Everest: From the First Attempt to the Final Victory* (New York: John Day, 1955). See also Galen Rowell, *In the Throne Room of the Mountain Gods* (San Francisco: Sierra Club Books, 1977).

5. See Adam Gruen, "The Port Unknown: A History of the Space Station Freedom Program," 1–2, unpublished manuscript, NASA History Office, NASA Headquarters,

Washington, D.C.; also Gruen, "The Port Unknown" (Ph.D. diss., Duke University, 1989).

6. Willy Ley, *Rockets: The Future of Travel beyond the Stratosphere* (New York: Viking Press, 1944), 223, 227; Fritz Sykora, "Guido von Pirquet: Austrian Pioneer of Astronautics," paper presented at the Fourth History Symposium of the International Academy of Astronautics, Constance, German Federal Republic, October 1970, NASA History Office; Willy Ley, *Rockets, Missiles, and Space Travel* (New York: Viking Press, 1951), 317; Willy Ley, *Rockets and Space Travel* (New York: Viking Press, 1948), 284. See also Cornelius Ryan, ed., *Man on the Moon* (London: Sidgwick and Jackson, 1953), 24.

7. See Oscar Schachter, "Who Owns the Universe?" *Collier's*, 22 March 1952, 36, 70–71; Senate Aeronautical and Space Sciences Committee, *Legal Problems of Space Exploration*, 87th Cong., 1st sess., 1961; Myres S. McDougal, Harold D. Lasswell, and Ivan A. Vlasic, *Law and Public Order in Space* (New Haven, Conn.: Yale University Press, 1963); and Clive Cussler, *Cyclops* (New York: Simon and Schuster, 1986).

8. "What Are We Waiting For?" *Collier's*, 22 March 1952, 23.

9. Hermann Oberth, *Man into Space: New Projects for Rocket and Space Travel*, trans. G. P. H. De Freville (London: Weidenfeld and Nicolson, 1957), 61, 107–8; Wernher von Braun, "Crossing the Last Frontier," *Collier's*, 22 March 1952.

10. See Curtis Peebles, "The Manned Orbiting Laboratory," a three-part article in *Spaceflight* 22 (April 1980): 155–60, 22 (June 1980): 248–53, and 24 (June 1982): 274–77.

11. See Paul B. Stares, *The Militarization of Space: U.S. Policy, 1945–1984* (Ithaca, N.Y.: Cornell University Press, 1985); and Curtis Peebles, *Battle for Space* (New York: Beaufort Books, 1983); Treaty on Principles Governing the Activities of States in the Exploration and Use of Outer Space, Including the Moon and Other Celestial Bodies (Outer Space Treaty), 10 October 1967.

12. Oberth, *Man into Space*, 73; Hermann Oberth, *Ways to Spaceflight*, NASA technical translation TT F-622 (Washington, D.C.: National Aeronautics and Space Administration, 1972), chap. 20; Ley, *Rockets*, 220; von Braun, "Crossing the Last Frontier," 28–29; Wernher von Braun, "Man on the Moon: The Journey," *Collier's*, 18 October 1952, 52; Robert Gilruth, "Manned Space Stations," *Spaceflight*, August 1969, 258; S. Fred Singer, *Manned Laboratories in Space* (New York: Springer-Verlag, 1969).

13. Hans Mark, *Space Station: A Personal Journey* (Durham, N.C.: Duke University Press, 1987), 50. See also Courtney G. Brooks, James M. Grimwood, and Loyd S. Swenson, *Chariots for Apollo: A History of Manned Lunar Spacecraft*, NASA SP-4205 (Washington, D.C.: National Aeronautics and Space Administration, 1979), chap. 3; and Wernher von Braun, "Concluding Remarks by Dr. Wernher von Braun about Mode Selection for the Lunar Landing Program," given to Dr. Joseph F. Shea, Deputy Director (Systems), Office of Manned Space Flight, 7 June 1962, NASA History Office.

14. Reinhold Messner, *Antarctica: Both Heaven and Earth* (Seattle: Mountaineers, 1991).

15. Testimony of George Mueller, Senate Aeronautical and Space Sciences Committee, *NASA Authorization for Fiscal Year 1970*, 91st cong., 1st sess., 1969; "What Are We Waiting For?" 23.

16. See Edward Everett Hale, *Brick Moon and Other Stories* (Freeport, N.Y.: Books for Libraries Press, 1970); Ward Kimball, "Man and the Moon" (Walt Disney, 1955) (also titled "Tomorrow the Moon"); Anatoly Andanov and Gennady Maximov, "Space Stations

of the Future—A Soviet View," *Spaceflight* 11 (August 1969): 264–65; Arthur C. Clarke, *Islands in the Sky* (New York: Holt, Rinehart, and Winston, 1954).

17. Von Braun, "Crossing the Last Frontier," 29, 72.

18. Willy Ley, "A Station in Space," *Collier's*, 22 March 1952, 30–31; Ward Kimball, "Man and the Moon" (Walt Disney, 1955).

19. The USS *Kitty Hawk*, for example, is 1,065 feet long and holds a crew of fifty-three hundred when the air wing is on board; Arthur C. Clarke, *2001: A Space Odyssey* (New York: New American Library, 1968).

20. Gene Roddenberry, *Star Trek: The Motion Picture* (Paramount, 1979). See also Gene Roddenberry, *Star Trek: The Motion Picture, a Novel* (New York: Simon and Schuster, 1979).

21. NASA, *Space Station: Key to the Future* (Washington, D.C.: GPO, n.d.); William Nromyle, "NASA Aims at 100–Man Station," *Aviation Week & Space Technology*, 24 February 1969, 16–17.

22. David Baker, "Space Station Situation Report—1: The North American Rockwell Proposal," *Spaceflight* 13 (September 1971): 318–34; Baker, "Space Station Situation Report—2: The McDonnell Douglas Proposal," *Spaceflight* 13 (September 1971): 344–51; McDonnell Douglas Astronautics Company, *Space Station*, MSFC-DRL-160, Executive Summary, Contract NAS8-25140, August 1970; Nieson S. Himmel, "Advanced Space Station Concepts," *Aviation Week & Space Technology*, 22 September 1969, 100–113; Irving Stone, "NASA Launches Space Station Task," *Air Force/Space Digest* 52 (July 1969): 79–82; John Logsdon, "Space Stations: A Policy History," prepared for the Johnson Space Center, NASA contract NAS9-16461, George Washington University, Washington, D.C., n.d.; Ray Hook, "Historical Review," *Journal of Engineering for Industry* 4 (November 1984): 276–86.

23. NASA, "NASA Announces Baseline Configuration for Space Station," *NASA News*, release 86-61, 14 May 1986.

24. NASA, *The Space Station: A Description of the Configuration Established at the Systems Requirements Review (SRR)* (Washington, D.C.: NASA Office of Space Station, 1986), 30.

25. Linda N. Ezell, *NASA Historical Data Book: Programs and Projects, 1969–1978*, NASA SP-4012 (Washington, D.C.: National Aeronautics and Space Administration, 1988), 94.

26. James M. Beggs, NASA Administrator, to Craig L. Fuller, Assistant to the President for Cabinet Affairs, 12 April 1984, NASA History Office; Reagan, "Address before a Joint Session of the Congress on the State of the Union," 25 January 1984, *Public Papers of the Presidents of the United States, 1984*, 90.

27. Michael J. Neufeld, "'Space Superiority': Wernher von Braun's Campaign for a Nuclear-Armed Space Station, 1946–1956," *Space Policy* 22 (February 2006): 52–62.

28. Von Braun, "Crossing the Last Frontier," 26.

29. Howard E. McCurdy, *The Space Station Decision: Incremental Politics and Technological Choice* (Baltimore: Johns Hopkins University Press, 1990), 149–50.

30. Robert W. Smith, *The Space Telescope: A Study of NASA, Science, Technology, and Politics* (New York: Cambridge University Press, 1989).

31. See Oberth, *Man into Space*, 67–71; NASA Space Station Task Force, *Program Description Document: Mission Description Document* (Washington, D.C.: National Aeronautics and Space Administration, 1984), bk. 2, sec. 3, 5–12.

32. Von Braun, "Crossing the Last Frontier," 72.

33. House Science and Technology Committee, Space Science and Applications Sub-

committee, *NASA's Space Station Activities*, 98th Cong., 1st sess., 1983, 6. See also Senate Commerce, Science, and Transportation Committee, Science, Technology, and Space Subcommittee, *Civil Space Station*, 98th Cong., 1st sess., 1983, 30; and McCurdy, *Space Station Decision*, 146–49.

34. "Ivory Tower in Space," *Nature* 307 (5 January 1984): 2.

35. Oberth, *Man into Space*, 63; Hermann Oberth, *Rockets in Planetary Space* (1923), NASA TT F-9227 (Washington, D.C.: National Aeronautics and Space Administration, 1964), 93–94. See also Space Science Board, Committee on Space Biology and Medicine, *A Strategy for Space Biology and Medical Science for the 1980s and 1990s* (Washington, D.C.: National Academy Press, 1987).

36. See John McLucas, *Space Commerce* (Cambridge, Mass.: Harvard University Press, 1991).

37. See McCurdy, *Space Station Decision*, 179–80; Craig Covault, "Reagan Briefed on Space Station," *Aviation Week & Space Technology*, 8 August 1983, 16–18; James R. Asker, "No Windfalls Yet, But Space Commerce Advances," *Aviation Week & Space Technology*, 19 April 1993, 26–27; and Camille M. Jernigan and Elizabeth Penetecost, *Space Industrialization Opportunities* (Park Ridge, N.J.: Noyes, 1985).

38. NASA, *Space Station*, 8–9.

39. McLucas, *Space Commerce*, 191.

40. Henry S. F. Cooper, "Annals of Space," *New Yorker*, 2 September 1991, 44–51.

41. Maxime A. Faget, Oral History Transcript, interviewed by Jim Slade, Houston, Texas, 18–19 June 1997, 89, Johnson Space Center Oral History Project, Houston, Texas; Eliot Marshall, "Space Stations in Lobbyland," *Air & Space*, December 1988–January 1989, 54–61; Gruen, "Port Unknown," 284–86; Kathy Sawyer, "Commercial Space Laboratory Not Needed, Expert Panel Says," *Washington Post*, 12 April 1989; T. A. Heppenheimer, "Son of Space Station," *Discover*, July 1988, 64–66.

42. NASA, "The Post-Apollo Space Program: Directions for the Future," summary of NASA's report to the President's Space Task Group, September 1969, 2, 9. See also Thomas Paine, Memorandum to the President, 26 February 1969, in *Exploring the Unknown: Selected Documents in the History of the U.S. Civil Space Program*, vol. 1, ed. John M. Logsdon, NASA SP-4407 (Washington, D.C.: National Aeronautics and Space Administration, 1995), 517–18.

43. Clarke Covington and Robert O. Piland, "Space Operations Center: Next Goal for Manned Space Flight?" *Astronautics & Aeronautics* 18 (September 1980): 30–37; NASA, *Space Operations Center: A Concept Analysis*, vol. 1, *Summary*, Johnson Space Center, 29 November 1979. See also John Noble Wilford, "Space Stations: NASA's Dream for Future of the Shuttle," *New York Times*, 16 March 1982.

44. Covington and Piland, "Space Operations Center," 33.

45. See McCurdy, *Space Station Decision*, 77–90.

46. John J. Madison and Howard E. McCurdy, "Spending without Results: Lessons from the Space Station Program," *Space Policy* 15 (1999): 213–21.

47. Senate Commerce, Science, and Transportation Committee, *Civil Space Station*, 27.

48. NASA, *Space Station*, 34; NASA, "NASA Announces Baseline Configuration for Space Station," *NASA News*, release 86-61, 14 May 1986.

49. NASA, *The Space Station*, 2, 34.

50. Ibid., 3, 36; Howard E. McCurdy, "Cost of Space Flight," *Space Policy* 10 (November 1994): 277–89.

51. NASA, "NASA Announces Baseline Configuration for Space Station," 2; NASA, *Space Station*.

52. House Committee on Science and Technology, *NASA's Space Station Activities*, 4; Mitchell Waldrop, "Space City: 2001 It's Not," *Science 83* 4 (October 1983): 60, 62, cover. See also John Noble Wilford, "When Man Has Stations in Space," *New York Times*, 19 October 1969.

53. NASA, "NASA Proceeding toward Space Station Development," *NASA News*, release 87-50, 3 April 1987; NASA, "NASA Issues Requests for Proposals for Space Station Development," *NASA News*, release 87-65, 24 April 1987. See also Andrew J. Stofan, "Space Station: A Step into the Future," NASA Headquarters, n.d.; Stofan, "Preparing for the Future," *Aerospace America* 25 (September 1987): 16–22.

54. Sally K. Ride, *Leadership and America's Future in Space*, Report to the Administrator (Washington, D.C.: National Aeronautics and Space Administration, 1987), 43.

55. Advisory Committee on the Future of the U.S. Space Program (Norman Augustine, chair), *Report of the Advisory Committee* (Washington, D.C.: GPO, 1990), 29; NASA, "Report to Congress on the Restructured Space Station," 20 March 1991; Technical and Administrative Services Corporation, "Space Station Freedom Media Handbook," May 1992, 23, TADCORPS, Washington, D.C.

56. NASA Space Station Redesign Team, "Final Report to the Advisory Committee on the Redesign of the Space Station," June 1993, 31, NASA History Office. See also Advisory Committee on the Redesign of the Space Station (Charles M. Vest, chair), *Final Report to the President* (Washington, D.C.: GPO, 1993); Warren E. Leary, "Clinton Plans to Ask Congress to Approve Smaller, Cheaper Space Station," *New York Times*, 18 June 1993; Daniel S. Goldin to John H. Gibbons, Cost Report for Space Station Alpha, 20 September 1993; all available in the NASA History Office.

57. NASA, "Alpha Station: Addendum to Program Implementation Plan," 1 November 1993; White House, Office of the Vice President, "Joint Statements on Space Cooperation, Aeronautics and Earth Observation," 2 September 1993; White House, Office of the Vice President, "United States—Russian Joint Commission on Energy and Space, Joint Statement on Cooperation in Space," 2 September 1993; Steven A. Holmes, "U.S. and Russians Join in New Plan for Space Station," *New York Times*, 3 September 1993; NASA, "Space Station Transition Status Report #1," 26 July 1993; NASA, "International Space Station: Creating a World-Class Orbiting Laboratory," January 1995; NASA, "International Space Station Fact Book," 1 June 1995; all but "International Space Station" available in the NASA History Office.

58. Advisory Committee on the Redesign of the Space Station, *Final Report to the President*, 39–43. See also Madison and McCurdy, "Spending without Results."

59. "What Are We Waiting For?" 23.

60. James M. Beggs to James A. Baker, 24 August 1983; Peggy Finarelli to OMB/Bart Borrasca, 8 September 1983; both in NASA History Office; John Hodge interview, 10 July 1985, in McCurdy, *Space Station Decision*, 171; NASA Office of Comptroller, "Aerospace Price Deflator," NASA Headquarters, 1994.

61. James C. Miller, memorandum for the President, 10 February 1987, NASA History Office. See also U.S. House Committee on Government Operations, Subcommittee

on Government Activities and Transportation, *Cost, Justification, and Benefits of NASA's Space Station*, 102nd Cong., 1st sess., 1991; and McCurdy, "Cost of Space Flight," 277–89.

62. U.S. Senate Commerce, Science, and Transportation Committee, Subcommittee on Science, Technology, and Space, *NASA Authorization for Fiscal Year 1984*, 98th Cong., 1st sess., 1983, 51.

63. Gruen, "The Port Unknown: A History," 131–33; Philip E. Culbertson to Neil B. Hutchinson, "Space Station Program Cost Estimates," 14 August 1985, NASA History Office.

64. Von Braun, "Crossing the Last Frontier," 24–25; Clarke, *2001*, 41–48.

65. Presidential Commission on the Space Shuttle Challenger Accident (William P. Rogers, chair), *Report to the President* (Washington, D.C.: GPO, 1986), 164.

66. NASA, "Report to Congress on the Restructured Space Station," 20 March 1991, 5.

67. NASA, *Space Station*, 26–27; Richard DeMeis, "Fleeing Freedom," *Aerospace America* 27 (May 1989): 38–41; Andrew Lawler, "NASA: No Permanent Station Crew without Escape Vehicle," *Space News*, 12 February 1990.

68. "What Are We Waiting For?" 23.

69. U.S. Senate Committee on Appropriations, *Department of Housing and Urban Development, and Certain Independent Agencies, Appropriations for Fiscal Year 1985*, 98th Cong., 2nd sess., 1984, esp. 1266.

70. Kathy Sawyer, "Astronauts Express Fears over Space Station," *Washington Post*, 19 July 1986; Craig Covault, "Launch Capability, EVA Concerns Force Space Station Redesign," *Aviation Week & Space Technology*, 21 July 1986, 18–20.

71. NASA Office of Space Station, "Proceedings of the Space Station Evolution Workshop, Williamsburg, Virginia, 10–13 September 1985," NASA History Office.

72. Internal NASA email from NASA Administrator Griffin, 18 August 2008, Space Ref.com, www.spaceref.com/news/viewsr.html?pid=29133 (accessed 19 February 2009); Review of U.S. Human Spaceflight Plans Committee (Norman R. Augustine, Chair), *Seeking a Human Spaceflight Program Worthy of a Great Nation*, October 2009, nasa.gov/offices/hsf/home/index.html (accessed 18 June 2010).

73. Synthesis Group on America's Space Exploration Initiative (Thomas P. Stafford, chair), *America at the Threshold* (Washington, D.C.: GPO, 1991), 6, 83; Advisory Committee on the Future of the U.S. Space Program, *Report of the Advisory Committee*, 29; John F. Connolly, "Constellation Program Overview," October 2006, Constellation Program Office, NASA Johnson Space Center.

74. NASA, *Reference Guide to the International Space Station*, NASA SP-2006-557 (Washington, D.C.: National Aeronautics and Space Administration, 2006).

75. McCurdy, *The Space Station Decision*, 47, 109–12. See also Space Task Group, *The Post-Apollo Space Program: Directions for the Future*, September 1969, NASA History Office.

CHAPTER 8: Spacecraft

Epigraph: Boyce Rensberger, "The Prophet in His Orbit," *Washington Post*, 7 November 1985, sec. C, p. 6.

1. Joseph J. Corn, *Winged Gospel: America's Romance with Aviation* (1983; Baltimore: Johns Hopkins University Press, 2002).

2. Stephen Pendo, *Aviation in the Cinema* (Metuchen, N.J.: Scarecrow Press, 1985).

3. Office of the White House Press Secretary, Press Conference of Dr. James Fletcher and George M. Low, 5 January 1972, San Clemente, California; NASA, "Space Shuttle," 1972; NASA, "Fact Sheet: The Economics of the Space Shuttle," July 1972, all in NASA History Office, NASA Headquarters, Washington, D.C. Also see House Committee on Science and Technology, Subcommittee on Space Science and Applications, *Operational Cost Estimates: Space Shuttle*, 94th Cong., 2nd sess., 1976.

4. President's Science Advisory Committee, "The Next Decade in Space," February 1970, 3, 50, NASA History Office.

5. See Tom Wolfe, *The Right Stuff* (New York: Farrar, Straus, Giroux, 1979); and Joseph D. Atkinson and Jay M. Shafritz, *The Real Stuff* (New York: Praeger, 1985).

6. Sylvia D. Fries, *NASA Engineers and the Age of Apollo*, NASA SP-4104 (Washington, D.C.: GPO, 1992); Howard E. McCurdy, *Inside NASA: High Technology and Organizational Change in the U.S. Space Program* (Baltimore: Johns Hopkins University Press, 1993), 78–89.

7. Mark Sullivan, *Our Times: The United States, 1900–1925* (New York: Charles Scribner's Sons, 1927), 2:556, 599. See also Roger E. Bilstein, *Flight in America: From the Wrights to the Astronauts*, rev. ed. (Baltimore: Johns Hopkins University Press, 1994); and Bilstein, "The Airplane, the Wrights, and the American Public," in *The Wright Brothers: Heirs of Prometheus*, ed. Richard P. Hallion (Washington, D.C.: Smithsonian Institution Press, 1978), 39–51.

8. Hallion, *Wright Brothers*, 75–87; Tom D. Crouch, *The Bishop's Boys: A Life of Wilbur and Orville Wright* (New York: W. W. Norton, 1986); Arthur G. Renstrom, *Wilbur and Orville Wright: A Chronology* (Washington, D.C.: Library of Congress, 1975).

9. Robert Scharff and Walter S. Taylor, *Over Land and Sea: A Biography of Glenn Hammond Curtiss* (New York: David McKay, 1968).

10. Corn, *Winged Gospel*, 12–13. See also Don Dwiggins, *The Barnstormers: Flying Daredevils of the Roaring Twenties* (New York: Grosset and Dunlap, 1968).

11. Orville Wright, "Future of the Aeroplane," *Country Life*, January 1909; Sullivan, *Our Times*, 2:558.

12. See Walter H. G. Armytage, *A Social History of Engineering* (Cambridge, Mass.: MIT Press, 1961), 268–70.

13. Bilstein, *Flight in America*.

14. Robert J. Serling, *Wrights to Wide-Bodies: The First Seventy-five Years* (Washington, D.C.: Air Transport Association, 1978).

15. "An Airplane in Every Garage?" *Scribner's*, September 1935, 179–82. See also Corn, *Winged Gospel*, chap. 5.

16. Douglas J. Ingells, *Tin Goose: The Fabulous Ford Trimotor* (Fallbook, Calif.: Aero Publishers, 1968); Fred E. Weick, "Development of the Ercoupe, an Airplane for Simplified Private Flying," *SAE Journal* 44 (December 1941): 520–31; Max Karant, "The Unbelievable Truth about Hammond's Experimental Plane," *Popular Aviation* 19 (October 1936): 56; Corn, *Winged Gospel*, chap. 5.

17. See Henry S. F. Cooper, "Annals of Space," *New Yorker*, 2 September 1991, 41–69; Stephan Wilkinson, "The Legacy of the Lifting Body," *Air & Space*, April–May 1991, 50–62.

18. Barton C. Hacker and James M. Grimwood, *On the Shoulders of Titans: A History*

of *Project Gemini*, NASA SP-4203 (Washington, D.C.: National Aeronautics and Space Administration, 1977), 139.

19. National Academy of Sciences, "Review of Project Mercury," *IG Bulletin* 80 (February 1964): 5.

20. Tom Wolfe, "Columbia's Landing Closes a Circle," *National Geographic*, October 1981, 475; Wolfe, "Everyman vs. Astropower," *Newsweek*, 10 February 1986, 41.

21. Hacker and Grimwood, *On the Shoulders of Titans*, 19–20, 123–25, 144–48, 170–73.

22. Bilstein, *Flight in America*, 24; Hallion, *Wright Brothers*, 80; Crouch, *Bishop's Boys*, 375–76, 434–35; Scharff and Taylor, *Over Land and Sea*, 215.

23. Bilstein, *Flight in America*, 57–58, 101.

24. Ibid.

25. H. G. Wells, *A Critical Edition of "The War of the Worlds,"* with introduction and notes by David Y. Hughes and Harry M. Geduld (Bloomington: Indiana University Press, 1993), 53, 200.

26. Jules Verne, *From the Earth to the Moon, and Round the Moon* (New York: Dodd, Mead, 1962). See also Ron Miller, "The Spaceship as Icon: Designs from Verne to the Early 1950s," in *Blueprint for Space: Science Fiction to Science Fact*, ed. Frederick I. Ordway and Randy Liebermann (Washington, D.C.: Smithsonian Institution Press, 1992), 51; Ron Miller, *The Dream Machines: An Illustrated History of the Spaceship in Art, Science, and Literature* (Malabar, Fla.: Keieger, 1993), 47–54.

27. Fritz Lang, *Frau im Mond* (1929; available through Foothill Video, Tujunga, Calif.).

28. Arthur C. Clarke, *2001: A Space Odyssey* (New York: New American Library, 1968), 46.

29. Ibid., 56–57.

30. Ibid., chap. 7; House Science and Astronautics Committee, Subcommittee on Manned Space Flight, *1974 NASA Authorization*, 93rd Cong., 1st sess., 1973, 1274.

31. Alex Raymond, *Flash Gordon* (New York: Nostalgia Press, 1974); Robert C. Dille, ed., *The Collected Works of Buck Rogers in the 25th Century* (New York: Chelsea House, 1969). See Miller, "Spaceship as Icon," 65.

32. Willy Ley, *Rockets, Missiles, and Men in Space* (New York: Viking Press, 1968), 506. See also Peter G. Cooksley, *Flying Bomb: The Story of Hitler's V-Weapons in World War II* (New York: Charles Scribner's Sons, 1979), 165.

33. U.S. Atomic Energy Commission, *Nuclear Propulsion for Space* (Oak Ridge, Tenn.: U.S. Atomic Energy Commission, 1967).

34. Synthesis Group on America's Space Exploration Initiative, *America at the Threshold* (Washington, D.C.: GPO, 1991), 66–68.

35. George Lucas, *Star Wars Episode IV: A New Hope* (Twentieth Century Fox, 1977). See also George Lucas, *Star Wars: A New Hope* (previously titled *Star Wars: The Adventures of Luke Skywalker*) (New York: Ballantine Books, 1976), 101, 110–11.

36. Lucas, *Star Wars*, 115.

37. See Peter Nicholls, *The Science in Science Fiction* (New York: Alfred A. Knopf, 1983).

38. E. E. "Doc" Smith, *Skylark of Space* (1928; New York: Pyramid Books, 1970).

39. George Mueller, "Space: The Future of Mankind," *Spaceflight* 27 (March 1985): 105; Mueller, "Antimatter & Distant Space Flight," *Spaceflight* 25 (May 1983): 207.

40. John H. Mauldin, *Prospects for Interstellar Travel*, Science and Technology Series,

vol. 80 (San Diego: American Astronautical Society, 1992); Brice N. Cassenti, "A Comparison of Interstellar Propulsion Methods," *Journal of the British Interplanetary Society* 35 (March 1982): 116–24; Alan Bond and Anthony Martin, "Project Daedalus—The Final Report on the BIS Starship Study," *Journal of the British Interplanetary Society,* supplement (1978): 5–8, 37–42; and L. D. Jaffe, C. Ivie, J. C. Lewis, R. Lipes, H. N. Norton, J. W. Stearns, L. D. Stimpson, and P. Weissman, "An Interstellar Precursor Mission," *Journal of the British Interplanetary Society* 33 (January 1980): 3–26.

41. Mueller, "Antimatter & Distant Space Flight"; Mauldin, *Prospects for Interstellar Travel*; Nicholls, *Science in Science Fiction*, 78–79.

42. Robert L. Forward, "Antimatter Propulsion," *Journal of the British Interplanetary Society* 35 (September 1982): 391–95.

43. Rick Sternbach and Michael Okuda, *Star Trek: The Next Generation Technical Manual* (New York: Pocket Books, 1991), 60–61, 67; E. F. Mallove, R. L. Forward, Z. Paprotny, and J. Lehmann, "Interstellar Travel and Communication: A Bibliography," *Journal of the British Interplanetary Society* 33 (June 1980): 201–48.

44. Sternbach and Okuda, *Star Trek*, 55.

45. Disneyworld, "Journey to Mars," opened 1975.

46. Mauldin, *Prospects for Interstellar Travel*, ix.

47. Space Task Group, *The Post-Apollo Space Program: Directions for the Future*, September 1969, 15, NASA History Office.

48. Wernher von Braun, "The Spaceplane That Can Put *You* in Orbit," *Popular Science,* July 1970, 37; Rick Gore, "When the Space Shuttle Finally Flies," *National Geographic,* March 1981, 317.

49. Michael Collins, "Orbiter Is First Spacecraft Designed for Shuttle Runs," *Smithsonian,* May 1977, 38.

50. Cooper, "Annals of Space," 64.

51. Quoted from Wolfe, "Everyman vs. Astropower," 41. See also James A. Michener, "Manifest Destiny," *Omni,* April 1981, 48–50, 102–4.

52. Quoted from McCurdy, *Inside NASA*, 87.

53. Thomas O. Paine, "Head of NASA Has a New Vision of 1984," *New York Times,* 17 July 1969.

54. President's Science Advisory Committee, "The Next Decade in Space," Executive Office of the President, Office of Science and Technology, March 1970, 38.

55. House Committee on Science and Technology, Subcommittee on Space Science and Applications, *Operational Cost Estimates: Space Shuttle*, 94th Cong., 2nd sess., 1976, 11; NASA, "Space Shuttle Economics Simplified," 26 January 1972, 3, NASA History Office; J. S. Butz, "The Coming Age of the Economy Flight into Space," *Air Force/Space Digest,* December 1969, 42.

56. Center for Aerospace Education Development, "Space Shuttle: A Space Transportation System Activities Book," Civil Air Patrol, U.S. Air Force, n.d.; James J. Haggerty, "Space Shuttle: Next Giant Step for Mankind," *Aerospace* 14 (December 1976): 3, 4. See also Florence S. Steinberg, *Aboard the Space Shuttle* (Washington, D.C.: NASA Division of Public Affairs, 1980); and Wernher von Braun, "The Reusable Space Transport," *American Scientist* 60 (November–December 1972): 730–38.

57. T. O. Paine to the President, 26 March 1970, NASA History Office; James C. Fletcher and William P. Clements, "NASA/DOD Memorandum of Understanding on

Management and Operation of Space Transportation System," 14 January 1977, 8, NASA History Office. See also Howard E. McCurdy, "The Costing Models of the Early 1970s and the Launching of the Space Shuttle Program," in *L'ambition technologique: Naissance d'Ariane*, ed. Emmanuel Chadeau (Paris: Institut d'Histoire de l'Industrie, 1995).

58. NASA, "Space Shuttle Economics" from "Space Shuttle: Appendix to Space Shuttle Fact Sheet," February 1972, NASA History Office; Klaus P. Heiss and Oskar Morgenstern, "Economic Analysis of the Space Shuttle System: Executive Summary," study prepared for NASA under contract NASW-2081, 31 January 1972.

59. See Brian O'Leary, "The Space Shuttle: NASA's White Elephant in the Sky," *Bulletin of the Atomic Scientists* 39 (February 1983): 36–43; John Logsdon, "Decision to Develop the Space Shuttle," *Space Policy* 2 (May 1986): 103–19; and Testimony of Ralph Lapp, Senate Aeronautical and Space Sciences Committee, *NASA Authorization for Fiscal Year 1973*, 92nd Cong., 2nd sess., 1972, 1069–86.

60. See Alex Roland, "The Shuttle: Triumph or Turkey?" *Discover*, November 1985, 14–24; John M. Logsdon, "The Space Shuttle Program: A Policy Failure?" *Science* 232 (30 May 1986): 1099–1105; Roger A. Pielke and Radford Byerly, "The Space Shuttle Program: Performance versus Promise," in *Space Policy Alternatives*, ed. Radford Byerly (Boulder, Colo.: Westview Press, 1992), 223–45.

61. NASA, Office of Space Flight, "Shuttle Launch Cost," 2 March 1983; Ed Campion to Lee Saegesser, 21 September 1990; both in NASA History Office; NASA, Office of Space Flight, "Budget Control Package in Support of FY 93 Budget to Congress: Shuttle Average Cost Per Flight," 1992; U.S. General Accounting Office, "Space Transportation: The Content and Uses of Shuttle Cost Estimates," GAO/NSIAD-93-115, January 1993.

62. See John Logsdon, "Space Shuttle Decision: Technological and Political Choice," *Journal of Contemporary Business* 7, no. 3 (1978): 13–30; Logsdon, "Decision to Develop the Space Shuttle."

63. W. R. Lucas, Program Development memorandum to Dr. Rees, 16 June 1970; NASA, "Space Shuttle," February 1972; both in NASA History Office.

64. NASA News, "Space Shuttle Decisions," release no. 72-61, 15 March 1972, NASA History Office.

65. See Cooper, "Annals of Space."

66. G. Harry Stine, "The Sky Is Going to Fall," *Analog Science Fiction/Science Fact*, August 1983, 75. See also Ben Bova, "The Shuttle, Yes," *New York Times*, 4 January 1982.

67. Rudy Abramson, "NASA to Study 2,000 Safety-Critical Parts," *Los Angeles Times*, 18 March 1986; Space Transportation System, "Return to Flight Status: Critical Item Waiver Status," Lyndon B. Johnson Space Center, August 1988; L Systems, Inc., "Risk Analysis of Space Transportation during the Space Station Era," prepared under contract NAS8-38076 for George C. Marshall Space Flight Center, 15 December 1989; Marcia Dunn, "Space Safety," Associated Press, 27 February 1995; all in NASA History Office. See also Kevin McKean, "They Fly in the Face of Danger," *Discover*, April 1986, 48–58.

68. Michael Collins, "Riding the Beast," *Washington Post*, 30 January 1986, sec. A, p. 25; Robert Block, "Griffin: Shuttle Is Dangerous but Options Are Limited," *The Write Stuff*, 4 September 2008, blogs.orlandosentinel.com/news_space_thewritestuff/ (accessed 18 June 2009).

69. R. P. Feynman, "Personal Observations on Reliability of Shuttle," appendix F, in Presidential Commission on the Space Shuttle Challenger Accident, *Report of the Presi-*

dential Commission (Washington, D.C.: GPO, 1986). See also *Report of the Presidential Commission*, chap. 8; Feynman, "An Outsider's View of the Challenger Inquiry," *Physics Today* 41 (February 1988): 26–37; Feynman, *What Do You Care What Other People Think?* as told to Ralph Leighton (New York: W. W. Norton, 1988); and Diane Vaughan, *The Challenger Launch Decision: Risky Technology, Culture, and Deviance at NASA* (Chicago: University of Chicago Press, 1996).

70. Space Shuttle Management Independent Review Team (Christopher C. Kraft, Chair), "Report of the Space Shuttle Management Independent Review Team," National Aeronautics and Space Administration, February 1995, vii, 3.

71. Columbia Accident Investigation Board, *Report*, vol. 1 (Washington, D.C.: GPO, August 2003), 106, 118; Parker V. Counts, "Space Shuttle Program 2020 Assessment," NASA, 21 August 2002.

72. Traci Watson, "NASA Administrator Says Space Shuttle Was a Mistake," *USA Today*, 27 September 2005.

73. Columbia Accident Investigation Board, *Report*, 23.

74. Andrew J. Butrica, "The X-33 History Project Home Page," 22 October 2004, NASA History Office, hq.nasa.gov/office/pao/History/x-33/home.htm (accessed 7 February 2009).

75. John C. Mankins, "The Maglifter: An Advanced Concept Using Electromagnetic Propulsion in Reducing the Cost of Space Launch," AIAA 30th Joint Propulsion Conference, 27–29 June 1994; George Pal, *When Worlds Collide* (Paramount, 1951).

76. "Why StarTram?" StarTram.com, n.d. (accessed May 13, 2010).

77. X Prize Foundation, "Ansari X Prize," space.xprize.org, 2009 (accessed 3 June 2009); Michael Dornheim, "Affordable Spaceship," *Aviation Week & Space Technology*, 21 April 2003, 64–73.

78. NASA, NASA Facts, "Constellation Program: America's Fleet of Next-Generation Launch Vehicles. The Ares I Crew Launch Vehicle," 2009; NASA, NASA Facts, "Constellation. Orion Crew Exploration Vehicle," 2008; Review of U.S. Human Spaceflight Committee (Norman R. Augustine, Chair), *Seeking a Human Spaceflight Program Worthy of a Great Nation*, October 2009, 90.

79. See for example Tariq Malik, "NASA's New Moon Plans: 'Apollo on Steroids,'" space.com, 19 September 2005 (accessed May 13, 2010).

80. The White House, Office of the Press Secretary, "Remarks by the President on Space Exploration in the 21st Century," 15 April 2010, John F. Kennedy Space Center, Merritt Island, Florida.

81. Statement for Charlie Bolden, National Press Club Event, 2 February 2010.

82. "Private Manned Rocket Reaches Space again for X-Prize," *South Florida Sun-Sentinel*, 4 October 2004; Peter Pae, "Rocket Takes 1st Prize of a New Space Race," *Los Angeles Times*, 5 October 2004.

CHAPTER 9: Robots

Epigraph: Joel Silver, Larry Wachowski, and Andy Wachowski, *The Matrix* (Warner Bros., 1999).

1. Arthur C. Clarke, "Extra-Terrestrial Relays: Can Rocket Stations Give World-Wide Radio Coverage?" *Wireless World* 51 (October 1945): 306; Wernher von Braun, "Crossing the Last Frontier," *Collier's*, 22 March 1952, 72.

2. Curtis Peebles, "The Manned Orbiting Laboratory," a three-part article in *Spaceflight* 22 (April 1980): 155–60, 22 (June 1980): 248–53, and 24 (June 1982): 274–77; Asif A. Siddiqi, "The Almaz Space Station Complex: A History, 1964–1992," *JBIS* 54 (November–December 2001): 389–416.

3. Kurt Neumann, *Rocketship X-M* (Lippert, 1950); Robert H. Goddard, *A Method of Reaching Extreme Altitudes*, Smithsonian Miscellaneous Collections, vol. 71, no. 2 (Washington, D.C.: Smithsonian Institution, 1919); and Maurice K. Hanson, "The Payload for the Lunar Trip," *Journal of the British Interplanetary Society* 5 (January 1939): 16.

4. See Roger D. Launius and Howard E. McCurdy, *Robots in Space: Technology, Evolution, and Interplanetary Travel* (Baltimore: Johns Hopkins University Press, 2008).

5. James A. Van Allen, "Space Station and Manned Flights Raise NASA Program Balance Issues," *Aviation Week & Space Technology*, 25 January 1988, 153; Alex Roland, "NASA's Manned-Space Nonsense," *New York Times*, 4 October 1987, sec. 4, p. 23.

6. Daniel S. Greenberg, "Robots in Space Are Less Costly," *Philadelphia Inquirer*, 18 September 1987; James A. Van Allen, "Space Science, Space Technology and the Space Station," *Scientific American* 254 (January 1986): 32, 37.

7. James A. Van Allen, "Is Human Space Flight Obsolete?" *Issues in Science and Technology* 20 (Summer 2004), www.issues.org/20.4/p_van_allen.html (accessed 3 July 2008).

8. Ibid.

9. Isaac Asimov, introduction and "Reason," in *Robot Visions* (New York: ROC, 1990), 10–11, 90. "Reason" originally appeared in the April 1941 issue of *Astounding Science-Fiction*.

10. Asimov, "Runaround," in *Robot Visions*, 126. A later modification of the laws required very advanced robots to protect and not injure humanity, termed the Zeroth Law.

11. Asimov, introduction to *Robot Visions*, 6–7.

12. Nicholas Nayfack, *Forbidden Planet* (Metro-Goldwyn-Mayer, 1956).

13. Irwin Allen, *Lost in Space* (Columbia Broadcasting System, 1965–68).

14. Douglas Trumbull, *Silent Running* (Universal Pictures, 1972).

15. George Lucas, *Star Wars Episode IV: A New Hope* (Twentieth Century Fox, 1977).

16. Frederick W. Taylor, *The Principles of Scientific Management* (New York: Harper and Brothers, 1911); Elton Mayo, *The Human Problems of an Industrial Civilization* (New York: Macmillan, 1933); Norbert Wiener, *Cybernetics; or, Control and Communication in the Animal and Machine* (New York: J. Wiley, 1948).

17. Peter Kussi, ed., *Toward the Radical Center: A Karel Capek Reader* (Highland Park, N.J.: Catbird Press, 1990).

18. Asimov, introduction and "Runaround," in *Robot Visions*, 7, 11, 126.

19. See Jim Oberg, "Space Explorers! The Evolution of Robotic Arms and Mobility in the Space Age," *Robot Magazine* 5 (Winter 2006): 20–24. See also www.jamesoberg.com/robots_on_planets.pdf (accessed 1 July 2010).

20. NASA, "The Vision for Space Exploration," February 2004, 3.

21. NASA, Office of Program Planning and Evaluation, "The Long Range Plan of the National Aeronautics and Space Administration," 16 December 1959; President's Science Advisory Committee, "Report of the Ad Hoc Panel on Man-in-Space," 16 December 1960. Both documents can be found in John M. Logsdon, ed., *Exploring the Unknown: Selected Documents in the History of the U.S. Civil Space Program*, vol. 1: *Organizing for Ex-*

ploration, NASA SP-4218 (Washington, D.C.:, National Aeronautics and Space Administration, 1995), 404, 410.

22. Oran W. Nicks, *The Far Travelers: NASA's Exploring Machines*, NASA SP-4102 (Washington, D.C.: National Aeronautics and Space Administration, 1982), 245–46.

23. Robert Glatzer, *The New Advertising* (New York: Citadel Press, 1970); Vance Packard, *The Hidden Persuaders* (New York: D. McKay, 1957).

24. Jane Stern and Michael Stern, *Auto Ads* (New York: Random House, 1979).

25. Robert B. Voas, "John Glenn's Three Orbits in *Friendship 7*," *National Geographic*, June 1962; Linda Neumann Ezell, *NASA Historical Data Book: Programs and Projects, 1959–1968*, NASA SP-4012 (Washington, D.C.: National Aeronautics and Space Administration 1988), 2:173; Loyd S. Swenson, James M. Grimwood, and Charles C. Alexander, *This New Ocean: A History of Project Mercury*, NASA SP-4201 (Washington, D.C.: National Aeronautics and Space Administration, 1966), 314, 575.

26. See Michael L. Smith, "Selling the Moon: The U.S. Manned Space Program and the Triumph of Commodity Scientism," in *The Culture of Consumption: Critical Essays in American History, 1880–1980*, ed. Richard W. Fox and T. J. Jackson Lears (New York: Pantheon Books, 1983), 195.

27. Nicks, *Far Travelers*, 88; Apollo 11 Crew Pre-Mission Press Conference, 5 July 1969, 2:00 P.M., NASA History Office, NASA Headquarters, Washington, D.C. See also John Noble Wilford, "Humans and Machines in Space: A Vision of Our Space Future," *Space Times*, March–April 1991, 15.

28. Advisory Committee on the Future of the U.S. Space Program (Norman Augustine, chair), *Report of the Advisory Committee* (Washington, D.C.: GPO, 1990), 6.

29. Van Allen, "Space Station and Manned Flights Raise NASA Program Balance Issues," 153.

30. Homer E. Newell, *The Mission of Man in Space* (Washington, D.C.: GPO, 1963), 5; James C. Fletcher to Edward M. Kennedy, 24 September 1971, NASA History Office.

31. Van Allen, "Space Science, Space Technology and the Space Station," 37; William D. McCann, "Mars Viewed as Last Stop for Manned Space Flights," *Plain Dealer*, 27 December 1969; Bill Green, "Earth to NASA," *New York Times*, 27 August 1989.

32. Robert Jastrow, "Man in Space or Chip in Space?" *New York Times Magazine*, 31 January 1971, 63.

33. James E. Tomayko, *Computers in Spaceflight: The NASA Experience*, NASA contractor report 182505, March 1988, NASA History Office; David A. Mindell, *Digital Apollo: Human and Machine in Spaceflight* (Cambridge, Mass.: MIT Press, 2008).

34. Gordon E. Moore, "Cramming More Components onto Integrated Circuits," *Electronics* 38 (19 April 1965): 114–17.

35. Irving John Good, "Speculations Concerning the First Ultraintelligent Machine," based on talks given in a Conference on the Conceptual Aspects of Biocommunications, Neuropsychiatric Institute, University of California, Los Angeles, October 1962; reprinted in *Advances in Computers* 6 (1965): 31–88.

36. Vernor Vinge, "The Coming Technological Singularity," Vision-21: Interdisciplinary Science & Engineering in the Era of CyberSpace, proceedings of a symposium held at NASA Lewis Research Center, NASA Conference Publication CP-10129, 30–31 March 1993.

37. Ray Kurzweil, *The Singularity Is Near: When Humans Transcend Biology* (New York:

Penguin Books, 2005), 7, 30. See also Kurzweil, *The Age of Intelligent Machines* (Cambridge, Mass.: MIT Press, 1990); Kurzweil, *The Age of Spiritual Machines: When Computers Exceed Human Intelligence* (New York: Viking, 1999).

38. Asimov, "The Bicentennial Man," in *Robot Visions*, 249.

39. Ibid., 287–88.

40. Isaac Asimov, *Gold* (New York: HarperPrism, 1995), 206.

41. Asimov, "The Bicentennial Man," in *Robot Visions*, 256.

42. Robert Scheerer, "The Measure of a Man," production 135, 13 February 1989, Star Trek Episode Archives, Paramount Pictures.

43. Stephen Hawking, "Science in the Next Millennium," White House Millennium Evenings, 6 March 1998, clinton4.nara.gov/textonly/Initiatives/Millennium/events.html (accessed 22 June 2009).

44. John Copeland, *Alien Planet* (Discovery Channel, 2005). See also Wayne Douglas Barlowe, *Expedition: Being an Account in Words and Artwork of the 2358 A.D. Voyage to Darwin IV* (New York: Workman, 1990).

45. Good, "Speculations Concerning the First Ultraintelligent Machine."

46. See Pamela Horn, *The Rise and Fall of the Victorian Servant* (New York: St. Martin's Press, 1975); and Frank E. Huggett, *Life below Stairs: Domestic Servants in England from Victorian Times* (New York: Charles Scribner's Sons, 1977).

47. See Candice Millard, *River of Doubt: Theodore Roosevelt's Darkest Journey* (New York: Doubleday, 2005); "Hillary of New Zealand and Tenzing Reach the Top," *Guardian*, 2 June 1953; Jamling Tenzing Norgay, *Touching My Father's Soul* (San Francisco, Calif.: HarperSanFrancisco, 2001).

48. Quoted from Anna Brendle, "Profile: African-American North Pole Explorer Matthew Henson," *National Geographic News*, 15 January 2003. See also Matthew A. Henson, *A Negro Explorer at the North Pole* (1912; reprint, New York: Arno Press, 1969).

49. See Wally Herbert, "Commander Robert E. Peary: Did He Reach the Pole?" *National Geographic* 124 (September 1988): 386–413.

50. Ridley Scott, *Blade Runner* (Warner Bros., 1982).

51. Philip K. Dick, *Do Androids Dream of Electric Sheep?* (New York: Del Rey, 1968), 17.

52. Silver, Wachowski, and Wachowski, *The Matrix*.

53. Stanley Kubrick, *2001: A Space Odyssey* (MGM, 1968).

54. Hans Moravec, *Mind Children: The Future of Robot and Human Intelligence* (Cambridge, Mass.: Harvard University Press, 1988), 100.

55. Bill Joy, "Why the Future Doesn't Need Us," *Wired* 8 (April 2000): 238–63. See also Hans Moravec, *Robot: Mere Machine to Transcendent Mind* (New York: Oxford University Press, 1999).

56. Kurzweil, *The Singularity Is Near*, 136, back cover.

57. Kurzweil, *The Age of Spiritual Machines*, 241.

58. Arthur C. Clarke, *3001: The Final Odyssey* (New York: Ballantine Books, 1997), 3; Clarke, *Rendezvous with Rama* (Orlando, Fla.: Harcourt Brace Jovanovich, 1972).

59. Kurzweil, *The Singularity Is Near*, 352.

60. Moravec, *Mind Children*, 101.

61. Manfred E. Clynes and Nathan S. Klein, "Cyborgs and Space," *Astronautics* 13 (September 1960): 27.

62. See Janice Hocker Rushing and Thomas S. Frentz, *Projecting the Shadow: The Cyborg Hero in American Film* (Chicago: University of Chicago Press, 1995).

63. Quoted from Swenson, Grimwood, and Alexander, *This New Ocean*, 194–95.

64. Ibid., 353, 359, 434; *Time*, 2 March 1962, 11. Also see Robert Holz, "Man in Space," *Aviation Week*, 5 March 1962, 13; "Cooperation in Space," *New Republic*, 5 March 1962, 3.

65. R. L. F. Boyd, "In Space: Instruments or Man?" *International Science and Technology*, May 1965, 70. Also see Swenson, Grimwood, and Alexander, *This New Ocean*, 428; and "Space: The New Ocean," *Time*, 2 March 1962, 12.

66. "Buran," *Encyclopedia Astronautica*, n.d., astronautix.com/craft/buran.htm (accessed 11 July 2008).

67. Quoted in David Whitehouse, "Robots to Rescue Hubble Telescope," *BBC News Online*, 2 June 2004, news.bbc.co.uk/2/hi/science/nature/3769445.stm (accessed June 18, 2010). See also Todd Halverson, "Canadian Robot Top Choice for Hubble Servicing Mission," *Florida Today*, 11 August 2004.

68. National Research Council, Division on Engineering and Physical Sciences, Space Studies Board and Aeronautics and Space Engineering Board, "News: Space Shuttle Should Conduct Final Servicing Mission To Hubble Space Telescope," 8 December 2004; National Research Council, *Assessment of Options for Extending the Life of the Hubble Space Telescope: Final Report* (Washington, D.C.: National Academies Press, 2005); U.S. Government Accountability Office, "Costs for Hubble Servicing Mission and Implementation of Safety Recommendations Not Yet Definitive," GAO-05-34, November 2004; Clinton Parks and Brian Berger, "NASA's Mission to Service Hubble in 2008 Will Cost $900 Million," space.com, 31 October 2006.

69. See David Mosher, "Space Station Ready for New Robot, Room," space.com, 7 March 2008; NASA, "STS-123, All Aboard: The Station Goes Global" (press kit), March 2008, 51–52.

CHAPTER 10: Space Commerce

Epigraph: *The National Aeronautics and Space Act*, Pub. L. No. 85-568: 72 Stat. 426 (29 July 1958) as amended.

1. See, for example, H. G. Wells, *The First Men in the Moon* (London: G. Newness, 1901); George Pal, *Destination Moon* (Eagle Lion, 1950); and Treaty on Principles Governing the Activities of States in the Exploration and Use of Outer Space, Including the Moon and Other Celestial Bodies (Outer Space Treaty), 10 October 1967.

2. See Lou Dobbs with H. P. Newquist, *Space: The Next Business Frontier* (New York: Pocket Books, 2001), 1; Jonathan N. Goodrich, *The Commercialization of Outer Space: Opportunities and Obstacles for American Business* (New York: Quorum Books, 1989); Tom Logsdon, *SpaceInc* (New York: Crown Publishers, 1988); John L. McLucas, *Space Commerce* (Cambridge, Mass.: Harvard University Press, 1991); D. V. Smitherman, ed., "New Space Industries for the Next Millennium," December 1998, NASA/CP-1988-209006, NASA Marshall Space Flight Center, Huntsville, Alabama; David P. Gump, *Space Enterprise: Beyond NASA* (New York: Praeger Publishers, 1990).

3. See Executive Office of the President of the United States, *Historical Tables: Budget of the United States Government, Fiscal year 2009* (Washington, D.C.: GPO, 2008): 49, 56.

4. In an offhand remark that revealed state secrets, Johnson noted that the knowledge gained from space reconnaissance technology was worth "ten times what the whole program has cost." By verifying the exact size of enemy forces and verifying arms agreements, spy satellites allowed political leaders to divert vast sums of money that would have otherwise been spent to defend against threats that did not exist. Evert Clark, "Satellite Spying Cited by Johnson," *New York Times*, 17 March 1967; quoted from Chalmers Roberts, *The Nuclear Years* (New York: Praeger, 1970), 87.

5. NASA Scientific and Technical Information (STI), "Apollo Spinoffs," 10 July 2007, www.sti.nasa.gov/tto/apollo.htm (accessed 25 August 2008); Daniel S. Goldin, "Space Station: Build It for America," *Washington Post*, 28 July 1992, sec. A, p. 19. See also NASA, *Spinoff* (Washington, D.C.: GPO, 1990, also 1994); and Tom Alexander, "The Unexpected Payoff of Project Apollo," *Fortune*, July 1969, 114–15.

6. J. D. Hunley, ed., *The Birth of NASA: The Diary of T. Keith Glennan*, NASA SP-4105 (Washington, D.C.: National Aeronautics and Space Administration, 1993).

7. Dobbs, *Space*, 1–2.

8. Howard E. McCurdy, *The Space Station Decision: Incremental Politics and Technical Choice* (Baltimore: Johns Hopkins University Press, 1990), 148, 180; Ronald Reagan, "Address before a Joint Session of the Congress on the State of the Union," 25 January 1984, *Public Papers of the Presidents of the United States, 1984* (Washington, D.C.: GPO, 1985), 1:90.

9. Stanley Kubrick, *2001: A Space Odyssey* (MGM, 1968); Isaac Asimov, "Reason," in *I, Robot* (New York: Bantam Books, 1950), 56–81; George Lucas, *Star Wars Episode IV: A New Hope* (Twentieth Century Fox, 1977); Gordon Carroll, David Giler, and Walter Hill, *Alien* (Brandywine-Shussell/Twentieth Century Fox, 1979).

10. Ben Bova, illustrations by Pat Rawlings, *Welcome to Moonbase* (New York: Ballantine Books, 1987); Pat Rawlings, "Leap of Faith," 2000, patrawlings.com (accessed 24 October 2008); NASA, Science@NASA, "Lunar Olympics," 2 August 2006; Paul Spudis, *The Once and Future Moon* (Washington, D.C.: Smithsonian Institution Press, 1996); Rick Tumlinson, ed., *Return to the Moon* (Burlington, Ont.: Apogee, 2005).

11. David G. Schrunk, Burton L. Sharpe, Bonnie L. Cooper, and Madhu Thangavelu, *The Moon: Resources, Future Development and Colonization*, Wiley-Prasix Series in Space Science and Technology (Chichester, West Sussex, England: Praxis Publishing, 1999); Harrison H. Schmitt, *Return to the Moon: Exploration, Enterprise, and Energy in the Human Settlement of Space* (New York: Praxis, 2006); Julie Wakefield, "Moon's Helium-3 Could Power Earth," space.com, 30 June 2000.

12. Robert Zubrin, *The Case for Mars: The Plan to Settle the Red Planet and Why We Must* (New York: Free Press, 1996), 219, 223–31.

13. National Commission on Space, *Pioneering the Space Frontier* (New York: Bantam Books, 1986), 84, 88, 89. See also Carl Sagan, *Pale Blue Dot: A Vision of the Human Future in Space* (New York: Random House, 1994); and John S. Lewis, *Mining the Sky* (Reading, Mass.: Helix Books, 1996).

14. See Commercial Space Transportation Study Alliance, *Commercial Space Transportation Study*, material provided to NASA HQ by the Boeing Company, 4 January 1997, hq.nasa.gov/webaccess/CommSpaceTrans/ (accessed 4 June 2009); G. Harry Stein, *Halfway to Anywhere: Achieving America's Destiny in Space* (New York: M. Evans, 1996); Gary C. Hudson, "Insanely Great: or Just Plain Insane?" *Wired* 4 (May 1996): 128–32.

15. Space Adventures, "Our Vision," 2008, spaceadventures.com (accessed 17 September 2008); Michael Belfiore, "The Five-Billion-Star Hotel," *Popular Science* 266 (March 2005): 50–57. See also NASA (D. O'Neill, compiler), "General Public Space Travel and Tourism," in vol. 1, *Executive Summary*, March 1998, NP-1998-3-11–MSFC, NASA Center for Aerospace Information, Linthicum Heights, Maryland.

16. George Orwell, *1984: A Novel* (New York: Harcourt, Brace, 1949); Aldous Huxley, *Brave New World* (New York: Harper and Brothers, 1946); Frederick A. Hayek, *The Road to Serfdom* (Chicago: University of Chicago Press, 1944).

17. Gail S. Davidson, "Packaging the New: Design and the American Consumer, 1925–1975," 8 February–14 August 1994, Cooper-Hewitt Museum of Design, New York.

18. David Gelernter, *1939: The Lost World of the Fair* (New York: Free Press, 1995).

19. Donald J. Bush, *The Streamlined Decade* (New York: George Braziller, 1975), 3.

20. Daniel Goldin, "Celebrating the Spirit of Columbus," in "America at 500: Pioneering the Space Frontier," *National Forum* 72 (Summer 1992): 8–9.

21. Bruce Gordon and David Mumford, "Tomorrowland, 1986: The Comet Returns," unpublished manuscript, Walt Disney Archives, Burbank, California. See also "Tomorrowland: Show World of Future," *Disneyland News*, July 1955; and "Disneyland's New Tomorrowland," *Vacationland*, Summer 1967.

22. Willie Ley, "Inside the Moon Ship," *Collier's* 18 October 1952, 56. See also Willy Ley, "Station in Space," *Colliers*, 22 March 1952, 30–31; Willy Ley, "Inside the Lunar Base," *Collier's*, 25 October 1952, 46–47.

23. See Michael L. Smith, "Selling the Moon: The U.S. Manned Space Program and the Triumph of Commodity Scientism," in *The Culture of Consumption: Critical Essays in American History, 1880–1980*, ed. Richard W. Fox and T. J. Jackson Lears (New York: Pantheon Books, 1983), 177–236.

24. Brooks Stevens, quoted in Vance Packard, *The Waste Makers* (New York: David McKay, 1960), 54.

25. Quoted from Howard E. McCurdy, *Inside NASA: High Technology and Organizational Change in the U.S. Space Program* (Baltimore: Johns Hopkins University Press, 1993), 73.

26. Futron, *Trends in Space Commerce* (Washington: U.S. Department of Commerce Office of Space Commercialization, [2000]).

27. Dobbs, *Space*, 2.

28. Christine Y. Chen, "Iridium: From Punch Line to Profit?" *Fortune*, 2 September 2002.

29. Corey Grice, "Iridium Owners Optimistic about New Satellite Focus," *Fortune* 12 December 2000.

30. Leonard David, "NASA Shuts Down X-33, X-34 Programs," space.com (1 March 2001).

31. Space Publications in collaboration with International Space Business Council, *State of the Space Industry, 1999* (Bethesda: International Space Business Council, 1999): 33; Futron, *Trends in Space Commerce*.

32. See Howard E. McCurdy, "The Cost of Space Flight," *Space Policy* 10, no. 4 (1994): 281; Mark Wade, "X-33," astronautix.com, n.d.

33. Michael Griffin, "NASA and the Business of Space," American Astronautical Society, 15 November 2005.

34. The White House, Office of the Press Secretary, "Remarks by the President on

Space Exploration in the 21st Century," John F. Kennedy Space Center, Merritt Island, Florida, 15 April 2010.

35. National Aeronautics and Space Administration, Fiscal Year 2011 Budget Estimates, 2010, 10.

36. Dobbs, *Space*, 136.

37. Futron, *Trends in Space Commerce*; Space Foundation, *The Space Report 2008: The Authoritative Guide to Global Space Activity* (Colorado Springs: Space Foundation, 2008).

38. U.S. White House, Office of Science and Technology Policy, National Security Council, "Fact Sheet: U.S. Global Positioning System Policy," 29 March 1996.

39. Futron, *Trends in Space Commerce*.

40. See Nathan C. Goldman, *Space Commerce* (Cambridge, Mass.: Ballinger, 1985); Roger Handberg, *The Future of the Space Industry: Private Enterprise and Public Policy* (Westport, Conn.: Quorum Books, 1995); U. S. Department of Transportation and Federal Aviation Administration, "Liability Risk Sharing Regime for U.S. Commercial Space Transportation: Study and Analysis," April 2002; Neil Dahlstrom, ed., "Commercial Space Policy in the 1980s: Proceedings of a Roundtable Discussion," July 2000, Space Business Archives, Alexandria, Virginia; Goodrich, *The Commercialization of Outer Space*.

CHAPTER 11: Back on Earth

Epigraph: NASA, *Why Man Explores*, symposium held at Bechman auditorium, California Institute of Technology, Pasadena, 2 July 1976 (Washington, D.C.: GPO, 1977), 11.

1. Joseph J. Corn, *Winged Gospel: America's Romance with Aviation* (1983; Baltimore: Johns Hopkins University Press, 2002), 88.

2. Quoted from W. E. Debnam, "Women's Place in Aviation as Seen by Endurance Fliers," *Southern Aviation* 4 (December 1932): 11. See also Claudia M. Oakes, *United States Women in Aviation, 1930–1939* (Washington, D.C.: Smithsonian Institution Press, 1985).

3. Mary S. Lovell, *Straight on Toward Morning: The Biography of Beryl Markham* (New York: St. Martin's Press, 1987); Beryl Markham, *West with the Night* (San Francisco: North Point Press, 1983).

4. Anne Morrow Lindbergh, *North to the Orient* (New York: Harcourt, Brace, 1935); Deborah G. Douglas, *United States Women in Aviation, 1940–1985*, Smithsonian Studies in Air and Space, no. 7 (Washington, D.C.: Smithsonian Institution Press, 1990).

5. Oakes, *United States Women in Aviation, 1930–1939*; see also Mary S. Lovell, *The Sound of Wings: The Life of Amelia Earhart* (New York: St. Martin's Press, 1989); and Doris L. Rich, *Amelia Earhart: A Biography* (Washington, D.C.: Smithsonian Institution, 1989).

6. Pamela Sargent, ed., *Women of Wonder: The Classic Years* (San Diego: Harcourt Brace, 1995).

7. Herbert E. Solow and Robert H. Justman, *Inside Star Trek* (New York: Pocket Books, 1996).

8. George Pal, *Conquest of Space* (Paramount, 1955); Nicholas Nayfack, *Forbidden Planet.* (Metro-Goldwyn-Mayer, 1956).

9. Speech by Wernher von Braun given at Mississippi State College, 19 November 1962, 6, NASA History Office, NASA Headquarters, Washington, D.C.; U.S. House Committee on Science and Astronautics, Special Subcommittee on the Selection of Astronauts, *Qualifications for Astronauts*, 87th Cong., 2nd sess., 1962, 5, 58.

10. See Hugh L. Dryden to Jacqueline Cochran, 18 June 1962, NASA History Office. See also Margaret A. Weitekamp, *Right Stuff, Wrong Sex: America's First Women in Space Program* (Baltimore: Johns Hopkins University Press, 2004).

11. U.S. House Committee on Science and Astronautics, *Qualifications for Astronauts*, 81; Jerrie Cobb with Jane Rieker, *Woman into Space: The Jerrie Cobb Story* (Englewood Cliffs, N.J.: Prentice-Hall, 1963), 149; "A Lady Proves She's Fit for Space Flight," *Life*, 29 August 1960, 72–76.

12. See paper presented by Jerrie Cobb, Aviation/Space Writers Association, twenty-third annual meeting and news conference, 1 May 1961, NASA History Office.

13. Weitekamp, *Right Stuff, Wrong Sex*, 126–28.

14. Lyndon B. Johnson to Miss Cobb, 23 April 1962; Jerrie Cobb to James E. Webb, 7 August 1962; Jerrie Cobb to the President, 10 February 1964; all in NASA History Office.

15. House Science and Astronautics Committee, *Qualifications for Astronauts*, 5.

16. Jacqueline Cochran to Jerrie Cobb, 23 March 1962. See also Cochran to James E. Webb, 14 June 1962; memo to James Webb and Hugh Dryden, 1 August 1962; and Joseph D. Atkinson and Jay M. Shafritz, *The Real Stuff: A History of NASA's Astronaut Recruitment Program* (New York: Praeger, 1985), chap. 5. Letters and memos in NASA History Office.

17. "13 Women Triumphing Vicariously," *New York Times*, 5 February 1995; "Collins Fulfills Dreams for Mercury 13 Women," *NASA Headquarter Bulletin*, 21 February 1995; Richard Paul, "Rocket Girls and Astro-nettes" (r.l.paul productions, 2010).

18. Elizabeth S. Bell, *Sisters of the Wind: Voices of Early Women Aviators* (Pasadena, Calif.: Trilogy Books, 1994); Margery Brown, "Flying Is Changing Women," *Pictorial Review*, June 1930.

19. Media General/Associate Press Public Opinion Poll, "The U.S. Space Program," poll no. 21, 22 June–2 July 1988. See also George Gallup, *Gallup Poll: Public Opinion, 1989* (Wilmington, Del.: Scholarly Resources, 1990), 171; George Gallup, *The Gallup Report*, no. 246 (March 1986): 11; Elizabeth H. Hastings and Philip K. Hastings, *Index to International Public Opinion, 1985–1986* (Westport, Conn.: Greenwood Press, 1987), 469–70.

20. Atkinson and Shafritz, *Real Stuff*.

21. Donna Shirley with Danelle Morton, *Managing Martians* (New York: Broadway Books, 1998), 81. See also Sylvia Fries, *NASA Engineers and the Age of Apollo*, NASA SP-4104 (Washington, D.C.: National Aeronautics and Space Administration, 1992).

22. See Robert Poole, *Earthrise: How Man First Saw the Earth* (New Haven, Conn.: Yale University Press, 2008).

23. Carl Sagan, *Pale Blue Dot: A Vision of the Human Future in Space* (New York: Random House, 1994), 6.

24. Archibald MacLeish, "A Reflection: Riders on Earth Together, Brothers in Eternal Cold," *New York Times*, 25 December 1968; Jimmy Carter, "Remarks at the Congressional Space Medal of Honor Awards Ceremony," Kennedy Space Center, *Weekly Compilation of Presidential Documents* 14, no. 39 (1 October 1978): 1685.

25. Barbara Ward and Rene Dubos, *Only One Earth: The Care and Maintenance of a Small Planet* (New York: Norton, 1972); Stewart Brand, *Whole Earth Catalog* (Menlo Park, Calif.: Portola Institute, Fall 1968); Davis Guggenheim, *An Inconvenient Truth* (Paramount, 2006).

26. Alvin Toffler, *The Third Wave* (New York: Bantam Books, 1980), 408. See also Toffler, "The Space Program's Impact on Society," in *Humans and Machines in Space: The Payoff*, ed. Paula Korn (San Diego: American Astronautical Society, 1992), 87.

27. Donella H. Meadows et al., *Limits to Growth: A Report for the Club of Rome's Project on the Predicament of Mankind* (New York: Universe Books, 1974), 24.

28. Ibid., 187. See also chap. 5; H. S. D. Cole, *Models of Doom: A Critique of the Limits to Growth* (New York: Universe Books, 1973).

29. See Lynn Margulis and Dorion Sagan, *Microcosmos: Four Billion Years of Microbial Evolution* (New York: Summit Books, 1986).

30. Lawrence E. Joseph, *Gaia: The Growth of an Idea* (New York: St. Martin's Press, 1990), 1–2. See also James Lovelock, *Ages of Gaia: A Biography of Our Living Earth* (New York: W. W. Norton, 1988).

31. NASA, "Terra: Flagship of the Earth Observing System," press kit, November 1999, 3; Sally K. Ride, *Leadership and America's Future in Space*, Report to the Administrator (Washington, D.C.: National Aeronautics and Space Administration, 1987), 23; Burton I. Edelson, "Mission to Planet Earth," *Science* 227 (25 January 1985): 367; Craig Covault, "Major Space Effort Mobilized to Blunt Environmental Threat," *Aviation Week & Space Technology*, 13 March 1989, 36–44; James R. Asker, "Earth Mission Faces Growing Pains," *Aviation Week & Space Technology*, 21 February 1994, 36; W. Henry Lambright, "Downsizing Big Science: Strategic Choices," *Public Administration Review* 58 (May–June 1998): 259–68.

32. J. C. Farman, B. G. Gardiner, and J. D. Shanklin, "Large Losses of Total Ozone in Antarctica Reveal Seasonal ClOx/NOx Interaction," *Nature* 315 (May 1985): 207–10; NASA Goddard Space Flight Center, Scientific Visualization Studio, Antarctic Ozone from TOMS: August 1, 2003 to September 23, 2003, animation number 2809, 2003.

33. Michael Meltzer, *Mission to Jupiter: A History of the Galileo Project*, NASA SP-2007-4231 (Washington, D.C.: National Aeronautics and Space Administration, 2007); Theresa M. Foley, "NASA Prepares for Protests over Nuclear System Launch on Shuttle in October," *Aviation Week & Space Technology*, 26 June 1989; "Court Rejects Activists' Bid to Halt Galileo/Shuttle Launch," *Aviation Week & Space Technology*, 16 October 1989, 21; Charles Perrow, "The Habit of Courting Disaster," *Nation*, 11 October 1986; and Perrow, *Normal Accidents: Living with High-Risk Technologies* (New York: Basic, 1984).

34. Scott Derrickson, *The Day the Earth Stood Still* (Twentieth Century Fox, 2008).

35. See Alvin Toffler, *Future Shock* (New York: Random House, 1970), chap. 4.

36. Sagan, *Pale Blue Dot*, 8–9.

CONCLUSION: Imagination and Culture

Epigraph: Sign in a store displaying Indian artifacts, author unknown, the Clotheshorse Trading Company, Seattle, Washington.

1. See Joseph J. Corn, *Winged Gospel: America's Romance with Aviation* (1983; Baltimore: Johns Hopkins University Press, 2002).

2. James E. Anderson, *Public Policymaking* (Boston: Houghton Mifflin, 1994).

3. See, for example, Roderick Nash, *Wilderness and the American Mind* (New Haven, Conn.: Yale University Press, 1967); Jackson Lears, *Fables of Abundance: A Cultural History of Advertising in America* (New York: Basic Books, 1994); Kristin Ross, *Fast Cars, Clean*

Bodies (Cambridge, Mass.: MIT Press, 1995); Nicholas B. Dirks, Geoff Eley, and Sherry B. Ortner, *Culture/Power/History* (Princeton, N.J.: Princeton University Press, 1994).

4. See Kenneth Clark, *Civilisation: A Personal View* (New York: Harper and Row, 1969), 322–23.

5. Charles Dickens, *Oliver Twist* (1838; reprint, New York: Bantam Books, 1982).

6. See Peter L. Berger and Thomas Luckmann, *The Social Construction of Reality: A Treatise in the Sociology of Knowledge* (Garden City, N.Y.: Anchor Books, 1966).

7. See Roderick Nash, *Wilderness and the American Mind*, 3rd ed. (New Haven, Conn.: Yale University Press, 1982).

8. See Alfred Runte, *National Parks* (Lincoln: University of Nebraska Press, 1987).

9. Anne R. Morand, Joni L. Kinsey, and Mary Panzer, *Splendors of the American West: Thomas Moran's Art of the Grand Canyon and Yellowstone* (Birmingham, Ala.: Birmingham Museum of Art, 1990). See also Barbara Novack, *Nature and Culture: American Landscape, 1825–1865* (New York: Oxford University Press, 1980).

10. See Paul Brooks, *Speaking for Nature: How Literary Naturalists from Henry Thoreau to Rachel Carson Have Shaped America* (Boston: Houghton Mifflin, 1980).

11. See G. Edward White, *The Eastern Establishment and the Western Experience* (New Haven, Conn.: Yale University Press, 1989).

12. See John Tytell, *Ezra Pound* (New York: Anchor Press, 1987); William Perlberg, *Miracle on 34th Street* (Fox, 1947); and J. D. Salinger, *Franny and Zooey* (Boston: Little, Brown, 1961).

13. Ken Kesey, *One Flew Over the Cuckoo's Nest* (New York: Viking Press, 1962).

14. Paul McHugh, "Psychiatric Misadventures," *American Scholar* 61 (Autumn 1992): 498. Also see Erving Goffman, *Asylums: Essays on the Social Situation of Mental Patients and Other Inmates* (Garden City, N.Y.: Doubleday Anchor, 1961), and Thomas S. Szasz, *The Myth of Mental Illness* (New York: Harper and Row, 1961).

15. See Myron Magnet, *The Dream and the Nightmare: The Sixties Legacy to the Underclass* (New York: William Morrow, 1993).

16. E. E. Schattschneider, *The Semi-Sovereign People* (New York: Holt, Rinehart and Winston, 1960), 3.

17. Frank R. Baumgartner and Bryan D. Jones, *Agendas and Instability in American Politics* (Chicago: University of Chicago Press, 1993); Stephen Jay Gould, *The Burgess Shale and the Nature of History* (New York: W. W. Norton, 1989).

18. Upton Sinclair, *The Jungle* (New York: New American Library, 1905).

19. Ibid., 349.

20. Gareth Morgan, *Imaginization: The Art of Creative Management* (Newbury Park, Calif.: Sage Publications, 1993).

21. See C. Northcote Parkinson, *Parkinson's Law and Other Studies in Administration* (Boston: Houghton Mifflin, 1957), and Anthony Downs, *Inside Bureaucracy* (Boston: Little, Brown, 1967). Also see Joe Queenan, "Evil Empire on the Potomac," *Washington Post*, 13 November 1994, sec. C, p. 5.

22. Alexis de Tocqueville, *Democracy in America* (New York: Schocken Books, 1835).

23. Mark Twain, *Huckleberry Finn* (New York: Harcourt, Brace & World, 1961).

24. See Scott L. Montgomery, *The Moon and Western Imagination* (Tucson: University of Arizona Press, 1999).

25. See Marshall McLuhan, *Understanding Media: The Extensions of Man* (New York: McGraw-Hill, 1964).

382   Notes to Pages 318–322

26. Wernher von Braun, "Man on the Moon: The Journey," *Collier's*, 18 October 1952, 54.

27. William Sims Bainbridge, *The Spaceflight Revolution: A Sociological Study* (New York: John Wiley and Sons, 1976), 13.

28. John Calvin Batchelor, in "What Is the Value of Space Exploration? A Symposium," 31, sponsored by the Mission from Planet Earth Study Office, Office of Space Science, NASA Headquarters, and the University of Maryland at College Park, 18–19 July 1994.

29. Paul Theroux, *Sailing through China* (Boston: Houghton Mifflin, 1984), 23. See also Dwayne A. Day, "Paradigm Lost," *Space Policy* 11 (August 1995): 153–59.

30. See Robert J. Samuelson, *The Good Life and Its Discontents* (New York: Times Books, 1995).

31. Lewis L. Strauss, "Remarks Prepared by Lewis L. Strauss, Chairman, United States Atomic Energy Commission, for Delivery at the Founders' Day Dinner, National Association of Science Writers," 16 September 1954, New York, Department of Energy History Division. Also see Brian Balogh, *Chain Reaction: Expert Debate and Public Participation in American Commercial Nuclear Power, 1945–1975* (New York: Cambridge University Press, 1991), 113.

32. See Corn, *Winged Gospel*, chap. 5.

33. Richard M. Nixon, "Space Shuttle Program," *Weekly Compilation of Presidential Documents* 8, no. 2 (5 January 1972): 27; Office of the White House Press Secretary, Press Conference of Dr. James Fletcher and George M. Low, 5 January 1972, 8, San Clemente, California.

34. Ryan A. Harmon, "Predicting the Future," *Disney News*, Fall 1991, 35.

35. David C. McClelland, *The Achieving Society* (Princeton, N.J.: Van Nostrand, 1961).

36. Glen A. Larson, Battlestar Galactica, battlestargalactica.com (accessed 14 June 2010); Joel Silver, *The Matrix* (Warner Brothers Pictures, 1999); James Cameron, *Avatar* (Twentieth Century Fox, 2009).

37. Damon Lindelof, J. J. Abrams, and Jeffrey Lieber, *Lost* (American Broadcasting Company, 2004–10).

38. See Jack Lemming, "The Future of Space Activities," *Spaceflight*, April 1994, 110–11; Henry C. Dethloff, *Suddenly, Tomorrow Came: A History of the Johnson Space Center*, NASA SP-4307 (Washington, D.C.: National Aeronautics and Space Administration, 1993), chap. 16.

39. See Leon Festinger, *When Prophecy Fails* (New York: Harper and Row, 1956); Festinger, *A Theory of Cognitive Dissonance* (Stanford, Calif.: Stanford University Press, 1957).

40. See Walter A. McDougall, *The Heavens and the Earth: A Political History of the Space Age* (New York: Random House, 1972); Robert A. Divine, *The Sputnik Challenge: Eisenhower's Response to the Soviet Satellite* (New York: Oxford University Press, 1993); John Logsdon, *The Decision to Go to the Moon* (Cambridge, Mass.: MIT Press, 1970).

41. U.S. Environmental Protection Agency, "Annual Performance Plan and Budget Overview of EPA (FY 2007)." Available from the EPA Web site, www.epa.gov/ocfo/budget/2007/2007bib.pdf (accessed November 14, 2007): Actual appropriation for FY 2007, $7.6 billion. U.S. Federal Bureau of Investigation, "About Us: Quick Facts," www.fbi.gov/quickfacts.htm (accessed November 14, 2007): Total budget, FY 2007, $6.04 bil-

lion. U. S. National Park Service: "NPS Overview." Available at NPS Web site, www.nps
.gov/pub_aff/refdesk/NPS_Overview.pdf (accessed November 14, 2007): Budget enacted
for FY 2006, $2.256 billion.

42. Psalm 116:6, New Living Translation; Hebrews 11:1, King James Bible.

43. See Benedict Anderson, *Imagined Communities*, rev. ed. (London: Verso, 1991);
Andrew Parker, "Bogeyman: Benedict Anderson's 'Derivative' Discourse," *Diacritics* 29,
no. 4 (1999): 40–57; David Thelen, "The Nation and Beyond: Transnational Perspectives
on United States History," *Journal of American History*, special issue, 86, no. 3 (December
1999): 965–75.

44. David Domke, Dhavan V. Shah, and Daniel B. Wackman. "Media Priming Effects:
Accessibility, Association, and Activation," *International Journal of Public Opinion Research*
10, no. 1 (1998): 51–74.

45. William Gamson, *Talking Politics* (Cambridge: Cambridge University Press,
1992), 6.

46. Walter Lippmann, *Public Opinion* (New York: Macmillan, 1922), 4–5.

47. William Gamson and Gadi Wolfsfeld, "Movements and Media as Interacting
Systems," *Annals of the American Academy of Political and Social Science* 528, no. 1 (1993):
114–25; Michael Newbury, "Celebrity Watching," *American Literary History* 12 (Spring–
Summer 2000): 272–83; Calvin Massey, "Civil Discourse and Cultural Transformation,"
Cardozo Studies in Law and Literature 12 (Summer, 2000): 193–215.

48. John Winthrop. "A Model of Christian Charity" (1630), religiousfreedom.lib
.virginia.edu/sacred/charity.htm (accessed 14 November 2007).

Index

Page numbers in boldface indicate illustrations.

Mars, **4**, **8**, **152**; base, **57**; canals, 123–26, **124**, 134–36; colonization, 165–67, 171–73; commerce, 272; exploration, 134–39, **156**; life on, 123–28, **128**, 134, 136–37; like the Moon, 134; Mars rock, 138–39, **147**; martian myth, 123–29, 139; terraforming, 174–75

"Mars and Beyond" (third Disney television program on space flight), 48, 127

Mars and Its Canals (Lowell), 125

Mars Declaration, 170–71

Mars Project, The (von Braun), 27

Mars Society, 171

Mars Underground, 173–74

Martian Chronicles, The (Bradbury), 80, 127, 166

mass-driver, 168

Matrix, The, 256–57, 321

Mauldin, John, 225

Mayo, Elton, 241

McAuliffe, Christa, 228

McCall, Robert T., art, 58, 175, **189**

McClelland, David, 320

McDougall, Walter, 33–34

McHugh, Paul, 313

McLuhan, Marshall, 302

McNamara, Robert, 108

media, influence of, 318, 323–24

Melville, Mike, 233

Men in Black, 145

Mercury project, 95–96; human control, 261–62; mishaps, **97**

Messner, Reinhold, 186

metaphors. *See* imagination

Method of Reaching Extreme Altitudes, A (Goddard), 20

Michaud, Michael, 167

Microwave Observing Project, 144–45, 148

Mikulski, Barbara, 148

military-industrial complex, 67, 69, 112

military reconnaissance satellites, 67

military significance of space, 73–74, 88

Millennium Falcon (imaginary spacecraft), 219–20

Miller, Ron, 215

Miller, William, 78

mining in space, 32, 168, 172–73, 272–73, **280**

minorities in astronaut corps, 299–300

Mission to Planet Earth, 305

Moon: commerce, 271–72; life on, 122; lunar bases, **160**, 165, 171–73; mining, **280**; as missile base, 72–73, **74**

Moon, The (Carpenter), 51

Moon hoax, 122

Moore, C. L. (Catherine), 294

Moore, Gordon E., Moore's law, 249

Moran, Thomas, 51, 312

Moravec, Hans, 257, 259

More, Thomas, 166

Morgan, Gareth, 317

Morrison, David, 89–90

Morrison, Samuel Eliot, 162

Morrow, Anne, 292–93

Mos Eisley Cantina, 133

mountaineering, 183–84

Mueller, George, 222

Munsey, Frank, 29, 35

Murray, Bruce, 170

NASA. *See* National Aeronautics and Space Administration

Nasmyth, James, 51

National Academy of Sciences, 195, 264

National Advisory Committee for Aeronautics (NACA), 55; attitudes toward space exploration, 87; Pilotless Aircraft Research Division, 86

National Aeronautics and Space Administration (NASA): female astronaut program, 295–300; field center spending, 201; Goett committee (Research Steering Committee on Manned Space Flight), 55–56; long-range plan, 11–12, 39, 56; minorities in astronaut corps, 299–300; Office of Exploration, 58, 173; policy on human and robotic flight, 243–46, 248, 264–65; separation of human and satellite flight centers, 69–70; Space Station Task Force, 192, 271

National Air and Space Museum, 105

National Commission on Space, 57–58, 157, 171–72, 272–73

National Geographic Society, 29

National Space Institute, 170

National Space Society, 170

Natural History (Pliny), 129

Naval Research Laboratory, 62–63, 96